T0189266

IFIP Advances in Information and Communication Technology 610

Editor-in-Chief

Kai Rannenberg, Goethe University Frankfurt, Germany

Editorial Board Members

TC 1 – Foundations of Computer Science
  Luís Soares Barbosa⬤, University of Minho, Braga, Portugal

TC 2 – Software: Theory and Practice
  Michael Goedicke, University of Duisburg-Essen, Germany

TC 3 – Education
  Arthur Tatnall⬤, Victoria University, Melbourne, Australia

TC 5 – Information Technology Applications
  Erich J. Neuhold, University of Vienna, Austria

TC 6 – Communication Systems
  Burkhard Stiller, University of Zurich, Zürich, Switzerland

TC 7 – System Modeling and Optimization
  Fredi Tröltzsch, TU Berlin, Germany

TC 8 – Information Systems
  Jan Pries-Heje, Roskilde University, Denmark

TC 9 – ICT and Society
  David Kreps⬤, National University of Ireland, Galway, Ireland

TC 10 – Computer Systems Technology
  Ricardo Reis⬤, Federal University of Rio Grande do Sul, Porto Alegre, Brazil

TC 11 – Security and Privacy Protection in Information Processing Systems
  Steven Furnell⬤, Plymouth University, UK

TC 12 – Artificial Intelligence
  Eunika Mercier-Laurent⬤, University of Reims Champagne-Ardenne, Reims, France

TC 13 – Human-Computer Interaction
  Marco Winckler⬤, University of Nice Sophia Antipolis, France

TC 14 – Entertainment Computing
  Rainer Malaka, University of Bremen, Germany

IFIP – The International Federation for Information Processing

IFIP was founded in 1960 under the auspices of UNESCO, following the first World Computer Congress held in Paris the previous year. A federation for societies working in information processing, IFIP's aim is two-fold: to support information processing in the countries of its members and to encourage technology transfer to developing nations. As its mission statement clearly states:

> *IFIP is the global non-profit federation of societies of ICT professionals that aims at achieving a worldwide professional and socially responsible development and application of information and communication technologies.*

IFIP is a non-profit-making organization, run almost solely by 2500 volunteers. It operates through a number of technical committees and working groups, which organize events and publications. IFIP's events range from large international open conferences to working conferences and local seminars.

The flagship event is the IFIP World Computer Congress, at which both invited and contributed papers are presented. Contributed papers are rigorously refereed and the rejection rate is high.

As with the Congress, participation in the open conferences is open to all and papers may be invited or submitted. Again, submitted papers are stringently refereed.

The working conferences are structured differently. They are usually run by a working group and attendance is generally smaller and occasionally by invitation only. Their purpose is to create an atmosphere conducive to innovation and development. Refereeing is also rigorous and papers are subjected to extensive group discussion.

Publications arising from IFIP events vary. The papers presented at the IFIP World Computer Congress and at open conferences are published as conference proceedings, while the results of the working conferences are often published as collections of selected and edited papers.

IFIP distinguishes three types of institutional membership: Country Representative Members, Members at Large, and Associate Members. The type of organization that can apply for membership is a wide variety and includes national or international societies of individual computer scientists/ICT professionals, associations or federations of such societies, government institutions/government related organizations, national or international research institutes or consortia, universities, academies of sciences, companies, national or international associations or federations of companies.

More information about this series at https://link.springer.com/bookseries/6102

Daryl John Powell · Erlend Alfnes ·
Marte D. Q. Holmemo ·
Eivind Reke (Eds.)

Learning in the Digital Era

7th European Lean Educator Conference, ELEC 2021
Trondheim, Norway, October 25–27, 2021
Proceedings

 Springer

Editors
Daryl John Powell
SINTEF
Trondheim, Norway

Marte D. Q. Holmemo
Norwegian University of Science
and Technology
Trondheim, Norway

Erlend Alfnes
Norwegian University of Science
and Technology
Trondheim, Norway

Eivind Reke
SINTEF
Trondheim, Norway

ISSN 1868-4238   ISSN 1868-422X (electronic)
IFIP Advances in Information and Communication Technology
ISBN 978-3-030-92936-7   ISBN 978-3-030-92934-3 (eBook)
https://doi.org/10.1007/978-3-030-92934-3

This Springer imprint is published by the registered company Springer Nature Switzerland AG
The registered company address is: Gewerbestrasse 11, 6330 Cham, Switzerland

Preface

In this book of proceedings, we share the papers that were presented at the 7th European Lean Educator Conference (ELEC 2021), which was hosted online from Trondheim, Norway, during October 25–27, 2021. This year's conference was special in that it was sponsored by the International Federation for Information Processing (IFIP) for the first time in its history, having been organized in close collaboration with the IFIP Working Group 5.7 Special Interest Group (SIG) on the Future of Lean Thinking and Practice.

The ELEC community is dedicated to fostering knowledge exchange within academia as well as between academic institutions and industry, which we believe is especially important. As such, ELEC 2021 provided a platform for professors, teachers, trainers, and coaches from academia, industry, and public sector organizations to share their knowledge and experiences and learn from one another.

ELEC conferences distinguish lean thinking as a broad management and operations improvement philosophy, fully grounded in science and supported by a continually growing set of methods, techniques, and tools. Given that this year's conference theme was 'learning in the digital era', ELEC 2021 made a special effort to highlight the role of digital technologies in the emerging digital lean manufacturing paradigm.

As such, this volume includes the 42 full papers that were presented at ELEC 2021. The papers were double-blind peer-reviewed to ensure quality. The topics which emerged at ELEC 2021 were:

- Learning Lean
- Teaching Lean in the Digital Era
- Lean and Digital
- Lean 4.0
- Lean Management
- Lean Coaching and Mentoring
- Skills and Knowledge Management
- Productivity and Performance Improvement
- New Perspectives on Lean

In addition to the technical program, ELEC 2021 featured four keynote talks:

- Torbjørn Netland, ETH Zurich, Switzerland, "Teaching Lean in the Digital Era".
- Eva Helen Rognskog, SATPOS, Horten, Norway, "Fail Fast, Learn Faster".
- Rose Heathcote, University of Buckingham, UK, "Forget business as usual. We've work to do".
- Torbjørn Gjerdevik, Laerdal Medical, Stavanger, Norway, "People First".

On behalf of the Conference Committee, We would like to thank all contributors for the high standard of work presented at ELEC 2021. This of course would not have been possible without the support of our International Scientific Committee and peer reviewers, so we would also like to thank these members for their efforts in reviewing and selecting the papers to be presented at ELEC 2021.

To conclude, We hope that ELEC 2021 has stimulated the exchange of both research results and practical experiences to enhance the state of the art of lean thinking and practice.

October 2021

Daryl John Powell
Eivind Reke

Organization

ELEC 2021 was organized by SINTEF Manufacturing AS with support from the Norwegian University of Science and Technology.

Conference Chair

Daryl John Powell   SINTEF Manufacturing AS, Horten, Norway

Program Chair

Eivind Reke   SINTEF Manufacturing AS, Trondheim, Norway

Conference Co-chairs

Erlend Alfnes   Norwegian University of Science and Technology,
                                     Norway
Marte Daae-Qvale   Norwegian University of Science and Technology,
   Holmemo   Norway

International Advisory Board

Constantin May   Ansbach University of Applied Sciences, Germany
Pia Anhede   Revere AB, Sweden
Joakim Hillberg   Revere AB, Sweden
John Bicheno   University of Buckingham, UK
Christoph Roser   University of Applied Sciences, Karlsruhe, Germany
Jannes Slomp   HAN University of Applied Sciences, The Netherlands
Dinis Carvalho   Universidade do Minho, Portugal
Monica Rossi   Politecnico di Milano, Italy
John Shook   Lean Global Network, Boston, USA

International Scientific Committee

Erlend Alfnes   Norwegian University of Science and Technology,
                                     Norway
Jos Benders   Norwegian University of Science and Technology,
                                     Norway
Dinis Carvalho   Universidade do Minho, Portugal
Eivind Arne Fauskanger   University of South-Eastern Norway, Norway
Peter Hines   Waterford Institute of Technology, Ireland
Poul Houman Andersen   Aalborg University, Denmark

Olivia McDermott	National University of Ireland, Galway, Ireland
Torbjørn Netland	ETH Zurich, Switzerland
Daryl John Powell	SINTEF Manufacturing AS, Horten, Norway
Eivind Reke	SINTEF Manufacturing AS, Trondheim, Norway
Christoph Roser	University of Applied Sciences, Karlsruhe, Germany
Matteo Rossini	Politecnico di Milano, Italy
Henrik Saabye	Velux, Denmark
José Carlos Sá	Instituto Superior de Engenharia do Porto, Portugal
Jannes Slomp	HAN University of Applied Sciences, The Netherlands

Peer Reviewers

Erlend Alfnes	Norwegian University of Science and Technology, Norway
Anabela Alves	University of Minho, Portugal
Jos Benders	Norwegian University of Science and Technology, Norway
Chris Buckell	University of Warwick, UK
Jenny Bäckstrand	Jonkoping University, Sweden
Federica Costa	Politecnico di Milano, Italy
Torgeir Dingsøyr	Norwegian University of Science and Technology, Norway
Jose Dinis-Carvalho	Universidade do Minho, Portugal
Heidi Dreyer	Norwegian University of Science and Technology, Norway
Eivind Arne Fauskanger	University of South-Eastern Norway, Norway
Torbjørn Hekneby	Storform, Norway
Peter Hines	Waterford Institute of Technology, Ireland
Marte Daae-Qvale Holmemo	Norwegian University of Science and Technology, Norway
Poul Houman Andersen	Aalborg University, Denmark
Natalia Iakymenko	SINTEF Manufacturing AS, Trondheim, Norway
Jonas Ingvaldsen	Norwegian University of Science and Technology, Norway
Bassel Kassem	Politecnico di Milano, Italy
Wilfred Knol	HAN University of Applied Sciences, The Netherlands
Alinda Kokkinou	Avans University of Applied Sciences, The Netherlands
Eirin Lodgaard	SINTEF Manufacturing AS, Trondheim, Norway
Olivia McDermott	National University of Ireland, Galway, Ireland
Maria Flavia Mogos	SINTEF Manufacturing AS, Trondheim, Norway
Torbjørn Netland	ETH Zurich, Switzerland
Mirco Peron	Norwegian University of Science and Technology, Norway
Bruno Pešec	Pesec Global, Norway
Daryl John Powell	SINTEF Manufacturing AS, Trondheim, Norway

Contents

Lean Management

Lean Coaching and Mentoring

Skills and Knowledge Management

Productivity and Performance Improvement

New Perspectives on Lean

Learning Lean

Sustaining Continuous Improvement Through Double Loop Learning

Chris Buckell[(⊠)] [iD] and Mairi Macintyre [iD]

University of Warwick, Coventry CV4 7AL, UK

Abstract. Public Service Organisations (PSOs) are facing continuing funding challenges and increased pressure to maintain and improve service delivery with fewer resources. One response, with the promise of improving efficiency rather than cutting services, has been to implement Continuous Improvement (CI) but success has been sporadic and unpredictable. The well documented CI methodologies, notably Lean and Six Sigma, have general agreement across practitioners and scholars alike, thus the reasons behind their potted success must lie elsewhere, in the culture or the environment perhaps? This work explores the wider contextual issues of CI implementation with the aim of providing guidelines to give a greater confidence of successful implementation. A structured literature review provided the initial conceptual framework that was further developed through a series of in-depth, semi-structured interviews carried out with industry experts. This is supported by a case study with a UK health sector organisation. The research shows that emphasis should be placed on addressing logic and mindsets at an individual and organisational level in order to re-focus CI efforts and achieve sustainable process improvement culture. Particular attention should be placed on the role of leaders. This research takes a unique approach to CI in the UK PSO context, providing insights into the achievement of sustainable CI and a theoretical framework for evaluating PSO logic. It establishes a theoretical foundation for the evaluation of organisational learning in relation to sustainable CI in UK PSOs. It also makes practical recommendations to support PSO to reveal, evaluate and address organisational principles through interactive workshops and a preliminary pilot study. Research should continue to focus on the critical role of organisational learning and governing variables in relation to addressing PSO logic for sustained CI.

Keywords: Public Service Organisation (PSO) · Continuous Improvement (CI) · Organisational learning · Sustainment

1 Introduction

This paper establishes preliminary work for the evaluation of thinking and behaviours in PSOs in relation to sustained CI implementation. The first part of this paper (Sects. 1, 2 and 3) discusses academic literature regarding the arrival and promulgation of CI methodologies in public sector, and organisational learning theory as a possible explanator of the current progress of CI in public sector to date. It concludes that conventional Public Service Organisations (PSOs) thinking, termed the PSO paradigm, is the root cause of the problem of unsustained CI in public services. To address this, PSOs need

© IFIP International Federation for Information Processing 2021
Published by Springer Nature Switzerland AG 2021
D. J. Powell et al. (Eds.): ELEC 2021, IFIP AICT 610, pp. 3–12, 2021.
https://doi.org/10.1007/978-3-030-92934-3_1

to re-evaluate fundamental principles and logic in relation to CI interventions. Section 4 presents an analysis of the current state of organisational learning in public sector by enhancing the literature with findings from semi-structured interviews conducted with CI professionals. Section five onwards advances the theoretical work further with a pilot study in a UK health sector organisation. The paper concludes that addressing organisational logics and mind sets is critical in achieving sustainable CI results.

2 The Need for CI in Public Services

In response to the global financial crisis in 2008 the UK public sector faced unprecedented austerity and budget cuts [1], increasing pressure to build a more efficient state, and do more with less. The UK Government policy between 2010–2015 echoed the political appetite for efficiency and cost savings. One response to this by Public Sector Organisations (PSOs) was to introduce Continuous Improvement (CI) initiatives from the late 1990s onwards. Before this time, a variety of CI methodologies had emerged from the manufacturing sector and were being applied in service organisations, promising process efficiency and removal of waste [2, 3]. Methodologies such as Lean and Six Sigma offered structured frameworks and tools which, was claimed, could be successfully transferred into service organisations. Consequently, manufacturing-originated CI methodologies became pervasive to service organisations and public sector alike [3, 4].

2.1 The Application of CI in Public Services

As CI became pervasive in public sector, a large body of knowledge was accumulated. Several studies identified critical success factors and barriers to success for CI change programmes in PSOs [5–9]. Despite this, PSOs have continued to report the same recurring problems for over a decade [10] and CI remains largely unsustained today [11].

Researchers such as Hines et al. [12, 13] and Radnor et al. [14, 15] began to recognise emerging problems with the sustainability of CI in the early 2000s, particularly methodologies which originated in the manufacturing sector. They encouraged adaptation of method with more emphasis on the socio-human and cultural elements of change; particularly the central role of leadership, staff empowerment and behaviours [16, 17]. Hines for example distinguished the 'visible' or tangible elements of tools, processes and technology from the 'enabling' intangible factors: strategy, alignment, leadership and engagement. Hines also placed importance on "the social norms" of the organisation in impacting the CI journey. Radnor et al.'s [14] 'House of Lean' developed a similar concept in a PSO-specific context, emphasising the importance of engagement and behaviours.

Despite this advancement in understanding, a cost reduction and tools-focus remained the predominant CI approach, largely in isolated or limited applications [2, 9, 10, 15, 16]. This has achieved cost efficiencies but has ultimately been unsustained (economically and socially). Radnor & Bateman argued more recently that CI should be considered a long-term endeavour that requires behavioural and cultural change in order to be sustained [10, 16].

2.2   Contemporary Thinking on the PSO Paradigm

Bateman's [11] review of the status of CI in ICiPS members provides the most recent comprehensive evaluation of CI in PSOs. Similar to Radnor's [14] report, it too focusses on strategy, training, techniques and barriers to implementation. Whilst still reporting the same barriers to implementation i.e. leadership, staff resistance, one stand-out point is insight into the learning that occurred. For example, one respondent from the study recognises a need "[not to be] hung up on methodology…but making it right for the problem rather than trying to get the problem to fit the tool".

However, there has been little discussion or evaluation of the tenets of PSO thinking, which this paper terms the PSO paradigm, and its relationship with methodology interpretation/deployment. Seddon & O'Donovan [18] argue in their critique of Lean that innovation in public services cannot be achieved until "a fundamental change in the mind-set of managers" occurs. Hines [19] raises a similar question his paper Lean: have we got it wrong? concluding that focus on waste cannot lead to sustainable Lean. Moreover, he notes "such a mindset is likely to become an obstacle in its own right". This presents an opportunity to consider how the current PSO paradigm can be addressed in order to unlock the sustainability (economic and social) issue.

Despite mounting questions regarding their efficacy and the way in which they were implemented, CI methodologies have been predominantly applied through a cost-reduction view [9]. On the one hand, PSOs have demonstrated some evidence of learning; regarding the well-established barriers [11] and the importance of leadership and employee engagement. However, Bateman et al.'s [10] recent editorial noted "a strong emphasis on tools" to reduce waste. This is an alarming situation as the same observations were reported by Radnor & Boaden 10 years earlier [20], despite numerous warnings originating back to the early 2000s. A critical point has now been reached where a fundamental review of the approach to CI initiatives is required. Failure to do so will result in re-occurring problems and worsening service delivery in the long term. To do this, urgent research should now be conducted to evaluate and re-assess existing PSO paradigm thinking. This is argued as necessary in order to allow already stretched public services to meet demand and ensure their survival going forward [10, 11, 21].

3   On Organisational Learning

The arguments laid out above point to a lack of learning or adaptation of CI principles since the introduction and promulgation of CI methodologies across public services since the 1990s. Given the body of research which highlights the chequered success of sustaining CI, particularly over the last decade, the consideration of organisational learning theory is presented in this section: specifically, in respect of revealing and replacing the underlying tenets of the PSO paradigm (cost reduction, internal efficiency, short term scope) to enable socially and economically sustainable CI.

Unlocking the PSO paradigm requires addressing deeply entrenched cognitive routines and norms (individual and organisational). Individuals must examine and re-evaluate their own behaviours, and the mental models that govern them. Becker [22] and Fiol & O'Connor [23, 24] might describe this type of "unlearning" as a necessary process in creating new mental models which enable learning to occur. This research draws on the single (SLL) and double (DLL) loop organisational learning theory of Argyris [25, 26] because it offers insights into addressing underlying thinking which drives behaviour and is a central theme of this paper.

3.1   Argyris on Organisational Learning

Argyris' research [25–28] highlights the importance of learning processes in problem solving and decision making. Argyris emphasises the importance of 'mental models' that influence reliable inquiry into organisations and their problems. According to Argyris, learning is achieved by comparing actions taken with "feedback from the environment" which in turn informs subsequent actions. Learning itself is defined as the "detection and correction of errors" [25] such that mismatches between the action taken and the desired outcome are identified. This is typically how organisations solve problems.

A shortfall occurs in most organisations as they solve problems by only correcting errors in the external environment without reflecting inwards [29]. This is defined as single loop learning. Of equal importance, is the need to change the way people "reason about their [individual and collective] behaviour". This is defined as double loop learning. To change behaviours, the cognitive processes used to identify and formulate actions need to be understood, unpacked and evaluated. Figure 1 below illustrates the processes of single and double loop learning.

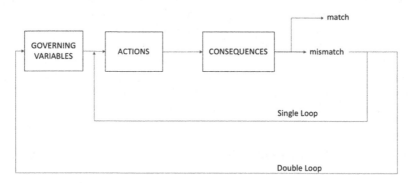

Fig. 1. The process of single and double loop learning

3.2   Tenets of Single Loop Learning (SLL)

Single loop learning occurs when a mismatch or unexpected consequence of an action is identified (first order error), then that action corrected. In this process, external errors are rectified, but the underlying 'governing variables' are not addressed (second order error). SLL relates to Model-I type reasoning, and is based on the following principles [25, 26];

- engage in defensive reasoning with others
- generate superficial single-loop responses which lead to single-loop solutions
- reinforce organisational routines
- inhibit genuine (double loop) organisational learning
- lack of awareness of unintended consequences of the status quo

At an individual level, members of an organisation rarely consider their own behaviours and unconsciously avoid coming to terms with difficult, potentially negative truths, or challenge the status quo. Consequently, this leads to 'maneuvers' [25] by individuals to interpret and suppress the information they give and receive to rationalise it against their theories-in-use. Argyris emphasises that organisational and individual mental models are often taken for granted [29]. Becker [22] and Fiol & O'Connor [23, 24] also emphasise the often-deep emotional and behavioural attachment to existing cognitive processes. Over time the theories-in-use in the organisation become less receptive to corrective feedback [25, 27]. Changing individual and collective thinking, therefore the PSO paradigm, must be recognised as a significant challenge.

3.3 The Importance of Double Loop Learning (DLL)

According to Argyris, "success in the marketplace depends on learning" [29], specifically sustained productive organisational learning [26]. Furthermore, Argyris argues that defensive routines of single loop learning which "preserve the status quo", must be disrupted in order for genuine learning to occur. What DLL provides, unlike SLL, is productive organisational 'inquiry' rather than unreflective corrective action. DLL involves reflection on values and logic in addition to outcomes. It illuminates the dilemmas that are otherwise suppressed and therefore allows genuine learning to occur (and the subsequent re-evaluation of governing variables). DLL learning relates to Model-II type reasoning, and is characterised by three principles [25, 26];

- Valid information- learning is enhanced by valid info
- Free and informed choice
- Internal commitment- including receptiveness for corrective feedback

DLL is an ideal, not an absolute state, because in a dynamic organisational environment the cycle of corrective action in response to valid information is continual. Enabling the principles of DLL fosters an environment where people can identify inconsistencies between espoused theories and theories-in-action (internally and externally), examine them through valid information, are free to take corrective and informed action, and are internally committed. When this reflection occurs, DLL can take place and the driving logic and mental model (governing variables) can be evaluated. This leads to continual organisational learning, and ultimately, sustained CI.

4 The Current State of CI in Public Sector

The principles of this part of the research and the development of the theoretical model are built on a series of in-depth interview with PSO CI professionals. A total of (8) interviews were conducted, lasting between 30 min and 1 h each subject to respondent availability. A total of approximately 10 h of interviews was conducted. The participants have been purposefully selected in view of their specialist knowledge relevant to addressing the research questions. The selected participants were a balance of experts of differing degrees of experience in CI in PSOs with backgrounds in research (4), practice (3) and executive education (1). The transcripts of the interviews were thematically coded according to criteria derived from Argyris' principles of SLL and DLL. Table 1 summarises the topics raised during the interviews relating to SLL behaviours.

Table 1. SLL behavioural themes in the interviews.

Item Raised During Interview	INTA1	INTA2	INTA3	INTA4	INTP1	INTP2	INTP3	INTE1	Total respondents	Total # references to topic
Indicators of SLL										
SLL indicators										
Unquestioning acceptance of status quo	x	x	x		x	x	x		6	8
Defensive reasoning			x			x	x		3	5
Espoused and in-use theories		x			x	x			3	4
Use of data	x				x	x			3	3
Superficial adoption of CI	x				x				2	4
Single loop solutions					x	x			2	4

The findings from this part of the research, although limited due to the small sample size of interview population, adds further validation of the arguments laid out in the literature review. A theoretical model (Fig. 2) was constructed, presenting a causal chain; originating with the external influences of central government, to the PSO paradigm, to its effects at a localised level in relation to SLL and CI in PSOs. The model incorporates the literature and interview findings with Argyris' theory, suggesting that the prevailing mindset in PSOs operate largely within a single loop model, whereby insufficient reflection on governing logic occurs. Therefore, genuine learning and re-evaluation of the principles behind CI interventions does not take place. As a result, CI eventually succumbs to recurring problems and is not sustained.

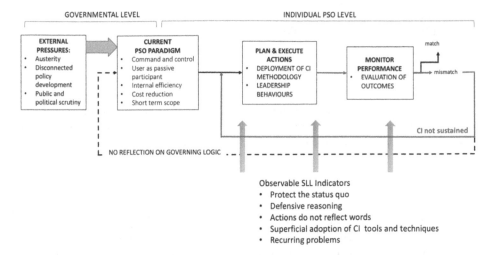

Fig. 2. Current state of organisational learning in PSOs

4.1   DLL as an Undervalued Perspective on CI

This paper has argued that the current PSO paradigm is pervasive in public sector and counterproductive to genuine organisational learning and sustainable CI. Furthermore, until the current paradigm changes and CI methodologies are re-interpreted, PSOs will only continue to achieve the same mixed results with the same recurring criticisms and unsustained service improvement. If CI cannot be sustainably established, it may lose momentum altogether and PSO service delivery will only continue to get worse. Double loop learning could offer a prescriptive solution for sustainable CI, but how this can be achieved however remains largely under-investigated in a PSO-specific context. There remains relatively little discussion or awareness of Argyris's theory in the academic literature or in practice, except a handful of studies of systems-thinking implementations [30].

This coincides with a developing body of knowledge which calls for a new approach to public service management (a new PSO paradigm) built on: co-production of service delivery between the PSO and the user, outside-in thinking, connected policy-operations policy development and adding value to the lives of citizens [21, 22] This research attempts to progress these ideas by offering theoretical and practical support to PSOs and enabling these principles to become realised.

5   Case Study Pilot of SLL and DLL Behaviours

To add further validation to the theoretical body of work, a case study was developed with a UK Health Research sector organisation, beginning March 2019. The case study began through the delivery of an interactive workshop, where a CI leadership network group were introduced to Argyris' theory and then invited to consider statements taken from the interviews in the initial research. Each statement related to an example of SLL or DLL respectively. The delegates were then asked to reflect on their own experiences and identify whether their organisation exhibited SLL and DLL-type behaviours. From there delegates recorded their own examples of SLL and DLL behaviours that they observed over a period of 6–8 weeks, as well as a review of organisational artefacts they encountered (such as processes, policies and procedures). Further analysis was then conducted to ascertain and expose themes in the underlying 'governing variables' in that organisation. An initial collection of examples of SLL and DLL behaviours have been gathered and themed (see Table 2 below). As a preliminary study, only the CI leadership network in the host organisation was included due to time and resource constraints.

5.1   Initial Findings

The analysis in Table 2 below shows evidence of both single loop learning (SLL) and double loop learning (DLL) behaviours in the organisation. The proportion of SLL and DLL behaviours was broadly in line with expectations and was validated further by similar findings from an internal CI maturity assessment in August 2019. This correlation suggests that there is significant value in continuing to take this work forward.

The analysis shows where the pilot organisation is demonstrating positive DLL behaviours; asking new questions and challenging the status quo, with 26% of the examples evidencing this. The number of examples regarding experimentation were relatively low (13%), while 23% of the examples evidenced the presence of deep-rooted organisational routines (SLL). This presents some significant opportunity to embed DLL behaviours (desirable), and address the SLL (undermine sustained CI). There were also examples indicating incongruence between espoused and in-use theories of action (15%). There were some double-loop learning examples of how we experiment and seek new information sources, for example from customers/stakeholders. However, there was also evidence of defensive reasoning (single-loop behaviour).

Table 2: Thematic analysis

Learning theme	Count	
	DLL	SLL
Asking new questions and challenging the status quo	16	
New information sources	7	
Experimentation	8	
Defensive reasoning		5
What we say and what we do are different		9
Deep-rooted organisational routines	1	14
Grand Total	32	28

The outlook of for the continuation of this pilot study is positive and will be supplemented with in-depth analysis of the themes and the implementation of targeted change activities, artefact, process and policy reviews, with a continual focus on leadership and employee engagement.

6 Conclusion

The aim of this research was not to 'reinvent the wheel' by creating entirely new CI frameworks or roles within PSOs. These would require significant up-front investment; generating awareness, training, developing tools etc. Rather, this attempts to re-focus and change emphasis on existing CI activity and methodologies, to recognise the central importance of behavioural and cognitive aspects of CI and organisational learning in PSOs and adapt method accordingly. This research offers a framework for organisations to expose their governing variables. The findings from the case study are limited due to the participant group, and further work should be conducted with an expanded population, including local CI practitioners and front-line staff in order to make broader conclusions.

However, the preliminary findings indicate there is good reason to be optimistic about the future of CI in PSOs. This research offers an undervalued theoretical perspective to consider the implementation of CI in PSOs. By instilling a new PSO paradigm, which puts the needs of the user at the heart of policy development and service delivery, PSOs can respond to the economic challenges now and in the future by designing and delivering sustainable public services. Future research should continue to build on the conceptual propositions of this study through observational testing and application of the illustrative models in different PSO contexts and continue to explore transitional strategies for enabling sustainable CI, as well as evaluation of the sustainability of 'new' DLL behaviours and practices.

References

1. Mawdsley, H., Lewis, D.: Lean and mean: how NPM facilitates the bullying of UK employees with long-term health conditions. Public Money Manag. 37(5), 317–324 (2017)
2. Gupta, S., Sharma, M., Sunder, M.V.: Lean services: a systematic review. Int. J. Product. Perform. Manag. 65(8), 1025–1056 (2016)
3. Samuel, D., Found, P., Williams, S.J.: How did the publication of the book the machine that changed the world change management thinking? Exploring 25 years of lean literature. Int. J. Oper. Prod. Manag. 35(10), 1386–1407 (2015)
4. Tortorella, G.L., Fogliatto, F.S., Anzanello, M., Marodin, G.A., Garcia, M., Reis Esteves, R.: Making the value flow: application of value stream mapping in a Brazilian public healthcare organisation. Total Qual. Manag. Bus. Excell. 28(13), 1544–1558 (2017)
5. Chay, T., Xu, Y., Tiwari, A., Chay, F.: Towards lean transformation: the analysis of lean implementation frameworks. J. Manuf. Technol. Manag. 26(7), 1031–1052 (2015)
6. Eaton, M.: Why change programmes fail. Train. J. 53–57 (2010)
7. Lucey, J., Bateman, N., Hines, P.: Why major lean transitions have not been sustained. Manag. Serv. 49(2), 9–13 (2005)
8. Masters, K.I.A.: From new public management to lean thinking : understanding and managing 'potentially avoidable failure induced demand' (2010)
9. Radnor, Z., Osborne, S.P.: Lean: A failed theory for public services? Public Manag. Rev. 15(2), 265–287 (2013)
10. Bateman, N., Radnor, Z., Glennon, R.: Editorial: the landscape of lean across public services. Public Money Manag. 38(1), 1–4 (2018)

11. Bateman, N., Buxton, R., Radnor, Z.: The status of continuous improvement in ICiPS members in 2015 (2017)
12. Hines, P., et al.: Staying Lean: Thriving, Not Just Surviving. 2nd edn. Productivity Press (2011b)
13. Hines, P., Lethbridge, S.: New development: creating a lean university. Public Money Manag. **28**(1), 53–56 (2008)
14. Radnor, Z., Bucci, G.: Evaluation of the lean programme in HMCS (2010)
15. Burgess, N., Radnor, Z.: Evaluating lean in healthcare. Int. J. Health Care Qual. Assur. **26** (3), 220–235 (2013)
16. Hines, P.: Service design and delivery. In: Macintyre, M., Parry, G., Angelis, J., Hines, P. (eds.) Staying Lean: Thriving, Not Just Surviving. 2nd edn. Springer Productivity Press (2011)
17. Lindsay, C.: Lean in healthcare: an evaluation of lean implementation in NHS lothian. Ph.D. Edinburgh Napier University (2016)
18. Seddon, J., O'Donovan, B.: An exploration into the failure of 'lean' (2015)
19. Hines, P.: The Lean Journey: have we got it wrong? Total Qual. Manag. Bus. Excellence **31**, (2018)
20. Radnor, Z., Boaden, R.: Editorial: lean in public services—panacea or paradox? Public Money Manag. **28**(1), 3–7 (2008)
21. Osborne, S.P.: From public service-dominant logic to public service logic: are public service organizations capable of co-production and value co-creation? Public Manag. Rev. **20**(2), 225–231 (2018)
22. Becker, K.: Organizational unlearning: time to expand our horizons? Learn. Organ. **25**(3), 180–189 (2018)
23. Fiol, C.M., O'Connor, E.J.: Unlearning established organizational routines – part II. Learn. Organ. **24**(2), 82–92 (2017)
24. Fiol, M., O'Connor, E.: Unlearning established organizational routines – part I. Learn. Organ. **24**(1), 13–29 (2017)
25. Argyris, C.: Single-loop and double-loop models in research on decision making. Adm. Sci. Q. **21**, 363–375 (1976)
26. Argyris, C.: On Organizational Learning, 2nd edn. Blackwell, Oxford (1999)
27. Argyris, C.: Good communication that blocks learning. Harvard Bus. Rev. **72**(4), 77–85 (1994)
28. Argyris, C.: Double-loop learning, teaching, and research. Acad. Manag. Learn. Educ. **1**(2), 206–218 (2002)
29. Argyris, C.: Teaching Smart People How to Learn. Harvard Business Press, Boston (1991)
30. Gibson, J., O'Donovan, B.: The vanguard method as applied to the design and management of english and welsh children's services departments. Syst. Pract. Action Res. **27**(1), 39–55 (2013)

Learning Spaces for Engineering Education: An Exploratory Research About the Role of Lean Thinking

Maira Callupe[✉] and Monica Rossi

Department of Management, Economics and Industrial Engineering,
Politecnico di Milano, Via Lambruschini 4/b, 20156 Milan, Italy
maira.callupe@polimi.it

Abstract. In the last decade, learning spaces have moved from the traditional classrooms and laboratories to sophisticated spaces that leverage on emerging technologies to facilitate and enhance active, social, and experiential learning. Engineering institutions around the globe are investing their resources in the creation of this spaces in order to provide students with a holistic training in line with the current demands in the job market. The present work identifies the main learning spaces implemented for engineering education and conducts and exploratory research about the role that Lean Thinking plays in their educational programs. The results suggest a clear distinction between Learning Factories and the group made of Fab Labs, Hackerspaces, and Makerspaces, which can be attributed to differences in governance and technical features. Learning Factories have successfully integrated Lean Thinking into their engineering curriculum, and while there is scarce literature concerning FLs, HSs, and MSs, there are elements in these spaces that can be considered lean enablers that could be exploited to integrate Lean Thinking into their research and educational activities.

Keywords: Learning space · Learning factory · Product development · Lean thinking

1 Introduction

The term "Learning space" refer to the physical setting where learning takes place, providing the context for the development of educational activities, the interaction of involved participants, and their use of available resources for the achievement of educational objectives. In the traditional sense, learning spaces include classrooms, libraries, laboratories, etc.; however, nowadays learning takes place in more diverse locations beyond these spaces. As more institutions are embracing experiential learning, an approach that recreates real-world situations and thinking, learning spaces are being re-conceptualized, with an increasing number of emerging models designed to facilitate active, social, and experiential learning [1]. Furthermore, the use of technology as an enhancer of the experience taking place within a learning space has been identified as conducive to active learning and having a significant impact on student learning [2]. In the particular case of engineering faculty and institutions, a variety of

© IFIP International Federation for Information Processing 2021
Published by Springer Nature Switzerland AG 2021
D. J. Powell et al. (Eds.): ELEC 2021, IFIP AICT 610, pp. 13–20, 2021.
https://doi.org/10.1007/978-3-030-92934-3_2

learning spaces are being established, leveraging heavily on modern technology to recreate spaces that enable hands-on experiential learning specific to this discipline. While individual learning spaces of this type are widely covered in literature, there are few publications that conduct comprehensive analysis of the spaces themselves, especially those with a thematic focus in particular. Given that hands-on experience in a realistic environment has been identified as one of the most suitable ways to interiorize lean concepts [3], this paper attempts to explore these facilities and to uncover the role that Lean Thinking plays in their educational programs, whether as a key component of their educational curricula, a philosophy behind the design of the latter, or others.

2 Research Objective and Methodology

Given the context presented in the previous section, the current work attempts to answer the following questions:

- RQ1. What are the learning spaces implemented in an engineering educational context?
- RQ2. What is the role of Lean Thinking in engineering learning spaces, if any?

Accordingly, the methodology adopted for this study is a systematic literature review consisting of two phases: first, the identification and characterization of the learning spaces relevant for the study, followed by the analysis of said learning spaces in the context of Lean Thinking.

2.1 Learning Spaces Identification and Characterization

The database used for sourcing the literature is Scopus, which was selected due to its rich metadata and relevance in the fields of education and engineering. The literature relevant for the first phase was retrieved using the keywords "learning space" and "engineering", resulting in a total of 382 articles, out of which 91 were excluded on the basis of scope and retrievability. The articles' metadata was mined and examined in search for mentions of learning spaces. Out of the nine learning spaces initially identified, five were excluded on the basis of scope and formalization. Formalization - in the context of this study- refers to the existence of a definition and a set of characteristics associated to the learning space that are well documented in academic literature. The learning spaces that fit these criteria are: i) Learning or Teaching Factories (TFs), ii) Fab Labs (FLs), iii) Hackerspaces (HSs), and iv) Makerspaces (MSs).

2.2 Analysis of the Role of Lean Thinking in Engineering Learning Spaces

A second set of literature was retrieved using the keywords "lean" and "learning", "teaching", or "education" in conjunction with the keywords corresponding to each of the four learning spaces identified. These search strings were composed considering the different known conventions in spelling and variations in the ending of the terms used to refer to these learning spaces. From the resulting 87 articles: 85 correspond to TFs, 2 correspond to MSs, while there were no articles related to FLs or HSs. Due to the low number of publications related to the latter 3 learning spaces, an internet search with the same keywords was conducted using the Google search engine, which resulted in additional 15 scientific articles and 9 websites associated to MSs, FLs, and HSs. All these sources are analyzed qualitatively with the objective of (i) extracting key characteristics of the learning spaces, (ii) identifying instances in which lean practices, methodologies, or tools are referenced, and (iii) extracting insights from observed trends.

3   Learning Spaces in Engineering

As a result of the first phase of the analysis, there were identified four learning spaces that are commonly implemented with the purpose of facilitating activities associated to engineering education. The following paragraphs include a brief summary of their definition and history, followed by their characterization based on features relevant for the study.

TFs originate as an approach to develop tools to recreate problems found in real industrial environments, which are addressed in an academic setting and result in the acquisition of competences. In terms of the physical setting, TFs are replicas -scaled down or actual size- of multiple phases of the value chain with a high degree of realism, grounded on a didactical concept with emphasis on active learning. Although the historical development of TFs goes back to the 80s, they have gained more prominence in recent times as they can be considered the response of academic institutions to the challenges posed by the Fourth Industrial Revolution. In the last decade several TFs have emerged in Europe and gained recognition from academia as well as industry [4]. The International Association of Learning Factories (IALF), which currently counts with 17 members, is the main existing network. On the other hand, the concept of FLs was initially developed in the early 2000's by the Massachusetts Institute of Technology (MIT) as a medium to explore the impact of personal fabrication in locations without access to it. Nowadays, FLs take the form of low-cost workshops equipped with computer power and simple tools for prototyping, facilitating entrepreneurship, research, innovation, and education through the collaboration and exchange of ideas among participants [5]. The Fab Lab Network has approximately 1500 FLs indexed in 90 countries all over the world [6].

In contrast to TFs and FLs, whose origins are closely tied to engineering faculty and institutions, HSs and MSs have emerged in different circumstances, notably outside of the university system. HSs emerged in Germany during the mid-1990 as a social club-like open space for social gathering and project development among computer enthusiasts. The movement was formalized through the publication of a document containing a set of general guidelines for the creation and organization of hackerspaces. Currently, there's a registry of 796 listed hackerspaces all over the world that consider themselves to be part of the hackerspace movement [7]. With a similar origin, the emergence of MSs is intertwined to the 'maker' movement which appeared around 2012 [8]. The movement brings together the DIY spirit and the sharing culture associated to the web and digital tools, initially just for children and later becoming more widespread. This is materialized in MSs with the inclusion of several pieces of equipment for prototyping, primarily 3D printers. As the cost of the technology associated decreased, MSs began to spread to more locations. However, there is no indication of the existence of an organized network at an international level.

Based on the information extracted from literature, there were identified eight features through which the identified learning spaces can be characterized. These features (Table 1) allow for a multi-dimensional understanding of these spaces as they describe the existence of an organized network and/or governance, as well as their environment, purpose, and technical features associated to products and processes. These features are defined as below:

Table 1. Characterization of the learning spaces included in the analysis.

Feature	Learning factory	Fab lab	Hackerspace	Makerspace
Main network	Int'l Assoc. of Learning Factories	Int'l Fab Lab Association The Fab Foundation	Hackerspaces. org	–
Governance	High	Medium	Low	Medium
Environment	Academic Non-academic	Academic	Non-academic	Academic Non-academic
Purpose	Education	Collaboration	Collaboration	Collaboration
	Research	Education	Education	Education
	Training	Entrepreneurship	Entrepreneurship	Entrepreneurship
		Innovation	Innovation	Innovation
		Research		
Entry barrier	High	Medium/Low	Medium/Low	Medium/Low
Product lifecycle	Full life cycle	Product development	Product development	Product development
Processes	Authentic Simulated	Authentic	Authentic	Authentic
Products	Selected for TF	Unlimited	Unlimited	Unlimited

- Main network. Existence of an organized and collaborative structure to which individual learning spaces can subscribe.
- Governance. Degree to which an underlying system controls the network and its operations.
- Environment. Setting in which the learning space is established.
- Purpose. Main purposes behind the establishment of the learning space.
- Entry barrier. Degree to which certain factors such as initial investment, required expertise or experience can prevent newcomers from establishing a learning space.

- Product life cycle. Phases of the life cycle reproduced at the learning space.
- Processes. Nature of the processes reproduced at the learning space.
- Products. Type of product that can be produced, manufactured, or assembled at the learning space.

Concerning the presence and/or role of Lean Thinking within these four learning spaces, two main trends were observed: i) Lean Thinking addressed as a learning topic, or ii) Lean Thinking implemented as an educational methodology. These trends are summarized into two categories called "Education for Lean" and "Lean for Education".

1. Education for Lean. Several examples of TFs showcase production lines dedicated to the teaching of the use and implementation of lean management tools and practices [9–11]. Lean management is often addressed with a holistic approach and has been one of the main topics in the educational and research agenda of learning factories in the last 10 years [12]. Furthermore, recently TFs have started to address the topic of Lean in conjunction with Industry 4.0 in terms of the competences required from the workforce [13]. A common approach observed in several TFs is the implementation of lean practices in a production line followed by the demonstration of the performance enhancements brought by Industry 4.0 [14].
2. Lean for Education. TFs and MSs make use of Lean concepts to create learning process methodologies such as problem-pull, theory-push, and reflection-first [15] or the implementation of the Lean Launchpad methodology for the development of the engineering curriculum [16].

4   Discussion

The analysis shows a clear difference between the presence and role of Lean Thinking between TFs and the group of learning spaces made of FLs, HSs and MS. This difference can be attributed to two factors linked to the features of these spaces: the level of governance and the technical features.

4.1   Level of Governance

The results show that there are more than a thousand FLs, HSs and MSs all over the world, and that the structure of the associated networks suggest a low level governance. While there are no exact figures about the number of MSs, literature suggests estimates of hundreds of such facilities that are organized in small regional clusters. In the case of FLs and HS, there is no centralized governance model that exerts control over the specific purposes, processes, techniques, methodologies, etc. implemented by each learning space node in the network. Furthermore, there are lists of technical equipment that is required for FLs or recommended for MSs in order for facilities to call themselves as such, but there are no further requirements or expectations in terms of their operations and activities. Nonetheless, TFs show a different type of network and governance. While the IALF counts with less than 20 TFs as members, literature suggests that currently there are around 100 TFs operating all over the world. While the

IALF does not set the research and educational goals for TFs, considering the low number of nodes within the network and the existence of an established cluster of TFs at its core, it can be observed from literature that the research activities of the TFs in the main network influence the activities of those within, in the form of internal collaborations, and those outside in the form of external collaborations. This difference in levels of governance is crucial to understand the degree to which the learning spaces included in the analysis are able to unlock and exploit the collaborative potential associated to organized networks.

TFs are learning spaces that address Lean Thinking as a core component of their research and educational agenda, making use of their production lines to showcase the implementation of lean tools and practices. The outcome of their activities is actively shared among the network community through an exchange of best practices, methodologies, potential research lines, etc. that collectively advances forward the body of research surrounding the topic and enhances the quality of the educational content delivered as well as the effectiveness of the learning experience. In contrast, the activities conducted at the large majority of FLs, HSs, and MSs are highly decentralized, with every individual space operating within its own confines or, at most, cooperating with a few other spaces. Furthermore, the educational activities imparted at these learning spaces often include introductory lessons about equipment use and safety measures, giving learners a high degree of autonomy to develop their own projects, hence the emphasis on innovation and entrepreneurship. While both formal and informal learning take place in these spaces, this configuration does not facilitate a smooth knowledge exchange and development such as the one taking place for TFs. Therefore, the lack of literature discussing in depth the implementation of Lean Thinking in these spaces might not necessarily be indicative of it absence but, instead, the absence of an organizational structure that enables its documentation and dissemination.

4.2 Technical Features

Concerning the technical features, the learning spaces included in the analysis show a clear divide between the recreated product lifecycle, with TFs focusing mostly on production, while FLs, HSs, and MSs are oriented towards product development. The TFs identified in the literature showcase a production line -simulated or authentic- built to manufacture or assemble a product in mind; therefore, the Lean Thinking curriculum includes the utilization of various known methods of the lean toolbox such as 5S, VSM, JIT, Kanban, supermarket, among others for the improvement of production processes. On the other hand, the facilities of FLs, HSs, and MSs are furnished with various equipment ranging from small 3D printers to large industrial machining centers such as precision measurement and laser cutting machines used for rapid prototyping. In some cases, these spaces operate in cooperation with incubators that nurture the development of startups. While there is abundant literature about the implementation of Lean Thinking in production -even outside the scope of this work- those addressing stages preceding production are less common. There are known methods and tools used in lean product development such as set-based engineering and rapid learning cycles; however, this research found no implementation of such methods at FLs, HSs, or MSs.

This absence could be attributed to the higher emphasis placed on production or also to the low governance of these spaces, although it could also be indicative of the lack of an structured and effective learning approach.

5   Conclusions and Future Works

The present work conducted an exploratory research about the role played by Lean Thinking in TFs, FLs, HSs, and MSs. The research shows that Lean Thinking has been successfully integrated into the engineering curriculum by a number of higher educational institutions through the implementation of TFs. TFs address Lean Thinking as a core component of their research and educational agenda, making use of their production lines to showcase the implementation of lean tools and practices. On the other hand, while there is scarce literature addressing Lean Thinking in FLs, HSs, and MSs, there are elements in these spaces that can be considered as lean enablers that could potentially be exploited in order to integrate lean methods and practices into their research and educational activities. For instance, the Lean Startup methodology could be adopted as a structural foundation for their activities such as rapid prototyping and entrepreneurship development. Therefore, while TFs offer an optimal setting for the teaching of Lean Manufacturing, FLs, HSs, and MSs have the potential to be used as settings for the teaching of Lean Product Development and the Lean Startup Methodology.

This exploratory research identified and characterized four main learning spaces implemented for engineering education. The results obtained from the analysis might be limited by the choice of a single scientific search engine -complemented by the use of an internet search engine- and the inclusion of works published in English language. Future studies to expand on this work, therefore, could include the obtention of a database that includes publications not indexed in Scopus and a wider grey literature such as theses or magazines in order to identify a larger sample of learning spaces, especially to account for the activities conducted by FLs, HSs, and MSs. Additionally, future research could attempt to study more in depth the reasons behind the emphasis placed by some learning spaces such as TFs on the production stages, and the pedagogic approaches for the teaching of Product Development within a wider scope of learning spaces. In the particular case of learning spaces for engineering education, to understand how to leverage on their technology, practices, environment, or any other feature to facilitate the teaching of Lean Product Development.

References

1. Oblinger, D.: Learning Spaces | EDUCAUSE e-book (2006)
2. Brooks, D.C.: Space matters: The impact of formal learning environments on student learning. Br. J. Educ. Technol. **42**(5), 719–726 (2011). https://doi.org/10.1111/j.1467-8535. 2010.01098.x

3. De Zan, G., De Toni, A.F., Fornasier, A., Battistella, C.: A methodology for the assessment of experiential learning lean: the lean experience factory case study. Eur. J. Train. Dev. **39**(4) (2015). https://doi.org/10.1108/EJTD-05-2014-0040

4. Callupe, M., Negri, E., Fumagalli, L.: An inclusive overview of learning factories around the globe. SSRN Electron. J. (2021). https://doi.org/10.2139/ssrn.3863491

5. Mikhak, B., Lyon, C., Gorton, T., Gershenfeld, N., Mcennis, C., Taylor, J.: Fab lab: an alternate model of ICT for development. Dev. Des. (2002)

6. The Fab Foundation: The Fab Foundation - Global Community. Fab Lab Network (2021). https://fabfoundation.org/global-community/

7. Hackerspace Wiki: Hackerspaces. List of Hackerspaces (2020). https://wiki.hackerspaces.org/List_of_Hacker_Spaces

8. Jensen, M.B., Semb, C.C.S., Vindal, S., Steinert, M.: State of the art of makerspaces - success criteria when designing makerspaces for norwegian industrial companies. Proc. CIRP **54**, 65–70 (2016). https://doi.org/10.1016/j.procir.2016.05.069

9. Rybski, C., Jochem, R.: Benefits of a learning factory in the context of lean management for the pharmaceutical industry. Proc. CIRP **54**, 31–34 (2016). https://doi.org/10.1016/j.procir.2016.05.106

10. Maheso, N., Mpofu, K., Ramatsetse, B.: A learning factory concept for skills enhancement in rail car manufacturing industries. In: Proc. Manuf. **31**, 187–193 (2019). https://doi.org/10.1016/j.promfg.2019.03.030. Research Conference on Learning Factories 2019 (CLF 2019), Braunschweig, Germany

11. Abele, E., et al.: Learning factories for research, education, and training. Proc. CIRP **32**, 1–6 (2015). https://doi.org/10.1016/j.procir.2015.02.187

12. Kreimeier, D., Morlock, F., Prinz, C., Krückhans, B., Bakir, D.C.: Holistic learning factories - a concept to train lean management, resource efficiency as well as management and organization improvement skills. Proc. CIRP **17**, 184–188 (2014). https://doi.org/10.1016/j.procir.2014.01.040

13. Enke, J., Glass, R., Kreß, A., Hambach, J., Tisch, M., Metternich, J.: Industrie 4.0 – competencies for a modern production system: a curriculum for learning factories. Proc. Manuf. **23**, 267–272 (2018). https://doi.org/10.1016/j.promfg.2018.04.028. Advanced Eng. Educ. Train. Manuf. Innov. CIRP Spons. Conf. Learn. Factories (CLF 2018)

14. Küsters, D., Praß, N., Gloy, Y.-S.: Textile learning factory 4.0 – preparing germany's textile industry for the digital future. Proc. Manuf. **9**, 214–221 (2017). https://doi.org/10.1016/j.promfg.2017.04.035. 7th Conf. Learn. Factories, CLF 2017

15. Tisch, M., Hertle, C., Abele, E., Metternich, J., Tenberg, R.: Learning factory design: a competency-oriented approach integrating three design levels. Int. J. Comput. Integr. Manuf. **29**(12), 1355–1375 (2016). https://doi.org/10.1080/0951192X.2015.1033017

16. Huang-Saad, A., Morton, C., Libarkin, J.: Unpacking the impact of engineering entrepreneurship education that leverages the lean launchpad curriculum. In: Proceedings - Frontiers in Education Conference, FIE, vol. 2016-November (2016). https://doi.org/10.1109/FIE.2016.7757373

The Learning Way to EBITDA Improvement

Torbjørn Hekneby[1]([✉]) and Daryl John Powell[2]

[1] School of Business and Law, University of Agder, Kristiansand, Norway
torbjorn@storform.no
[2] Norwegian University of Science and Technology, Trondheim, Norway

Abstract. Multi-national firms pursue enhanced marked positioning by production performance, profit realization and cost reduction. As such, a key strategy is to apply standardized management concepts such as lean and Toyota Production System, or more recently, to develop a Company-Specific production System (XPS). However, a fundamental challenge is how to document the financial impact of such programs. The promise of cost reduction is frequently discussed, often hidden behind other organizational initiatives, such as downsizing and restructuring. This study investigates a Norwegian multi-national in Process industry, producing silicon to the global market. The company has developed, implemented, and institutionalized its own XPS since 1991 and claims that this has directly contributed to extensive cost reduction and significantly strengthened competitive position. The company claims a cost-reduction of 5–7% of total production cost, year-on-year since 2013. This is supposedly the result of planned change activities related to the XPS implementation. We challenge this claim, assuming that such a significant and sustained cost reduction must be explained by other variables. Our findings, however, indicate that the XPS first created institutionalized learning and secondly that this 'learning capability' managed to link continuous improvement work directly to improving the cost level of the organization. Our data were controlled against downsizing, marked change, exchange rates, new investments, new technology and other contingency factors. The findings have implications for how firms might pursue business improvement. By using an XPS as catalyst for organizational learning, continuous improvement work might be linked more directly to financial performance for the company.

Keywords: Lean · Company specific production system (XPS) · Financial results · Organizational learning

1 Introduction

Multi-national corporations (MNCs) are constantly seeking enhanced efficiency and business performance through technological and organizational development. This "race for efficiency" has been a driving force for MNCs since the birth of industrial capitalism, with different best-practice organizational concepts [1, 2] becoming roadmaps for organizational implementation. Since then, lean has been presented as a "superior" global management concept [3] and is one of the most popular organizational paradigms of our time - that builds on the knowledge of the Toyota Production

© IFIP International Federation for Information Processing 2021
Published by Springer Nature Switzerland AG 2021
D. J. Powell et al. (Eds.): ELEC 2021, IFIP AICT 610, pp. 21–31, 2021.
https://doi.org/10.1007/978-3-030-92934-3_3

System (TPS). Since the late 1980s, several companies from different industries and organizations have made significant attempts to implement TPS and lean as models for best practice, aiming for production performance and enhanced financial outcome. However, it turned out that copying TPS and lean was more challenging than initially expected [4]. Facing different contextual environments, many MNCs argue that the lean statement of "universal applicability anywhere by anyone" [3] did not fulfill its promise. A significant number of studies in the last 30 years have documented the gap between the promised universality and practical reality [5–8].

Popular organizational concepts such as lean have two key characteristics: room for interpretation, and the promise of performance improvement [9, 10]. An organizational concept is usually presented with a set of principles, methods, and tools. These principles are easy to understand, but also ambiguous and imprecise, which allows for different interpretations. Such "interpretive viability" [11] makes it possible for different consumers (e.g., managers, consultants) to adapt the concept to different local conditions in their own organizations [11]. This in turn gives the concept more applicability and increases the field of distribution, because consumers use the elements that are most beneficial to their own interests. A prerequisite for popularity is thus the concept of "ambiguity", and it is this possibility of providing one's "own interpretation" that defines the "interpretative space" [11, 12].

Arguably, lean is a good example of a concept with high ambiguity. As suggested by Womack and Jones (1996), lean can be defined according to five main principles:

1. Define customer value
2. Identify the value stream(s) for each customer
3. Create flow
4. Secure pull
5. Pursue perfection through continuous improvement (*kaizen*).

With such a level of abstraction, lean has significant scope for interpretation, which, in turn, allows consumers, whether intentionally or not, to choose components that they find appropriate for their own context. This explains the long and ongoing discussion about lean and the content of the concept [13]. lean might be viewed thorough different 'lenses' [14] i.e., as a system [15], as a philosophy [16], as a set of tools and practices [17], as a 'soft lean' version [18], as a start-up program for new businesses [19], as a management concept [20], as an organizational learning system [21], or as a concept for cost cutting termed "hard lean" [22]. Hence, without a common understanding of the content, any attempt to measure the success of implementation and expected financial outcome relies on how the concept is interpreted. Consequently, the discussion of successful implementation is based on how the concept is defined. This explains how and why lean often is used to camouflage extensive downsizing in an organization, directly affecting the financial outcome regardless of enhanced production performance [23].

In response to the challenge of implementing 'best practice' concepts, MNCs began to develop new strategies to secure business performance in their network, with a focus on adjusting and tailoring lean (and other concepts) to fit the company's uniqueness [24]. This "own-best-way" approach to the "one-best-way" phenomenon suggests that companies should adjust and tailor the principles and concept of lean to their contextual environment. Such adjustment is supposed to be made at the corporate level, with the new concept being implemented in the corporate network to ensure standardization and homogeneity among its subsidiaries. This company-specific production system (XPS) carries the company name, where the "X" represents the name of the company [25].

An XPS is also portrayed as a "multi-plant improvement program" [26] and a "corporate lean program" [27, 28]. The main resemblance is to the "own-best-way" approach, according to which the corporate level uses different concepts to create their own improvement program. This distinguishes it from other (global) lean programs by referring to a coordinated initiative at the corporate level whereby a tailored program is created and implemented among the company's subsidiaries.

The XPS phenomenon seems to be a growing trend among MNCs [24]. Hence, adjusting organizational concepts at a corporate level with the aim of network standardization implies a significant strategic initiative followed by a vital change process for an MNC [25]. Consequently, the phenomenon has academic and practical relevance. Despite the growing interest and strategic impact, research on XPS is very limited. In particular, the XPS phenomenon lacks empirical documentation on causal relations between XPS implementation and financial effects. Except for the work of Netland [24–26] and Ostermann [15, 29], and studies on corporate lean programs [21, 28], we have discovered no other empirical material describing either the creation and implementation process of an XPS or direct financial outcome of such implementation. This claim is based on an extensive search of the scientific literature. Thus, there is a need for more empirical data to examine the effect on production performance and financial outcomes when implementing an XPS in an MNC.

2 Research Design

This article built on the empirical material taken from the first author's recently completed doctoral project at agder business school and post doctor data collection spring 2021. The first author examined the creation, implementation, and institutionalization of an XPC in a Norwegian MNC within process industry (from now called *Norwegian Chemical Company NCC)* [30, 31]. Data were obtained in the period 2017 to 2020. (See Table 1 for an overview of data material).

In this previous work, it was discovered that one division of NCC claimed to have developed the XPS to secure direct financial impact in its plants. Based on the knowledge of lean implementation and scarce XPS documentation, we set out to challenge this claim, hypothesizing that direct cost reduction must be explained by other variables than simply the implementation of an XPS. We collected data from one of the plants in the NCC division (from now called NCC Plant). This enabled a precise examination of possible financial outcomes and contextual variables regardless of variation within the NCC division's other plants (See Table 1). Data was collected based on interviews and archival data from the plant. The plant was also visited during the data collection. Data was then sorted and analyzed using the reflexive methodology, emphasizing careful interpretation, challenging the reflection of data [32].

Table 1. Data material Norwegian Chemical Company (NCC)

Plant visits	Observations	Interviews	Archival data
Four NCC plants, Norway	One week observation shop floor level	Top managers 11 interviews	XPS written material
NCC plant, Brazil	One week observation shop floor level	Managers 32 interviews	Assessment written material
NCC plant, China	One week observation shop floor level	Shopfloor interviews 30 interviews	NCC performance data
NCC plant, Norway	One week assessment program observation One week NCC university observation	Other 12 interviews with former CEO/managers 8 interviews with managers NCC plant, Norway	

3 Findings

Our findings indicate that NCC Plant had implemented the XPS to an extent that continuous improvement had been institutionalized [33] with extensive operator involvement [34, 35]. In the plant, 80 out of 105 of the operators had voluntarily participated in 'critical process groups', working systematically to reduce the furnaces' instability [34]. We also found that the operators reported a strong motivation for this job, stating that the continuous improvement work was one of the most important activities for improving performance at the plant.

> "Because... you know, if the furnace is good, my job is good. The group work has been very important for my daily work". (Furnace operator NCC Plant).

We found the XPS was initiated and created from corporate level in 1994 to 2006 [36] and that the content of the XPS implied two important organizational choices: first, to show 'respect for people' and involve everyone in continuous problem-solving, and, second, to decentralize decision-making in autonomous work teams and remove the position of the team supervisor. Extensive empowerment should then lead to better problem-solving and higher employee motivation. In the words of the CEO of the NCC:

> "I had to understand the [people] dimension and how strong it is. The enormous energy you can release through the organization when people are properly trained and are made responsible... and your decisions are decentralized".

Since the beginning of the XPS creation process, the NCC started to implement and institutionalize the XPS in its global network [36]. Different 'best practice' concepts from different organizational traditions (i.e., lean, TPS, Socio-technical System theory) influenced the creation and later development of the XPS but the core has remained from its initial consolidation in 1999. A dual emphasis was placed on improving both technology and human resources, as illustrated in the 'the double integrated value chain' (see Fig. 1). Technological and human development should be equally emphasized, as specified in the statement: to create world-class production, we need world-class operators.

Fig. 1. The double integrated value chain (from NCC company presentation)

In developing the XPS, we found that one important concept was implemented early in the process: Critical Process Management (CPM). Metallurgical upstream processes have an extensive number of variables that influence the output. To ensure stable production, these variables must first be defined, then constantly measured, and monitored. Central to this is the organization's ability to ensure that the variables are identical every time (to stabilize them) and then to develop and improve the process (to make it capable). In 2001, the NCC started to implement CPM more extensively. At that time, the NCC Plant was struggling to get the furnaces under control. Division managers then started hiring specialists to further develop the knowledge within the division and in the NCC Plant. Finally, in 2006, CPM were blended into the NCC's XPS.

We found that CPM knowledge was to have a decisive impact on the NCC's production [36] and major consequences both for the furnaces and for production performance [34]. The quality of the production increased considerably, and fewer resources were required, but, perhaps most importantly, the operators and engineers experienced the furnaces becoming more stable and less unpredictable. This directly affected working conditions on the shifts, which gradually became calmer and more controlled. We found that this 'learning from direct experience' [37] contributed to the institutionalization of continuous improvement, later to be important for further improvements of plant performance.

We found that a new concept was in 2013 rolled out under the XPS umbrella. Cost Road Map (CRM) was launched as a strategic program in the overall XPS, aimed to secure not only production performance, but also the best possible cost position among its competitors. The CRM was directly designed with purpose of connecting the organisation and the improvement work to the financials. Defined KPI's had to document causality between the action taken and its result and lead to a direct improvement of 'Fully Absorbed Cost' (FAC) which is the company's total production unit cost, including all costing such as raw material, labor, capital cost and credits, FAC was measured against the production of tonnage produced product. All improvement had to be demonstrated in FAC reduction, and the priority of projects was crucial. As an example, time efficiency not related to direct cost reduction (i.e., 1 min saved per employee per day) was not defined as CRM initiatives, whereas increased load, yield improvements, reduced power or raw material consumption represented examples of defined CRM projects. The CRM improvement was measured against a predefined plant specific baseline established in cooperation with the plants finance team. Such a predefined baseline was anchored in the actual cost position from last year's fourth quarter, aiming to secure a precise measure of actual cost reduction for the plant. Target was to reduce actual cost every year by 7% measured against baseline from previous year.

Anchored in the overall XPS philosophy and the 'double integrated value chain' (see Fig. 1), we found that operators, employees, engineers and managers in the different departments were regularly involved in 'brainstorm'-sessions, initiating possible CRM projects for the existing and next year's project portfolio. Incoming suggestions were reviewed by top management and prioritized. Then 'defined projects' was distributed back to the operators and departments for execution. An important and extensive CRM standard for the NCC division was established. The responsibility for the CRM development was not outsourced to the financial department. It was a direct responsibility for the top manager and his chain of command, only to be supported from the financial department and the plants' help chains.

In the NCC Plant, we observed several projects related directly to the CRM program. Operators, managers, technical staff were participating, using typical lean production tools for improvement (i.e., A3, visual mapping, CPM, 5S, VSA etc.). To secure that the projects initiated for the CRM target had sustainability, the financial team did not report project successful until 3-month sustainability of improvement on financial report had been established. The projects initiated for CRM was spread over all parts of the plant's value stream and involved the whole organization from raw material to final product deliverance.

We found that the XPS combined with the new integrated CRM program have had significant effect on financial outcome in the NCC Plant. Since 2013 the plant has managed to reduce its cost position every year, measured in Euro cost pr tonnage produced silicon pr year. FAC has been reduced real term with approximately 7% every year, according to target (see Fig. 2).

Fig. 2. Cost reduction NCC plant 2013–2021

Some fluctuations occur in the period, also described as 'bad news and good news' among plant management (see Fig. 3). This relates to external factors like raw material price, currency etc. I. e. in the period 2014 to 2018 'good news' resulted in nominal cost development beyond target of 7%, whereas in 2019 and 2020 'bad news' due to e.g. increased price on raw materials resulted in nominal cost development below target. Importantly, we found that this fluctuation was not related to the actual improvement work in the NCC Plant. Fluctuation was founded in external factors influencing the cost position regardless of actual underlying improvement.

Fig. 3. Fluctuations in cost reduction NCC plant 2013–2021

We also tried to find other explanations for the financial performance in the plant, not related to the XPS. We found that the numbers of employees were stable in the period, indicating no downsizing or change in numbers of operators. Management were also stable in the period, indicating no major change in leadership or management of the plant.

We found no significant investments in new technology that alone had resulted in increases in production volume, process knowledge or the like, and thus significantly reduced the cost position. For example, we found no major investments in technology for expansions and increased capacity on plant furnaces. We only found that the NCC Plant had invested in equipment aimed for enhanced flexibility and increased process knowledge (i.e., constantly new measuring points), related to daily operation and continuous improvement efforts.

4 Discussion

Two main topics emerge from our findings. (1) The topic of sustainability and how the institutionalization of learning might generate financial results in an organization [33] and (2) the use of 'best practice' concepts (i.e., lean) as catalyst for organizational learning [28].

Within conceptual organizational theory, institutionalization involves the long-term persistence of change, indicating the sustainability of new practices in an organization [33]. In this stage of the change process, new practices and concepts become shared norms, values, and knowledge, and a normative consensus has been reached [38]. There is a distinct notion that continuous improvement brings evidence of lean being institutionalized in an organization [8, 18, 28, 39–42]. Continuous improvement has its roots in the evolution of TPS and the formalization of improvement work during the 1950s and 1960s. Institutionalized practice became one of the main experiences when Toyota and General Motors (GM) established their joint venture, New United Motor Manufacturing Inc. (NUMMI) starting to distribute TPS to the Western world [43, 44]. In the NUMMI project, continuous improvement was claimed to be institutionalized among the operators in GE's Freemont plant, constantly developing performance in the production line [44]. This was later re-established and documented by Fujimoto (1999) who explained the logic behind Toyota's "manufacturing learning capability" and how this institutionalized learning explained the success of the company. Womack and Jones (1996) further emphasized this in the fifth principle of lean: conducting "continuous kaizen" by improving the standards in the flow [45]. Many companies deciding to implement lean have, therefore, considered lean means to establish continuous improvement and create a learning organization.

A significant number of studies have reported problems with institutionalizing continuous improvement in an organization [7, 46]. Studies have shown that organizations tend to establish short-term projects with external or internal consultants responsible for the implementation process [6, 18]. Such 'outsourcing' of the implementation process often results in short-term effects, leaving the organization with no shared assumptions or consensus about the deeper principles of the concept [7, 46, 47]. In the case of the NCC, institutionalization of continuous improvement began as early as 1999 when the XPS was consolidated [36]. Significant resources were allocated to secure the long-term sustainability of the XPS. By creating a global university, the corporate XPS department aimed to coordinate all improvement activities, assessment programs, cross country learning initiatives, leadership training, XPS coordination, global reporting system etc. As such, the NCC's global culture was systematically developed into a learning organization, made possible by constant re-examination of basic assumptions [48] in the global network [34, 35]. The NCC 'long time initiative' from 1999 explains the institutionalized learning capability in the organization, representing the core of the XPS [35].

Secondly, the NCC did not use 'best practice' concepts as standard for the creation and implementation of their XPS. The NCC used lean and other management concepts (i.e., Socio-Technical System Theory) [49] as catalysts for organizational learning [36]. Resembling the original concept of TPS, the NCC experimented with different concepts, and extracted the learning into its own XPS. This process lasted for 15 years [36]. Then the XPS was distributed and institutionalized, constantly being developed by experimenting with new concepts. In 2006, CPM was formally consolidated into the XPS [35]. And, as we have seen, the NCC started also to deploy CRM as the direct cost improvement initiative in 2013. The constant development and integration of new concepts demonstrates the NCC's learning capability and might explain the success of connecting organizational learning directly to its financial outcomes. Financial results have traditionally been hard to demonstrate due to lean implementation. For example, [5] refers to the Wall Street analyst Cliff Ransom estimating that only 1–2% of firms that implement lean do so effectively enough to see the results financially [5]. In the case of the NCC, financial results have been linked directly to its organizational learning initiative, securing sustainable cost reduction and EBITDA improvement steadily since 2013.

5 Conclusion and Implications

We found that NCC managed to link organizational learning directly to the improvement of financial results in their organization. The XPS created institutionalized learning in the organization and this 'learning capability' managed to link continuous improvement work directly to improving the cost level of the value chain. The XPS was developed resembling the creation of TPS, where different concepts and ideas were used for experimentation, constantly extracting new knowledge to the overall XPS. The evolving XPS concept, with its associated departments, values, and practices, served as a repository for the organization's accumulated experience. By codifying the lessons learned, the XPS practices functioned analogously to how standard operational procedures (SOP) should function in a learning shop-floor environment [42, 50].

What are the implications of this study for managers wanting to link improvement directly to financial results? First, they should appreciate that institutionalized learning requires significant time, attention, support, and dedication. The duration of the XPS development in the NCC implies that there must be consistency in top-management support. This may be easily endangered when there are changes in top-management positions. Second, top managers should stimulate the organization to pick up new ideas and actively build a network for external learning. Third, top managers should allow the organization to experiment with different concepts before the final content of the XPS is consolidated. Finally, top managers need to realize the importance of allocating resources for institutionalization of the XPS. The creation process is an opportunity for building shared norms in the organization, later to be used directly for cost reduction and financial improvement.

Acknowledgements. The authors would like to give a special thanks to contributors in the NCC team for providing data and important knowledge to this paper.

References

1. Bodrožić, Z., Adler, P.S.: The evolution of management models: a neo-schumpeterian theory. Adm. Sci. Q. **63**(1), 85–129 (2018)
2. Sturdy, A., et al.: The Oxford Handbook of Management Ideas. Oxford University Press, Oxford (2019)
3. Womack, J.P., Jones, D.T., Roos, D.: The Machine that Changed the World. Free Press (1990)
4. Hines, P., Taylor, D., Walsh, A.: The Lean journey: have we got it wrong? Total Qual. Manag. Bus. Excell. **31**(3–4), 389–406 (2020)
5. Hopp, W.J.: Positive lean: merging the science of efficiency with the psychology of work. Int. J. Prod. Res. **56**(1–2), 398–413 (2018)
6. Holmemo, M.D.-Q., Ingvaldsen, J.A.: Bypassing the dinosaurs?–How middle managers become the missing link in lean implementation. Total Qual. Manag. Bus. Excell. **27**(11–12), 1332–1345 (2016)
7. McLean, R.S., Antony, J., Dahlgaard, J.J.: Failure of continuous improvement initiatives in manufacturing environments: a systematic review of the evidence. Total Qual. Manag. Bus. Excell. **28**(3–4), 219–237 (2017)
8. Bhasin, S.: An appropriate change strategy for lean success. Manag. Decis. **50**(3), 439–458 (2012)
9. Benders, J., Van Grinsven, M., Ingvaldsen, J.: The persistence of management ideas. In: The Oxford Handbook of Management Ideas, pp. 270–285. Oxford University Press, Oxford (2019)
10. Røvik, K.A.: Trender og translasjoner: ideer som former det 21. århundrets organisasjon. Universitetsforl (2007)
11. Benders, J., Van Veen, K.: What's in a fashion? Interpretative viability and management fashions. Organization **8**(1), 33–53 (2001)
12. Hines, P., Holweg, M., Rich, N.: Learning to evolve: a review of contemporary lean thinking. Int. J. Oper. Prod. Manag. **24**(10), 994–1011 (2004)
13. Osterman, C.: Defining gaps in Lean: increasing the ability to solve problems in a production system. Mälardalen University (2020)
14. Hopp, W.J., Spearman, M.S.: The lenses of lean: visioning the science and practice of efficiency. J. Oper. Manag. **67**, 610–626 (2021). https://doi.org/10.1002/joom.1115
15. Osterman, C., Fundin, A.: A systems theory for lean describing natural connections in an XPS. TQM J. **32**(6) (2020)
16. Shah, R., Ward, P.T.: Defining and developing measures of lean production. J. Oper. Manag. **25**(4), 785–805 (2007)
17. Dennis, P.: Lean Production Simplified: A Plain-Language Guide to the World's Most Powerful Production System. CRC Press (2017)
18. Holmemo, M.D.-Q., Rolfsen, M., Ingvaldsen, J.A.: Lean thinking: outside-in, bottom-up? The paradox of contemporary soft lean and consultant-driven lean implementation. Total Qual. Manag. Bus. Excell. **29**(1–2), 148–160 (2018)
19. Reis, E.: The Lean Startup, p. 27. Crown Business, New York (2011)
20. Liker, J.K.: The toyota way. Esensi (2005)
21. Powell, D.J., Coughlan, P.: Rethinking lean supplier development as a learning system. Int. J. Oper. Prod. Manag. **40**(7–8) (2020)
22. Holmemo, M.D.-Q.: Lean implementation in Norwegian public service sector (2017)
23. Kinnie, N., Hutchinson, S., Purcell, J.: Downsizing: is it always lean and mean? Pers. Rev. **27**(4) (1998)

24. Netland, T.: Exploring the phenomenon of company-specific production systems: one-best-way or own-best-way? Int. J. Prod. Res. **51**(4), 1084–1097 (2013)
25. Netland, T.: Coordinating production improvement in international production networks: what's new? In: Johansen, J., Farooq, S., Cheng, Y. (eds.) International Operations Networks, pp. 119–132. Springer, London (2014). https://doi.org/10.1007/978-1-4471-5646-8_8
26. Netland, T.H., Aspelund, A.: Multi-plant improvement programmes: a literature review and research agenda. Int. J. Oper. Prod. Manag. **34**(3), 390–418 (2014)
27. Netland, T.: Critical success factors for implementing lean production: the effect of contingencies. Int. J. Prod. Res. **54**(8), 2433–2448 (2016)
28. Powell, D., Coughlan, P.: Corporate Lean programs: practical insights and implications for learning and continuous improvement. Proc. CIRP **93**, 820–825 (2020)
29. Osterman, C., Fundin, A.: Understanding company specific Lean production systems. Is Lean getting lost in translation? In: 25th Annual EurOMA Conference EurOMA, 24 June 2018, Budapest, Hungary (2018)
30. Eisenhardt, K.M., Graebner, M.E.: Theory building from cases: opportunities and challenges. Acad. Manag. J. **50**(1), 25–32 (2007)
31. Langley, A.: Strategies for theorizing from process data. Acad. Manag. Rev. **24**(4), 691–710 (1999)
32. Alvesson, M., Skjøldberg, K.: Reflexive Methodology, New Vistas for Qualitative Research. Sage, London (2018)
33. Buchanan, D., et al.: No going back: a review of the literature on sustaining organizational change. Int. J. Manag. Rev. **7**(3), 189–205 (2005)
34. Hekneby, T., Benders, J., Ingvaldsen, J.A.: Not so different altogether: putting Lean and sociotechnical design into practice in a process industry (2021)
35. Hekneby, T., Ingvaldsen, J.A., Benders, J.: Managing adoption by cultural development: exploring the plant level effect of a 'Company Specific Production System'(XPS) in a Norwegian multinational company. J. Ind. Eng. Manag. **13**(2), 402–416 (2020)
36. Hekneby, T., Ingvaldsen, J.A., Benders, J.: Orchestrated learning: creating a company-specific production system (XPS). Int. J. Lean Six Sigma (2021). https://www.emerald.com/insight/2040-4166.htm
37. Levitt, B., March, J.G.: Organizational learning. Ann. Rev. Sociol. **14**(1), 319–338 (1988)
38. Cummings, T.G., Worley, C.G.: Organization Development and Change. Cengage Learning (2014)
39. Besser, T.L.: Team Toyota: Transplanting the Toyota Culture to the Camry Plant in Kentucky. SUNY Press (1996)
40. Liker, J.K., et al.: Toyota Culture. McGraw-Hill Publishing, New York (2008)
41. Marodin, G.A., Saurin, T.A.: Implementing lean production systems: research areas and opportunities for future studies. Int. J. Prod. Res. **51**(22), 6663–6680 (2013)
42. Spear, S.J.: Learning to lead at Toyota. Harv. Bus. Rev. **82**(5), 78–91 (2004)
43. Holweg, M.: The genealogy of lean production. J. Oper. Manag. **25**(2), 420–437 (2007)
44. Adler, P.S.: Time-and-motion regained. Harv. Bus. Rev. **71**(1/2), 97–108 (1993)
45. Womack, J.P., Jones, D.T.: Lean Thinking. Free Press, New York (1996)
46. Lagrosen, Y., Lagrosen, S.: Creating a culture for sustainability and quality–a lean-inspired way of working. Total Qual. Manag. (2019). https://doi.org/10.1080/14783363.2019.1575199
47. Asif, M.: Lean Six Sigma institutionalization and knowledge creation: towards developing theory. Total Qual. Manag. (2019). https://doi.org/10.1080/14783363.2019.1640598
48. Schein, E.H.: Organizational Culture and Leadership, vol. 2. Wiley, Hoboken (2010)
49. Thorsrud, E., Emery, F.E.: Mot en ny bedriftsorganisasjon: eksperimenter i industrielt demokrati, vol. 2. Tanum (1969)
50. Adler, P.S., Cole, R.: Designed for Learning: A Tale of Two Auto Plants (1993)

Top Down or Bottom Up

Perspectives on Critical Success Factors of Lean in Institutes of Higher Education

Alinda Kokkinou$^{(\boxtimes)}$ (ID) and Ton van Kollenburg (ID)

AVANS University of Applied Sciences, Breda, The Netherlands
{a.kokkinou, ajc.vankollenburg}@avans.nl

Abstract. The application of continuous improvement initiatives such as Lean in Higher Education Institutes is an emerging topic for research, as these organizations are increasingly adopting the tools and methods to improve their quality practices. Nevertheless, Institutes of Higher Education differ significantly from business organizations, which limits the applicability of previous research findings. Using Q-methodology, the present study examines the prevailing perspectives on critical success factors of Lean at Dutch and Belgian Institutes of Higher Education. Findings show that Lean implementation at Institutes of Higher Education takes place bottom-up, with relatively little management involvement and commitment, and mostly involves supporting processes. This impedes the organizational culture change that needs to take place for Lean implementation to be sustainable in the long term, as successes are less visible to management, leading to less management involvement. However, as this is due to structural difference of Higher Education from other industries, it requires a different approach than the conventional, top-down approach prescribed in the literature. A bottom-up implementation of Lean is recommended, centered on improving university-wide supporting processes, promoting cross-departmental cooperation, and overcoming the silo mentality.

Keywords: Lean implementation · Critical success factors · Higher education

1 Introduction

Across the world, institutes of Higher Education (HE) have been increasingly embracing continuous improvement initiatives, and Lean management in particular, to improve their academic and administrative operations [1, 2]. Changes in student enrollment, reductions in national or local funding, increased competition, and a rise in student expectations are pressuring institutes of HE to do more with less [1, 3, 4]. While Total Quality Management (TQM) was initially the programme of choice, it has steadily given way to Lean management, Six Sigma, or a combination of both [2].

Lean management uses a customer perspective to identify and eliminate non-value-added activities [5]. The simplicity of its approach and tools fueled its popularity and it has now been applied to a variety of industries beyond the automotive industry, including service industries. As not all implementations have been successful,

© IFIP International Federation for Information Processing 2021
Published by Springer Nature Switzerland AG 2021
D. J. Powell et al. (Eds.): ELEC 2021, IFIP AICT 610, pp. 32–41, 2021.
https://doi.org/10.1007/978-3-030-92934-3_4

extensive academic attention has been devoted to Critical Success Factors (CSFs) of Lean and Six Sigma implementation [6–8]. Generally accepted CSFs have included management involvement and commitment, cultural change, communication, organization infrastructure, training, project management skills, project prioritization and selection, amongst others [9].

Institutes of HE share a number of characteristics that make the implementation of such programmes less evident. First, institutes of HE rarely have a distinct and recognizable strategy that easily translates to metrics. Second, there is significant complexity in HE in defining customers [8, 10], value, and defects [11]. Third, senior leadership lacks process thinking and clarity regarding how to incorporate Lean thinking in strategy, tactics and operations [10].

Despite these issues, there is consensus that Institutes of HE could significantly benefit from continuous improvement programmes [8, 10–12]. Therefore, the purpose of this study was to examine CSFs of Lean at HEs. The findings of the study are used to formulate recommendations to Institutes of HE seeking to use Lean to improve their academic and administrative operations.

2   Lean in Higher Education

2.1   Lean in Higher Education

Academic attention has been drawn to the issue of successfully implementing Lean, Six Sigma, or Lean Six Sigma (LSS) in HE, leading to two streams of research. The first stream consists of conceptual articles, drawing from evidence of successful LSS implementation in other industries in combination with the authors' personal experience in HE, to ascertain the relevance and benefits of LSS to HE [10]; to examine readiness factors for the implementation of LSS in HE [8, 10] and to formulate frameworks for deploying LSS in HE [11, 13]. The second research stream consists of the empirical investigation of Lean, Six Sigma, and LSS implementations in HE, mostly in the form of single case studies, oftentimes drawing on the authors' personal experience in HE [12, 14].

A review of these publications on Lean in HE shows that many describe single departmental initiatives [1, 2]. Typically, a single individual, or a small group of colleagues uses Lean tools to improve a specific sub-process. This may concern a single or a small number of departments and is caused by the silos that are often characteristic of HE [10]. Focusing on sub-processes decreases the need for coordination and makes for easier appropriation [1]. This approach, referred to as bottom-up, is characteristic of a lack of leadership or broader institutional support [2].

This contrasts with the prescribed top-down implementation approach recommended to Institutes of HE implementing Lean or Lean Six Sigma [8, 11, 12]. This approach, also coined 'institution-wide Lean in HE' [2], advocates first building top-level commitment, and focusing on cultural change in the organization. Several authors argue that the integration of Lean and Six Sigma is most appropriate for HE [10–12], as the Lean approach allows for the tackling of low hanging fruit, and Six Sigma can thereafter be used to reduce variation in processes [10]. Yet, few academic papers have documented the successful implementation of this top-down, integrated approach [1, 12]. A possible explanation for this is that Institutes of HE are structurally different from other industries and thus continuous improvement methodologies need to be adapted to account for these differences [15].

2.2 Success Factors of Lean in Higher Education

An extensive body of research has examined the CSFs of Lean Six Sigma across industries [6, 7]. The consensus is that leadership and management involvement and commitment, linking LSS to the business strategy, and customer orientation are the most important CSFs for organizations implementing Lean Six Sigma. In the context of HE, these same CSFs, also called readiness factors, have been recognized, namely (i) leadership and vision, (ii) management involvement, commitment and resources, (iii) link between LSS and strategy, and (iv) customer focus [8]. However, there is a stark contrast between these CSFs that assume a top-down approach and documented implementations of Lean that show a bottom-up approach.

2.3 Perspectives on CSFs Using Q-Methodology

Traditionally, research on CSFs of Lean has employed a quantitative approach, using surveys requiring participants to rate the importance of a set of CSFs using 5-point Likert scales [6, 16]. This approach has two limitations. First, this approach allows respondents to rate many, or all CSFs highly, and thus does not discriminate between CSFs that are more important than others. For example, in Antony's [6] survey of UK service enterprises, six of thirteen CSFs had a mean rating above 4, making the interpretation of which CSFs are truly important quite arbitrary. Second, this approach assumes that there is consensus about which CSFs are important and does allow for multiple viewpoints. Yet, there is sufficient evidence to suggest that CSFs are context specific. The relative importance of CSFs may depend on the industry [15], organizational culture [17], national culture [18], or stage of implementation of the Lean programme.

Q-Methodology, a qualitative approach that seeks to objectively and scientifically observe subjectivity [19], can overcome these two limitations. A Q-methodology study starts with compiling a set of statements that participants are asked to sort according to their viewpoint or preference, following a prescribed normal distribution. In the context of CSFs, this implies that, while a participant may believe them all to be important, he or she may will still have to rate some as more important than others. This research approach thus supports discriminating between more or less important CSFs [20].

Q-methodology "employs a by-person factor analysis in order to identify groups of participants who make sense of a pool of items in comparable ways" [21]. In other words, Q-methodology helps identify patterns in individuals' subjective viewpoints about a particular topic [22]. These different perspectives can be linked to organizational, cultural and other characteristics, leading to new theoretical insights and better tailored practical recommendations [20].

3   Methodology

For the purpose of investigating CSFs of Lean in HE, the conventional steps of Q-methodology were followed. First a set of statements about CSFs of Lean Six Sigma were compiled from previous research on the topic [7, 9, 23]. This resulted in 42 statements that each included a statement about the importance of a single CSF. These CSFs were purposefully diverse and encompassed the subjects of leadership, training, resources, rewards etc. In a second step, participants to the study were asked to sort the 42 statements according to their agreement as to whether the particular CSF was more or less important, according to a forced normal distribution. The output thereof is called a Q-sort. In a third step, each participant's Q-sort was converted to numerical data for subsequent analysis. In this step, the two most important statements were assigned a score of +4, the next three most important statements were assigned a score of +3, all the way to the two least important statements which were assigned a score of −4.

Participants were recruited from the network of Lean HE Netherlands and Belgium. Lean HE is "the peer led community of practice for people working to apply lean and similar approaches in Higher Education" [24]. Lean HE Netherlands and Belgium, the local division of the global network allowed access to their network, an active group of practitioners involved in implementing or executing Lean at their home institution. In total, 28 participants, representing 15 institutes of HE in the Netherlands (12) and Belgium (2) participated in the study.

Each participant received an e-mail with instructions on how to complete the Q-sort, a personalized link to the online platform Miro, and a link to post-sort survey in Qualtrics. The online platform Miro was used to facilitate the sorting procedure. Participants first read and pre-sorted the statements by dragging them to three areas on the board representing, agree, neutral, and disagree. In a second step, they could sort each group of statements in a pre-formatted grid. This two-step reduced the cognitive complexity of the task. Participants were also asked by means of a survey in Qualtrics to provide information about themselves (training and experience with Lean), information about their home institution (type of implementation, time since implementation started), and provide some clarification about the choices they made during the Q-sort. Finally, in-depth interviews with a sub-sample of participants were used to add context to the quantitative findings. The interviews were conducted online and recorded. The study findings were presented to the Lean HE Netherlands and Belgium network in March and June 2021.

4   Findings

The survey findings, Q-sorts, and interview transcripts were analyzed separately. The Qualtrics survey was used to collect data about participants and their home institution. The 28 participants represented 15 Institutes of HE in the Netherlands and Belgium. Of the 15 Institutes of HE surveyed, 14 had been implementing Lean or an equivalent continuous improvement programme for less than 5 years. For six Institutes of HE, the implementation was qualified as structured, while for 12 Institutes of HE it was described as a Bottom-Up approach. A third of the Institutes of Higher Education defined their continuous improvement programme as Pure Lean. Another third defined it as Lean with some or many Six Sigma influences. The remaining third defined it as an own amalgamation of various continuous improvement programmes, oftentimes encompassing lean tools.

Individual participants similarly exhibited varied experiences with Lean. Consistent with the relatively short duration of implementation of Lean in the Institutes of HE sampled, two thirds of participants had 5 or less years of experience with Lean and other continuous improvement programmes. Almost all had at least a Lean or Lean Six Sigma Green Belt, with eight participants indicating they had a Lean or Lean Six Sigma Black Belt.

4.1 Quantitative Findings

To analyze the Q-sorts, the procedures as described by Zabala were used [25]. Q-methodology does not have strict guidelines and thus there is no 'right' number of factors. Instead, several quantitative criteria (such as eigenvalues and number of Q-sorts loading on each factor) and qualitative criteria (factor interpretation) are used to compare different solutions [21, 22]. Using these criteria, a three-factor solution was identified as most suitable (see Table 1). Specifically, the eigenvalue exceeded one for each factor. Each factor represented the viewpoint of at least four respondents. The total variance explained by the three-factor solution was 51.85%.

The next step was to examine which statements distinguished each perspective from the others. For this, the z-score of each perspective was compared to the z-scores of the other perspectives. Figure 1 compares selected statements per perspective. The z-scores were converted back to the original Q-sort values (ranging from −4 to +4) for better interpretation. Thee three perspectives could then be described based on their distinguishing statements.

Table 1. Factor characteristics

	Customer	Top-down	Bottom-up
Average reliability coefficient	0.8	0.8	0.8
Number of loading Q-sorts	14	7	5
Eigenvalues	7.03	4.33	3.16
Percentage of explained variance	25.10	15.45	11.30
Composite reliability	0.98	0.97	0.95
Standard error of factor scores	0.13	0.19	0.22

The first perspective, representing the views of fourteen participants, was named the customer-driven perspective as according to this perspective, it is important to consult customers often, and LSS projects should be linked to what is important to the customer. Participants in this perspective also placed a lot of importance on top management empowering employees.

The second perspective, named Top-Down, represented the views of seven participants. This perspective mirrored the customer-orientation of the first perspective, but also placed great importance to projects being aligned with the business strategy. Similarly, while underwriting the importance of empowerment, the Top-Down perspective also considered it important that top management take responsibility for quality performance, and that middle managers participate in the execution of projects.

The third perspective, representing the views of five participants, was conversely named the Bottom-Up approach and represented a much more internal focus. Participants in this perspective rejected the notion that customers had to be consulted often and did not consider it important that projects be linked to what the customer wanted. Instead, in this perspective, stronger emphasis was given to project leaders' project management skills and ensuring that employees understood how LSS worked. In this perspective, the role of top management was limited to providing financial resources.

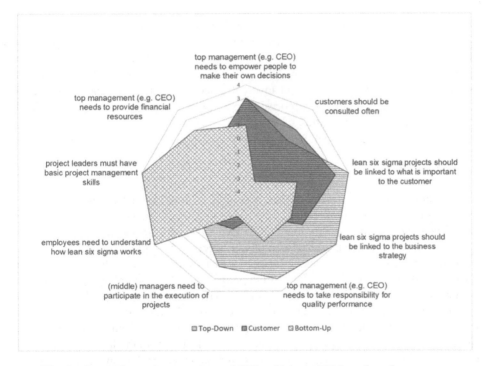

Fig. 1. Comparison of perspectives of CFSs of lean in HE for selected statements

4.2   Qualitative Findings

The in-depth interviews conducted with participants of the Q-sort helped sketch a picture of the organizational context of Institutes of HE in which Lean was being implemented. Three main topics that recurred across interviews were the lack of involvement from (top) management, the lack of process ownership impeding the improvement of end-to-end processes, and the difficulty of applying lean to the primary educational and research processes.

The leadership and top management at Institutes of Higher Education in the Netherlands was described as neither enthusiastic nor skeptical about the added-value of Lean for their institution. Instead, respondents described HE leaders that did not have a process mindset and thus had difficulty appreciating how improved processes could lead to improved outcomes for stakeholders. A related issue was the lack of recognizable strategy to link Lean to. In other words, interviewees found their institution's strategy too vague to be able to link clear performance metrics to.

Instead, respondents described Lean implementations that were initiated by an individual or a group of colleagues, by applying Lean tools to improve a departmental or inter-departmental process. However, as the processes tackled became bigger and more stakeholders were involved, the lack of process ownership became an impediment, as no one was able or willing to take responsibility of changes that may affect more than one department.

All process improvement projects discussed during the interviews concerned supporting or administrative processes. These processes were easier to observe, and thus also better suited to the application of Lean tools. Interviewees expressed the desire to apply continuous improvement methods to the primary processes of research and teaching but were finding it difficult to involve lecturers and researchers. They also found the education and research processes to be more complex and less tangible, and thus less suited to the application of Lean.

5   Discussion

The present study used Q-methodology, a combination of qualitative and quantitative research methods, to identify and describe various perspectives on the CSFs of Lean in HE. Three perspectives were identified. The first two perspectives, "Customer-Oriented" and "Top-Down" reflect commonly accepted best practices of Lean implementation in business organizations, namely a focus on customer value, and the need for support from the top [10]. The second perspective "Top-Down" is also consistent with conceptually based LSS implementation in HE frameworks [8, 10, 11] that argue that, for Lean to succeed in HE, visionary leadership and management commitment are imperative and thus the focus should be on cultural change, starting at the top [13]. The second, "Top-Down," perspective identified through the Q-Methodology is therefore an idealistic view of how Lean should be implemented in HE, with only limited documented applications in HE [2]. This viewpoint does not do justice to the structural differences of the HE domain [15] and stands in stark contrast with documented implementations in HE, which follow a bottom-up approach.

Instead, our findings support the existence of a third perspective on CSFs of Lean in HE, namely a "Bottom-Up" perspective, that advocates for top management to provide resources, but then step aside and let project managers and employees take the lead in improving processes. According to this perspective, top management is perceived as lacking the process mindset needed to appreciate Lean [13]. This also hinders the process of assimilating Lean in the organizational strategy of institutes of HE. The "Bottom-Up" approach is characterized by processes being improved locally, within a single department, or between two departments. As HE is characterized by silos [10], this implies that a sub-process can be made more efficient, while decreasing overall efficiency for the organization. This is where the lack of leadership becomes a hindrance, as the lack of process ownership for end-to-end processes means requires management to get involved.

Our findings also showed that, contrary to common wisdom about applying Lean, Lean implementation in HE primarily involves supporting processes. While these supporting processes such as course enrolment and exam may be visible and important to customers and other stakeholders, teaching and conducting research are the primary activities of HE. There are several possible explanations for the lack of Lean projects concerning these primary activities. First, as Lean is not typically incorporated in HE organizations' strategy, project selection will not be either. This is linked to the second explanation, namely that in the context of HE it is difficult to define who the customers are and what their requirements are [1]. Third, education processes, and teaching in particular, are co-creation processes, requiring the input and interaction of two parties: teacher and student [15].

6 Recommendations for Practitioners

Our findings underscore the startling gap between empirical and conceptual studies of Lean implementation on HE, and support Wiegel and Hadzialic's [15] position that the structural differences between the domains in which Lean was developed on the one hand, and HE on the other hand, require Lean to be adapted for use in HE. Therefore, we propose that the focus of early-stage Lean implementation in HE should be on how to replicate the local departmental successes across the institution, with limited involvement from senior leadership and management.

For organizations of HE having initiated Lean implementation locally, it is recommended to keep the project selection to supporting processes, but remove silos [13] by improving key end-to-end supporting processes such as student enrolment. While ensuring that processes will not only be improved locally, this will also highlight areas with a lack of process ownership. This will furthermore create an opportunity to involve more senior management as their input will be needed to resolve this lack of ownership. To improve these end-to-end processes, a multi-disciplinary, multi-departmental team will be needed. This will create bridges across departments, further reducing silos across the organization. The successes of these projects should be brought to the attention of senior leadership, as a evidence-based way to pique their interest further.

7 Limitations and Further Research

The present study employed Q-Methodology, a combination of qualitative and quantitative methods to identify viewpoints, or perspectives of CSFs in HE. This method's results are primarily qualitative and descriptive in nature and cannot be generalized across a population. Furthermore, the sample used for this study consisted of employees at institutes of HE in the Netherlands and Belgium. While the use of this combination of method and sample was consistent with the purpose of this study, future research should investigate whether the various perspectives identified are also relevant in other cultures, and whether their prevalence differs from our study results.

An important topic to emerge from this study was the perceived difficulty of applying Lean management principles to the primary processes of HE, namely teaching and research. A possible reason for this, meriting further investigation, is the co-creation characteristic of these processes [15]. Future research should therefore examine whether Lean can also be applied to processes that are heavily reliant on co-creation, such as diagnostic evaluations by healthcare providers and teaching at HEs.

Acknowledgments. We thank the Taskforce for Applied Research SIA for funding this project. We also would like to thank the steering committee of Lean in HE – Netherlands and Belgium for their commitment and support of this project.

References

1. Nadeau, S.: Lean, six sigma and Lean six sigma in higher education: a review of experiences around the world. AJIBM **07**, 591–603 (2017)
2. Balzer, W.K., Francis, D.E., Krehbiel, T.C., Shea, N.: A review and perspective on Lean in higher education. QAE **24**, 442–462 (2016)
3. Davidson, J.M., Price, O.M., Pepper, M.: Lean Six Sigma and quality frameworks in higher education – a review of literature. IJLSS **11**, 1005–1018 (2020)
4. Brookes, M., Becket, N.: Quality management in higher education: a review of international issues and practice. Int. J. Qual. Stand. **1**, 1–37 (2007)
5. Koning, H.D., Does, R.J.M.M., Bisgaard, S.: Lean six sigma in financial services. IJSSCA **4**, 1 (2008)
6. Antony, J.: Six Sigma in the UK service organisations: results from a pilot survey. Manag. Audit. J. **19**, 1006–1013 (2004)
7. Antony, J., Desai, D.A.: Assessing the status of six sigma implementation in the Indian industry: results from an exploratory empirical study. Manag. Res. News **32**, 413–423 (2009)
8. Antony, J.: Readiness factors for the Lean Six Sigma journey in the higher education sector. Int. J. Product. Perform. Manag. **63**, 257–264 (2014)
9. Banuelas Coronado, R., Antony, J.: Critical success factors for the successful implementation of six sigma projects in organisations. TQM Mag **14**, 92–99 (2002)
10. Antony, J., Krishan, N., Cullen, D., Kumar, M.: Lean Six Sigma for higher education institutions (HEIs): challenges, barriers, success factors, tools/techniques. Int J Prod. Perform. Manag. **61**, 940–948 (2012)
11. Sunder, M.V., Antony, J.: A conceptual Lean Six Sigma framework for quality excellence in higher education institutions. IJQRM **35**, 857–874 (2018)
12. Svensson, C., Antony, J., Ba-Essa, M., Bakhsh, M., Albliwi, S.: A Lean Six Sigma program in higher education. Int. J. Qual. Reliab. Manag. **32**, 951–969 (2015)
13. Balzer, W.K., Brodke, M.H., Thomas Kizhakethalackal, E.: Lean higher education: successes, challenges, and realizing potential. Int. J. Qual. Reliab. Manag. **32**, 924–933 (2015)
14. Höfer, S., Naeve, J.: The application of lean management in higher education. IJCM **16**, 63–80 (2017)

15. Wiegel, V., Hadzialic, L.B.: Lessons from higher education: adapting Lean Six Sigma to account for structural differences in application domains. IJSSCA **9**, 72 (2015)
16. Desai, D.A., Antony, J., Patel, M.B.: An assessment of the critical success factors for Six Sigma implementation in Indian industries. Int J Prod. Perform. Manag. **61**, 426–444 (2012)
17. Bortolotti, T., Boscari, S., Danese, P.: Successful lean implementation: organizational culture and soft lean practices. Int. J. Prod. Econ. **160**, 182–201 (2015)
18. Kokkinou, A., van Kollenburg, T.: An exploration of the interplay between national culture and the successful implementation of Lean Six Sigma in international companies. In: Rossi, M., Rossini, M., Terzi, S. (eds.) ELEC 2019. LNNS, vol. 122, pp. 179–188. Springer, Cham (2020). https://doi.org/10.1007/978-3-030-41429-0_18
19. Ramlo, S.E., Newman, I.: Q methodology and its position in the mixed-methods continuum (2011)
20. Kokkinou, A., van Kollenburg, T., Touw, P.: The role of training in the implementation of Lean Six Sigma. In: EurOMA proceedings, Berlin (Forthcoming)
21. Watts, S., Stenner, P.: Doing Q ethodology: theory, method and interpretation. Qual. Res. Psychol. **2**, 67–91 (2005)
22. Ponsignon, F., Maull, R.S., Smart, P.A.: Four archetypes of process improvement: a Q-methodological study. Int. J. Prod. Res. **52**, 4507–4525 (2014)
23. Zu, X., Robbins, T.L., Fredendall, L.D.: Mapping the critical links between organizational culture and TQM/Six Sigma practices. Int. J. Prod. Econ. **123**, 86–106 (2010)
24. Lean, H.E.: Lean HE - What is Lean HE? https://www.leanhe.org/lean-he. Accessed 11 June 2021
25. Zabala, A.: qmethod: a package to explore human perspectives using Q methodology. R J. **6**, 163–173 (2014)

The Production Cultural Biorhythm as a New LEAN Learning Process

Bernd Langer[1(✉)], Bernd Gems[2], Yannik A. Langer[3], Maik Mussler[1],
Christoph Roser[1], and Timo Schäfer[4]

[1] Karlsruhe University of Applied Sciences, 76133 Karlsruhe, Germany
bernd.langer@h-ka.de
[2] accirrus.de, 79194 Heuweiler, Germany
[3] Heilbronn University of Applied Sciences, 74081 Heilbronn, Germany
[4] Agilent Technologies, 76337 Waldbronn, Germany

Abstract. Bottlenecks limit value streams, extend lead times and thus cause high costs. ADaM24 provides a new approach to bottleneck problems by offering Advanced Data Management over 24 h for individual and organizational learning to eliminate bottlenecks (Langer et al. 2021). The basic principle to this new approach is to determine a standard day by recording it in minute intervals which are averaged over a longer period of time.

When ADaM24 is deployed at a company, it reveals six previously undiscovered and characteristic patterns of waste. These are MURA patterns of human-machine interaction, which we call "ProductionCultural Biorhythm" or PCB due to their company-related specificity. Patterns and the expression PCB were first established by Langer et al. (2021) [1] and can be demonstrated across different industries and company sizes. The processes measured at companies show that the maximum possible capacity is actually never used, in particular at bottlenecks. ADaM24's clear and easy to grasp graphical representations open up new opportunities for learning about how to make desirable behavioral changes. Moreover, it provides managers with new intervention options. Now time slots within 24 h can be selected for optimization projects, during which not only the radicality of an intervention, but also its improvement dynamics and implementation stability can be measured.

Overall, the PCB topology offers a wealth of new approaches, such as the integration of AI or even the ability to continuously identify bottlenecks "now and next" in dynamic value streams.

Keywords: Mura · Key figures · Bottleneck

1 Making Unexpected Potentials Easily Visible

ADaM24 analyzes and visualizes the results and effects of a production in a completely different complementary way, independent of the chosen production control methods or used IT systems. It is the central point of view how a workforce and the management team deals with bottlenecks in principle = culture at the bottleneck.

© IFIP International Federation for Information Processing 2021
Published by Springer Nature Switzerland AG 2021
D. J. Powell et al. (Eds.): ELEC 2021, IFIP AICT 610, pp. 42–49, 2021.
https://doi.org/10.1007/978-3-030-92934-3_5

1.1   Idea and Basic Principle Behind ADaM24 and PCB

In the LEAN context, daily shop floor management is one of the essential ways of integrating the benefits of the LEAN world into the day-to-day work of the relevant workforce. Up to now, key figures have been entered into monthly charts on a daily basis at fixed points in time (Fig. 1 left). It is thus possible to identify deviations from the nominal value, spot problems and initiate appropriate activities. The disadvantage of the monthly charts typically used is that each month is recorded in a separate template so that trends cannot be identified. The collected data is rarely presented in the form of key figures in a chart with a continuous time dimension (Fig. 1, center).

As part of ADaM24, key figures are continuously recorded based on short time intervals, typically five minutes, over the entire day or shift - instead of a single measured value per day or shift. In the next step, these measured values are averaged over a longer period of time, for example three months. This is how the profile of a standard day is created (Fig. 1 right). The higher the temporal resolution and the more representative the selected averaging period, the more meaningful information is included in the profile. This data collection method is the core element of ADaM24, i.e. Advanced Data Management over 24 h. It helps visualize the so-called Produc-tionCultural Biorhythm PCB in the form of the ROP (Real Output Profile, Fig. 1 right). These profiles allow users to identify inefficiencies, and thus provide the basis for valuable learning experiences for both staff and management. In addition, this method used to determine the standard day is not dependent on product mix or employees and is therefore considered uncritical by employees' representatives.

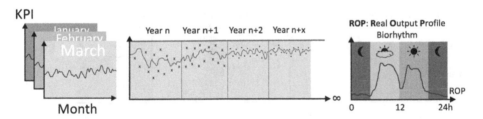

Fig. 1. Left: monthly charts; center: time series toward infinity; right: profile of a standard day (Real Output Profile ROP)

1.2   Data Collection in the Digital Age

Actual data collection in the digital age is still often laborious and performed manually. Ideally, the desired data is automatically collected from the machine controls or sensor technology. However, this state of technology has not yet been broadly implemented in the SME sector. Alternatively, data can be collected via existing MES (Manufacturing Execution Systems) or through logging operational data, although data quality remains a critical factor.

To overcome this problem, our research group is planning to develop a so-called "MuraBox". The aim is to create a plug-and-play solution, which automatically records the tact of a machine or a workstation and visualizes the ROP. In addition, it will automatically determine and present the potential analyses. This helps automate data collection in the digital age, independent of interfaces, and provides an easy-to-use solution for small and medium-sized businesses. Furthermore, data analysis is simplified by means of AI. Data will be collected using the following collection matrix:

- a single machine or a single workstation, a line or an entire plant
- manual, partially or fully automated processes
- production, logistics or indirect areas.

Interesting values are, for example: output in pieces, power consumption in kW, first time through, which refers to the percentage of faultless products in the first run, logistical activities or ERP bookings.

2 In 6 Learning Steps to LEAN Success: The ADaM24 Procedure Model According to Langer and Mussler

As a structured standard method, ADaM24 offers the possibility to use collective learning for the implementation of precise change projects in order to turn them into strong habits. The implementation of the current improvement process automatically becomes the new, familiar reference and is visualized accordingly. In doing so, we place a new layer of habit on top of the previously familiar operational levels when developing new solutions in an unfamiliar environment. This corresponds to the idea of a meta-KATA. Figure 2 illustrates the process.

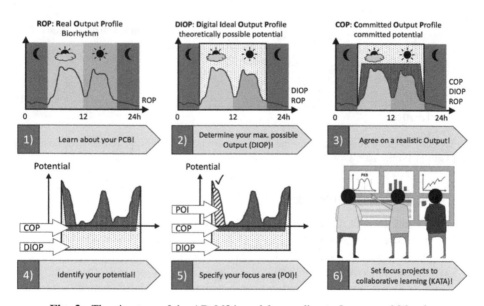

Fig. 2. The six steps of the ADaM24 model according to Langer and Mussler

1. At a bottleneck machine (according to Goldratt *"Step 1: Identify the System's Constraints"*) the ROP is determined and visualized. Previously hidden or easy to conceal inefficiencies become visible at first glance. This new form of transparency reveals phases of low productivity both in terms of shifts and in the form of deviations spread out over the day in a variety of ways. Hiding singular events proves to be advantageous. Measured phenomena, typical of the production process being examined, are shown. These are the production-cultural effects.

2. The DIOP (Digital Ideal Output Profile) visualizes what would be expected from machines, facilities and workstations according to the investment. The delta between ROP and DIOP is the potential that can be assessed using ADaM24. According to Goldratt, this corresponds to *"Step 2: The bottleneck has to be utilized to the maximum"*.

3. The COP describes the part of the potential that management and the workforce are confident of realizing. At the same time, the radicality $R_V = COP/DIOP$ of the change becomes measurable for the first time.

4. The DIOP potential view is the difference between DIOP and ROP. It makes the maximum potential of improvement obvious. The COP potential is the difference between COP and ROP and the focus of the activities that will follow. The information can be given e.g. in pieces, EUR, capacity gain or also throughput optimization at the bottleneck. It reveals the need for action and thus has a motivating effect on all those involved.

5. As we know from Change Management, it makes sense to start at one point and achieve initial success there. For this purpose, we define a Point Of Interest (POI) timed over the course of the day on a COP potential basis as part of the ADaM24 method. The choice of this point is a good opportunity to turn those affected into actors: the workforce determines this starting point itself, which activates self-esteem, autonomy motive and self-commitment. An approach that is congruent with the motive significantly increases the chances of success.

6. Optimizations are now carried out mono-causally through POI-focused projects – this can be done by using KATA (Rother 2013 [2]), for example. After reaching the COP at the POI, additional POIs can be set according to step 5 until the total COP is reached.

The ROP approaches the DIOP as part of the ongoing optimization progress. The dynamics of change can be visualized by the dynamic PCB, while sustainability can be ensured by the static PCB. When going through the six steps, the superior and his/her team are able to jointly and consciously enter the terrain of unknown potential. A collective learning process is thus fostered, lessons learned increase the organizational knowledge base.

3   ADaM24 in Action Against Bottlenecks

It makes sense to prioritize mainly constraints for ADaM24 projects. To this end, it is required to know these bottlenecks. Detecting them is a challenge due to the shifting of the bottlenecks. There are numerous different bottleneck methods available, but only few of them work reliably with shifting bottlenecks (Lima et al. 2008 [3]). The bottleneck walk is recommended for shifting bottlenecks for most real-world systems (Roser et al. 2015 [4]). Another solution is the active period method for highly digitized systems or simulation-based experiments (Roser et al. 2001 [5]). The bottleneck walk does not detect the bottleneck directly, but instead analyzes which inventory or process waits for other processes in order to determine the bottleneck. The active period method defines the bottleneck at any given time as the process covering the longest period without waiting for another process (the active period) at that time.

System performance is defined by the bottleneck, or in most real-world production systems, by the interaction and change of shifting bottlenecks over time. Following the biological biorhythm, bottlenecks can be seen as constrictions of the arteries of the production system. The PCB of a production system also reflects the behavior at the bottleneck.

Figure 3 shows that when using the PCB, it is helpful to first have a look at the dominant bottleneck machine (M3 on the left) and possibly reach the DIOP (M3 on the right). This helps increase the throughput at the critical spot and thus resolves the bottleneck. The M4 machine becomes the new bottleneck. It is not advisable to start using ADaM24 for non-bottleneck machines; this would result in more push throughput, growing inventories and even longer lead times, as known from value stream mapping.

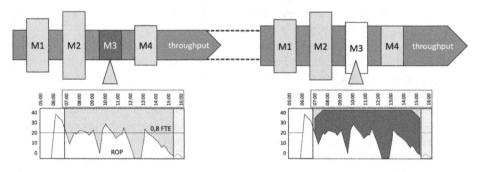

Fig. 3. Increase throughput based on POI projects

A broader view helps to identify patterns over time intervals (e.g. per week, month or quarter). As a result, we can identify phenomena such as bottlenecks that occur at different locations/points on monday morning than on other days of the week, for example.

The previously mentioned MuraBox is used to match the ADaM24 and the work-in-process inventories, so that the current bottleneck and the bottleneck that follows after unblocking the previous constraint (now and next) can be determined predictively.

4   PCB in the Context of LEAN Thinking and Industry 4.0

LEAN managers are always on the lookout for new innovative ways to optimize and increase efficiency. ADaM24 allows them to visualize company-specific MURA patterns and discover a new perspective on industrial production processes. Within the scope of LEAN consulting mandates, project works, bachelor's and master's theses, approx. 280 profiles have been created and examined for PCB patterns.

Figure 4 shows a selection of ROPs for different supply levels (OEM, Tier 1 - Tier n), industries and company sizes. The ProductionCultural Biorhythm reveals weaknesses in production, logistics, operations and so on.

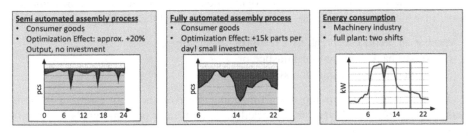

Fig. 4. ROP examples in industry

In most cases, superiors and employees are equally surprised by the patterns of the PCB effect. During interventions where ADaM24 was used to help change human behavior and human-machine interaction at bottleneck machines, significant improvements of up to 83% could be achieved.

5   Cybernetics and PCB in Learning Organizations

The PCB methodology opens up a new perspective on managing complex systems. It applies cybernetics to control and regulate a system according to its purpose, whereby various problems of system control are to be solved by the manager (Steckelberg and Harrer 2021 [6]). In doing so, it is essential that managers lower the degrees of freedom and/or reduce complex issues to merely complicated ones.

Measurement of the PCB does not initially reduce the degrees of freedom of the system, nor does it restrict the state of workers, machines, lines or plants. However, the PCB does represent a crucial and novel variable that is provided as feedback to system managers. This enables them to make targeted interventions within complex systems. This system approach is also described in models of the learning organization according to Senge 2021 [7]. The PCB suggests that teams and managers look into the following questions:

- Which of our cultural behaviors are causal to the ROP identified?
- What level of radicality R_V do we want to aim for in change projects?
- How do we achieve general organizational, technology-related and cultural learning through the PCB?

These questions, stimulated by the PCB, enable learning in the process of change. The PCB thus provides cybernetic system access at a previously inaccessible, cultural level of production units.

6   The Psychology of Learning in Use for the Implementation

How hard can it be? Why can't they just do it? How can I get my people on the right track? Experienced LEAN managers are all too familiar with these types of questions. There is no universally valid answer to these types of questions. The answer always depends on the personality traits, relationships and relationship histories of the people involved, as well as on the traits of the organization. We have picked three possible use cases, in which answers become transparent through the PCB: Employees.

- receive an additional break as a result of wrongdoing,
- get to pay back experienced injustice or
- feel clever by subverting "the system".

When managers try to stop this wrongdoing, they are actually asking a lot from these employees: "Employees are expected to give up their personal goals in favor of the goals of the manager".

From the employees' point of view, there have to be very good reasons to do so. Many change management procedures fall short here, because they focus on replacing one process by another that is perceived as equivalent. Changing habits in this sense proves difficult enough - but getting employees to adopt an inner attitude that involves giving up their personal goals is a far greater challenge. If this is achieved, so-called "extra-productive behaviour" sets in - and LEAN principles are followed in accordance with a deep, inner attitude.

There is no standard solution when it comes to implementing the PCB. We know from Dalal's (2005 [8]) meta-analysis of 45 studies with a total of 16,721 respondents that the difference between counterproductive and extra-productive behavior is significantly determined by the personality trait conscientiousness as well as by job satisfaction, commitment, and perceived fairness.

Determining a concrete procedure requires a business-psychological and organizational diagnosis supported by ADaM24.

7   PCB Topology Shows Concrete Lines of Action

The PCB topology depicts the ADaM24 and PCB action map. In addition to the procedure model, other interesting research, action and transfer strands are derived:

- To date, six basic MURA patterns have been identified that are independent of industry and organization. Further patterns are to be discovered by means of AI in order to develop generic, industrial solution approaches.
- Radicality R_V is examined as an evaluation measure for change processes.
- Determination of the POI or COP/DIOP potential is to be automated, initially at machine or workstation level, and later for the entire value stream.
- It is now possible to show the sustainability (stability) of change activities as a measure via the static PCB.
- The change dynamics of change processes can now also be made visible as a measure via the dynamic PCB.

- In addition, the robustness of processes can be determined by the static PCB and subsequently by the dynamic PCB.
- The visualization of dynamic bottlenecks (now + next) on shop floor level is a powerful management tool that motivates employees and steers them in the right direction.
- New indicator lamps with a count- and sum-up display show "now + next" downtimes right at the bottleneck machines.

From all this we may conclude that ADaM24 and the PCB make a significant positive contribution to change activities in organizations and consulting business when it comes to bottleneck optimization.

References

1. Langer, B., Gems, B., Wenger, D., Schäfer, T., Roser, C.: Auch Maschinen haben einen Biorhythmus [Machines also show a biorhythm]. ZFO Zeitschrift Führung + Organisation **90** (2), 74–82 (2021)
2. Rother, M.: Die Kata des Weltmarktführers: Toyotas Erfolgsmethoden [The Kata of the Global Market Leader: Toyota's Success Principles]. 2nd edn. Campus, Frankfurt/Main (2013)
3. Lima, E., Chwif, L., Pereira Barreto, M.R.: Methodology for selecting the best suitable bottleneck detection method. In: 2008 Winter Simulation Conference, WSC 2008, pp. 1746–51 (2008). https://doi.org/10.1109/WSC.2008.4736262
4. Roser, C., Lorentzen, K., Deuse, J.: Reliable shop floor bottleneck detection for flow lines through process and inventory observation: the bottleneck walk. Logist. Res. **8**, Article No. 7 (2015). http://link.springer.com/article/10.1007/s12159-015-0127-2
5. Roser, C., Nakano, M., Tanaka, M.: A practical bottleneck detection method. In: Peters, B.A., Smith, J.S., Medeiros, D.J., Rohrer, M.W. (eds.) Proceedings of the Winter Simulation Conference, vol. 2, pp. 949–53. Institute of Electrical and Electronics Engineers, Arlington (2001)
6. Steckelberg, A.V., Harrer, C. (eds.): Leadership & Management. Ganzheitliche Betrachtung von Führung und ästhetikbasierte Qualifizierung von (Nachwuchs-)Führungskräften. 1st ed. Springer Fachmedien Wiesbaden GmbH; Springer Gabler, Wiesbaden (2021)
7. Senge, P.M.: Die fünfte Disziplin. Kunst und Praxis der lernenden Organisation. [The Fifth Discipline: The Art and Practice of the Learning Organization], 11th edn. Schäffer-Poeschel, Freiburg (2021). (Systemisches Management) [Systemic Management]
8. Dalal, R.S.: A meta-analysis of the relationship between organizational citizenship behavior and counterproductive work behavior. J. Appl. Psychol. **90**, 1241–1255 (2005)

Proposing VSM as a Tool to Compare Synchronic Online Teaching and Face-to-Face Teaching

Felipe Martinez[(✉)] [ID]

Prague University of Economics and Business, Prague, Czech Republic
`felipe.martinez@vse.cz`

Abstract. The C19 forces university teachers to turn into the digital world in a short time. Although online teaching has been proven to be a successful approach for learning, the rapid change that we were subjected to make did not leave space for each of us to assess the impact of this change on the student. All that we knew was: We have to teach online. There is no other possible way right now! However, is it ok to teach online?

Many educators had already faced this question and provided answers for it from different perspectives. However, the C19 did not give time to find answers. We needed to teach online from scratch, and we had little time to learn technologies and prepare or adapt the lessons online.

Nevertheless, after a year of online teaching, there has been time to learn technologies and prepare online lessons. In addition, each of us can collect our data to compare both learning approaches from personal perspectives.

Thus, this aims to purpose the Value Stream Mapping (VSM) as a lecturer tool to assess educational process from the lecturer's perspective in both synchronic situations (online vs. face-to-face).

The paper explains VSM as a lecturer's self-assessment tool implementing an example in a specific teaching process. This process belongs to the teaching activities of the author. Therefore, the paper avoids generalizing possible answers to the research question. However, further research will require the results from the implemented tool to obtain a sufficient data set for generalizations.

Keywords: Lean management · Teaching process · Process value added

1 Introduction

Social distancing restrictions due to the pandemic directly affect the way classes have to be taught (Obrad 2020). When the C19 hits our reality, not just university teachers had to turn into the digital world in a short time (Watermeyer et al. 2021). Online teaching has been proven to be a successful approach for learning besides some certain obstacles (Sun et al. 2008). However, the rapid change in the teaching method that had to be done did not leave space to evaluate the impact of this change from the perspective of the student, teachers, courses, technology, design, and the environment (Sun et al. 2008). All that we knew was: We have to teach online. There is no other possible way right now!

© IFIP International Federation for Information Processing 2021
Published by Springer Nature Switzerland AG 2021
D. J. Powell et al. (Eds.): ELEC 2021, IFIP AICT 610, pp. 50–59, 2021.
https://doi.org/10.1007/978-3-030-92934-3_6

As expected, educators and no educators raise questions about the efficacy or validity of the provided education (Obrad 2020). Or, in simple words, is it ok to teach online? Many educators had already faced this question and provided answers for it from different perspectives such as course environment, students' outcomes & characteristics, and institutional together with the organizational factors (Tallent-Runnels et al. 2006). The phenomena of Web 2.0 provide new perspectives on student participation and creativity, and online identity formation (Greenhow et al. 2009). The assessment is an online environment that requires formative feedback from the lecturer (Gikandi et al. 2011). Furthermore, massive open online courses (MOOCs) have been proven to be a good alternative for learning (Liyanagunawardena et al. 2013). The relationship between the student and teacher also provides insights into the learning process quality (Tormey 2021).

However, in the framework of C19, there was no time to dedicate to find answers to this question. We needed to teach online from scratch, and we had little time to learn technologies and prepare or adapt the lessons to these platforms (Watermeyer et al. 2021).

Nevertheless, after a year of online teaching, there has been time to learn technologies and prepare online lessons, and there has been at least one entire semester of synchronic online teaching as a replacement for face-to-face teaching. Therefore, each of us can collect our data to compare both learning approaches from personal perspectives. But how about the Value for the student? Does synchronic online teaching provide the exact Value to students as face-to-face teaching?

2 Teaching Process as a Value Stream

The value stream mapping (VSM) is a strong Lean Management tool to understand how the process flows and creates Value (Abdulmalek and Rajgopal 2007). The VSM tool has implementations in manufacturing and services (Sundar et al. 2014), including education. For example, the tool has been implemented to improve academic curriculum creation (Zighan and EL-Qasem 2021) or to develop strategies to reduce work stress in primary education. Thus, the VSM is a versatile tool with a framework to improve a process or assess the Value and other variables.

The development of the VSM requires an understanding of the activities that provides Value to the customer. Therefore, it is necessary to determine the customer and the output of the teaching process. Teaching is a complex set of knowledge transfer activities involving at least one teacher and one student (Eshchar and Fragaszy 2015). Since the teacher provides the transfer, it is possible to determine the student as the customer of this knowledge transaction. Thus, the output of the process of teaching is the acquired knowledge by the student.

The long list of activities related to the teaching process includes setting learning goals, preparing for conferences and seminars, determining case studies, searching for teaching materials, and many more (Eshchar and Fragaszy 2015). From the Lean/VSM perspective, it is necessary to determine the source of Value in the process. In this case, these are the activities that create Value for the customer (student). Nevertheless, it is also necessary to determine the activities that are not providing Value. Some of these non-value activities are necessary to develop the process, and others are entirely waste (Chowdhury et al. 2016).

The source of Value in an e-learning course is among the teaching activities within six dimensions: Student, teachers, courses, technology, design, and environment (Sun et al. 2008). From this perspective, the main factors influencing the Value in the process are the student computer anxiety, the teacher attitude toward e-Learning, e-Learning course flexibility, e-Learning course quality, perceived usefulness, perceived ease of use, and diversity in assessments (Sun et al. 2008). A different set of factors influencing teaching courses online are course environment, students' outcomes, students' characteristics, and institutional factors and organizational factors (Tallent-Runnels et al. 2006). In both cases, the value activities are related to the students acquiring knowledge, mainly happening at the sessions in synchronic teaching. While it is understood that asynchronous communication tends to facilitate deeper communication, it is not much more than in traditional classes (Tallent-Runnels et al. 2006). Therefore, the means of communication through which the sessions are held is necessary but not decisive. Thus, the online teaching process factors are related to the technological possibilities rather than the knowledge transfer itself (Eshchar and Fragaszy 2015; Tallent-Runnels et al. 2006).

3   Methodology

This paper approaches the comparison of synchronic online teaching and face-to-face teaching implementing the Value Stream Mapping (VSM) and Lean principles. Despite the multiple research approaches implemented to assess distance education, this paper explores the VSM possibilities as a personal tool to assess educational process from the lecturer's perspective in both synchronic situations (online vs. face-to-face). Therefore, this paper proposes a self-assessment tool for teaching processes using Value as the assessment parameter.

The first step is to determine the process in which the tool will be implemented. Then, it is necessary to determine the activities that provide Value, and finally, it is necessary to calculate the Value of these activities as a percentage of their time divided by the total time of the process (Chowdhury et al. 2016).

The chosen process is a "5S Methodology training". It is a four-hour learning activity with lecturing and workshops for twelve participants. One training was delivered in October 2019 before the C19 and as face-to-face training in a classroom. The training participants are employees from different manufacturer organizations related to quality, manufacturing, warehouse, and other similar activities. The second training was delivered on January 2021, in the middle of the lockdown due to C19. Thus, it was delivered online for workers of the same company related to similar quality, manufacturing, warehouse, and others. This specific training allows the comparison since both teaching experiences have a similar number of people, similar profile of participants, similar positive assessment and it is promoted by the same organization. The mentioned training assessment is performed by the same organization as a feedback and continues improvement strategy. Thus, the learning outcome is fulfilled similarly, and the only significant change is the obligation to do the training online due to the pandemic lockdown. The implementation of the SIPOC tool helps to illustrate the process.

The implementation of the activities diagram tool facilitates listing the process activities and their classification as Value or non-value activity. This tool has a list of activities, the typology of activities (Operation, Transport, Checking, Delay, Storage), the time of each activity, the distance (if needed), and the number of workers developing the activity (Greasley 2013). First, the value-added (VA) activities are identified, and their times are selected as VA time. Then, the other activities are identified as non-value-added activities (NonVA). This permits the calculation of the percentage of the VA in the process in both situations (synchronic online vs. face-to-face). Then, the graphical representation of both VSMs displays the processes to determine comparison.

Additionally, the lists of times represent a set of time data of the process. Thus, the implementation of the 2-Sample t Test of the Mean and the 2-Sample Standard Deviation Test provide insights to determine the comparison of both trainings.

4   Findings

The chosen process to illustrate the implementation of the tool is a "5S Methodology training". The synchronic online (Synch) version and the face-to-face (F-T-F) version of the training can be summarized in the same SIPOC diagram (Fig. 1).

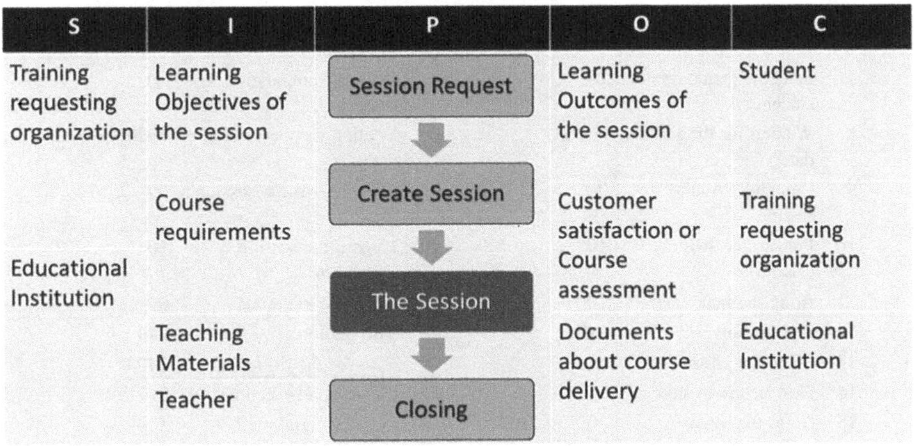

Fig. 1. SIPOC - The teaching process of a specific training

The four-hour training is similar in both situations since the only change is the environment where the educational process takes place. The activities before the start of the training are the same in both situations. Also, the activities after the end of the training are the same. These activities include conversations via email or telephone, sending documents, and evaluations. From the customer's point of view (student), the session is the only source of change between both trainings. Thus, the VA time is only 240 min.

The implementation of the activity diagram of the face-to-face (F-T-F) training shows that there are activities before the session such as understanding of the session requirements, determining the session flow, the workshops, etc. The assumption is that the teacher prepares the session two weeks before the session starts. In addition, the F-T-F training has the specifics of transportation. In this case, the transportation shows the distance from the University to the training site. Also, there is the preparation of the classroom before the session and bringing the materials for the workshops (Table 1).

Table 1. Process activities F-T-F and Synch

VSM	\<Face To Face (F-T-F)\>				\<Synchronic online (Synch)\>			
	No	Activity	Time (min)	Distance (Meters)	No	Activity	Time (min)	Distance (Meters)
A	1	Receive the requirement	5		1	Receive the requirement	5	
	2	Understand the learning objectives	60		2	Understand the learning objectives	60	
B	3	Determine the session flow	30		3	Determine the session flow	30	
	4	Determine the workshops	60		4	Determine the workshops	180	
	5	Ask for the materials for the workshops	60		5	Ask for software or apps for the workshops	60	
	6	Create the presentation	60		6	Create the presentation	180	
	7	Send the preparation to students	20		7	Send the preparation to students	20	
C	8	Waiting for the session date	19200		8	Waiting for the session date	19200	
	9	Commute to the session	30	2300	9	Commute to the session	0	0
	10	Prepare the session classroom	10		10	Prepare the session classroom	10	
	11	Bring the materials	10	15	11	Bring the materials	0	0
	12	**The session**	**240**		**12**	**The session**	**240**	
D	13	Clean the classroom	5		*16*	*Wait for the feedback*	*240*	
	14	Pick lecturers' stuff	5		13	Clean the classroom	0	
	15	Go to the office	30	2300	14	Pick lecturers' stuff	5	
	16	*Wait for the feedback*	*960*		15	Go to the office	0	0
	17	Lessons learned	30		17	Lessons learned	30	
	Total		**20815**	4615	Total		**20260**	0

After the session, the activities are cleaning and organizing the classroom before leaving. Also, the transportation to return to the office at the University is included. Finally, the time required to obtain the course's feedback to develop the lessons learned for subsequent sessions.

The activity diagram of the synchronic online (Synch) version of the training shows that the time creating the session's slides is longer since the lectures and workshops have to be more frequent to have a better relationship between the student and teacher (Tormey 2021). The difference in this preparation time is observed in the 88-slide (34 MB) presentation of the Synch training compared to the 22-slide (5 MB) presentation of the F-T-F training. Likewise, the number of workshops in Synch training is more significant, and therefore the time to prepare them increases. Activity number five has the same time but changes its objective. Since the session is online, then materials for workshops are changed by platforms for teaching. The most significant change in the Synch training is the absence of transportation. The lecturer does not require to commute to the training site. Thus, all the transportation activities have null time and distance.

Similarly, the activities after the Synch session related to transportation or classroom have null time and distance. Moreover, the activity waiting for feedback is happening faster and right after the end of the session. Since everything is online, students deliver their assessment of the course immediately or at least the same day, while in the F-T-F version, it might take up to two days if they are doing online or even longer for paper-based assessments.

The consolidation of the activities in a four-step VSM shows that the F-T-F training has 500 min of value-added activities, 20315 min of non-value activities, and a VA% of 2,4% (Fig. 2).

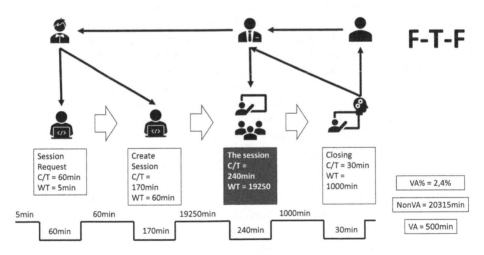

Fig. 2. VSM of the F-T-F teaching process

The Synch training the VSM illustrates a higher VA time of 740 min, a lower Non-VA of 19510 min, and a higher VA% of 3,65%.

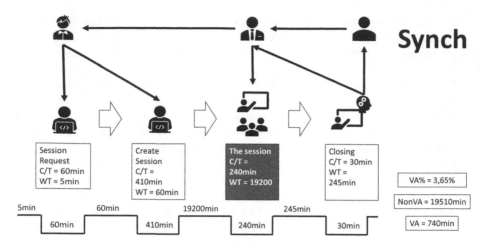

Fig. 3. VSM of the Synch teaching process

The additional analysis of the 2-Sample t Test of the Mean of Synch and F-T-F determine that the means of both data sets do not differ (P = 0,673). Similarly, the 2-Sample Standard Deviation Test for Synch and F-T-F reveals that the standard deviation of both data sets does not differ (P = 0,654) (Fig. 3).

5   Discussion

There are some important differences between the two teaching processes presented. First, the list of activities for the Synch training is lower than F-T-F by three activities (Activities 9, 11, and 15). These are the activities related to transportation, which losses their sense within this process because there is no transportation on the Synch training. Then, the fifth activity changes its purpose since physical materials are not required, but platforms, software, and applications are required to develop the workshops online but synchronously. Also, the flow of the process changes with activity 16 since this is related to the session's feedback, which in the online session can be done almost immediately. However, these differences are not necessarily averse to the process since both online, and face-to-face have constraints, consequences, and success factors (Obrad 2020; Eshchar and Fragaszy 2015; Tallent-Runnels et al. 2006).

The exploration of the value-added times and non-value-added times in both processes reveals that the differences are relatively small. The F-T-F process is longer than Synch, has less VA time, more NonVA time, so then less VA% in comparison with the Synch process. Thus, from this VA perspective, synchronous online teaching is better than face-to-face teaching. However, the factors that influence the learning experience are broader (Eshchar and Fragaszy 2015; Sun et al. 2008; Tallent-Runnels et al. 2006). Thus, this paper exposes the teaching experiences rather than the learning experience. The teaching preparation for the online sessions requires different and more preparation than the face-to-face sessions. While in the classroom, the teacher-student interaction flows during the session allowing the teacher to navigate the content and activities within a prepared framework; at the online session, the lecturing and work-shops must be detailed designed on specific timing to guarantee good teacher-student interaction (Tormey 2021). Then, the time saved in transfers and transportation is required to develop a much more interactive session, with more short workshops and fewer long lectures.

The VA analysis takes the preparation of the classes as valuable time. However, from a strict perspective of the concept of Value towards the student, there are only 240 min that the student is learning with the teacher or the session time. In this case, the VA% for F-T-F is 1,15%, while the Synch is 1,18%. This confirms that the difference between both processes is relatively small. Additionally, the results of the 2-Sample t Test of the Mean and the 2-Sample Standard Deviation Test provide more evidence to argue that the difference between both processes is minor.

The similarity of both trainings allows the presented comparison. This similarity also includes the positive assessment of both trainings. Thus, the outputs from both processes are similar. Moreover, further research should investigate the impact of the changes within the 240 min session taking in consideration that the customer of a training has less chance to take the same course twice. Thus, the improvements based on the current customer's feedback will be applied to a different customer and therefore the new feedback is not coming from the same customer (Eshchar and Fragaszy 2015; Gikandi et al. 2011). Nevertheless, the teaching/learning process effectiveness during these 4 h session requires an in-dept analysis for further discussions.

The effectiveness of the VSM in representing and evaluating educational processes has been well described in the literature (Zighan and EL-Qasem 2021; Sundar et al. 2014; Abdulmalek and Rajgopal 2007). Furthermore, this paper also implements SIPOC as a tool to determine the scope of the process in this analysis. However, there are other tools that might be considered for the development of a similar analysis. This paper implements VSM since the tool provides information about the value flow, but tools such as flowchart might provide insights on sequential or parallel activities as well as other interactions (Damelio 2011).

6   Conclusion

The VSM has proven again that it is an excellent tool to assess the Value of any process (Zighan and EL-Qasem 2021; Sundar et al. 2014; Abdulmalek and Rajgopal 2007). In this case, the VSM helps compare the teaching process of a face-to-face session and an online synchronic session. Listing the activities performed by the teacher with the time to develop them allows the teacher to develop a simple basic assessment of the effort to develop the session and the provided Value to the student. Although some activities change their flow or composition, the idea is to keep the VA of the process as similar as possible.

Moreover, the VA analysis of the teaching process collects insights from the tea-cher's perspective rather than the learning experience. Education is much more com-plex than the teacher's effort to provide knowledge by delivering a session. Nevertheless, the teacher needs to review how the VA of the entire process changes due to the change of the teaching means. This paper shows the comparison of one specific training delivered both online and face-to-face. The analysis of the VA change from both processes provides the teacher with insights on how these teaching methods change the VA provided to the students. By following the teaching process assessment principles of this comparison, the teacher can create a self-assessment tool to verify the change in the provided VA.

The implementation of the VA analysis of the teaching process using the VSM perspective requires further research. However, it is expected that other teachers implement the tool to enhance its characteristics and usability.

References

Abdulmalek, F.A., Rajgopal, J.: Analyzing the benefits of lean manufacturing and value stream mapping via simulation: a process sector case study. Int. J. Prod. Econ. **107**(1), 223–236 (2007). https://doi.org/10.1016/j.ijpe.2006.09.009

Damelio, R.: The Basics of Process Mapping. CRC Press, Boca Raton (2011)

Eshchar, Y., Fragaszy, D.: What is teaching? A clear, integrative, operational definition for teaching is still needed. Behav. Brain Sci. **38** (2015). https://doi.org/10.1017/S0140525X14 000661

Gikandi, J.W., Morrow, D., Davis, N.E.: Online formative assessment in higher education: a review of the literature. Comput. Educ. **57**(4), 2333–2351 (2011). https://doi.org/10.1016/j. compedu.2011.06.004

Greasley, A.: Operations Management. John Wiley & Sons Inc., New York (2013)

Greenhow, C., Robelia, B., Hughes, J.E.: Learning, teaching, and scholarship in a digital age web 2.0 and classroom research: what path should we take now? Educ. Res. **38**(4), 246–259 (2009). https://doi.org/10.3102/0013189X09336671

Chowdhury, A.H., Shahriar, S., Hossen, T., Mahmud, P.: Reduction of process lead time using lean tool - value stream mapping (VSM). Appl. Mech. Mater. **860**, 74–80 (2016). https://doi. org/10.4028/www.scientific.net/AMM.860.74

Liyanagunawardena, T.R., Adams, A.A., Williams, S.A.: MOOCs: a systematic study of the published literature 2008–2012. Int. Rev. Res. Open Distrib. Learn. **14**(3), 202–227 (2013). https://doi.org/10.19173/irrodl.v14i3.1455

Obrad, C.: Constraints and consequences of online teaching. Sustainability **12**(17) (2020). https:// doi.org/10.3390/su12176982

Sun, P.-C., Tsai, R.J., Finger, G., Chen, Y.-Y., Yeh, D.: What drives a successful e-learning? An empirical investigation of the critical factors influencing learner satisfaction. Comput. Educ. **50**(4), 1183–1202 (2008). https://doi.org/10.1016/j.compedu.2006.11.007

Sundar, R., Balaji, A.N., SatheeshKumar, R.M.: A review on lean manufacturing implementation techniques. In: 12TH Global Congress on Manufacturing and Management (GCMM - 2014) (stránky 1875–1885). VIT University, Vellore, INDIA: VIT University, School of Mechanical and Building Science; Queensland University Technology (2014). https://doi.org/10.1016/j. proeng.2014.12.341

Tallent-Runnels, M.K., et al.: Teaching courses online: a review of the research. Rev. Educ. Res. **76**(1), 93–135 (2006). https://doi.org/10.3102/00346543076001093

Tormey, R.: Rethinking student-teacher relationships in higher education: a multidimensional approach. High. Educ. **82**(5), 993–1011 (2021). https://doi.org/10.1007/s10734-021-00711-w

Watermeyer, R., Crick, T., Knight, C., Goodall, J.: COVID-19 and digital disruption in UK universities: afflictions and affordances of emergency online migration. High. Educ. **81**(3), 623–641 (2021). https://doi.org/10.1007/s10734-020-00561-y

Zighan, S., EL-Qasem, A.: Lean thinking and higher education management: revaluing the business school programme management. Int. J. Product. Perform. Manag. **70**(3), 675–703 (2021). https://doi.org/10.1108/IJPPM-05-2019-0215

Hybrid Learning Factories for Lean Education: Approach and Morphology for Competency-Oriented Design of Suitable Virtual Reality Learning Environments

Thomas Riemann[✉], Antonio Kreß, Liane Klassen,
and Joachim Metternich

Technical University of Darmstadt, Otto-Berndt-Straße 2,
64785 Darmstadt, Germany
T.Riemann@ptw.tu-darmstadt.de

Abstract. In recent years, learning factories have proven to be an effective instrument for developing competencies, especially in lean production and digitization. The concept of learning factories has been enriched in the recent past by elements and training units in virtual reality (VR). This enrichment allows an expansion of the mapping abilities of different training environments and value streams in the context of lean education. Nevertheless, learning factory developers are faced with the challenge of selecting suitable scenarios in terms of content and scope. An approach for the competency-oriented and structured design of such scenarios will be presented in this publication and illustrated by means of an application example of the research project PortaL (Virtual action tasks for personalized, adaptive learning).

Keywords: Virtual reality · Learning factory · Learning environment · Lean education

1 Introduction

The achievement of strategic goal, innovative capabilities and finally profitability depend strongly on the ability to build up the relevant competencies and to use and further develop the existing competencies of employees efficiently. Companies and other institutions are measured by how well they are able to leverage their knowledge to create value [1]. Learning factories have proven to be an effective tool to develop competencies. By using VR, the mapping capability of learning factories can be extended [2]. If a learning factory applies both physical and virtual elements, it is called hybrid. The goal of this publication is to present a procedure for the development of virtual training scenarios in hybrid learning factories.

2 Virtual Reality and Learning Factories

Learning factories provide a reality-conform production environment for learning [3]. Virtual learning factories are basically used for training in the same areas as traditional physical learning factories. Mostly, the tasks of the trainings originate from the

© IFIP International Federation for Information Processing 2021
Published by Springer Nature Switzerland AG 2021
D. J. Powell et al. (Eds.): ELEC 2021, IFIP AICT 610, pp. 60–67, 2021.
https://doi.org/10.1007/978-3-030-92934-3_7

planning or simulation background [4]. Several learning factories address lean education topics. They can be seen as an extension of digital learning factories, as they provide visual software tools and infrastructures such as user interfaces to enable the visualization of digital models [5]. Virtual learning factories enrich physical learning experience by a higher degree of flexibility and expanded opportunities. Therefore, it is important that the virtual learning factory represents the real learning factory in all its relevant processes, activities and resources [5]. Using VR for the development of competencies in virtual learning factories thus offers several advantages. Due to adaptable virtual environments, the learning scenarios can be personalized specifically to the learner [6]. Additionally, acquired knowledge can be tested in the virtual environment without generating economic damage. To be able to design a constructive learning environment, the focus of competency development should be on the application of the learning content, not on the reproduction of knowledge. The learner should thus be given authentic tasks that correspond to the intended competencies. The prerequisite for competency development in a virtual environment is, on the one hand, the use of the interaction possibilities and, on the other hand, the link to a learning outcome [7]. Finally, it is important to develop and consolidate the necessary competencies with the help of transfer-based action tasks.

3   Methodology

The primary goal of the methodology is to develop virtual learning scenarios that are able to develop the intended competencies while being adaptive and individualizable. Thus, the advantages of VR technology can be implemented, and competency development can be individualized and personalized.

3.1   Development of the VR Scenarios

The development process involves the inclusion of three steps, as shown simplified in Fig. 1. The development of a scenario for a virtual learning factory is close to the development of a physical learning factory. Based on the design approach by Tisch for physical learning factories [8], the first step is to clarify the organizational requirements (step 1). This includes the organizational environment and targets as well as the target group. In the second step, the learning targets and intended competencies are deduced (step 2). Finally, the configuration or design of the learning scenario is developed based on the previous steps. This includes the design of the product and processes, which should also be based on didactic principles (step 3).

The first two steps are identical to the procedure for a physical learning factory scenario. In the first didactical transformation the intended competencies are derived from the organizational requirements or environment, the organizational targets and the target group. In the second didactical transformation the socio-technical infrastructure and didactical aspects are derived from the intended competencies [9]. This approach can also be applied when a virtual learning factory scenario needs to be designed, but additional (e.g., maximum time in VR) or changed requirements (e.g., space for VR tracking) need to be observed.

Once the organizational requirements and learning targets have been defined (these two steps are not in focus of this paper), the concrete configuration (infrastructure and didactics) of the scenarios takes place within the third step, the design phase. A learning factory has two central characteristics that should be represented in the infrastructural and didactical aspects of the virtual scenarios: On the one hand, the high contextualization of the learning environment (i.e., a factory environment that is close to reality) and, on the other hand, the possibility of the learners' first practical experiences. However, the contextualization should not be too detailed, firstly because of the high complexity of the implementation in VR and secondly because of the risk of overtaxing the participants. The scenario should contain interaction possibilities which allow the user to perform the competency-relevant action tasks and it should facilitate the necessary thought processes for problem solving. The implemented actions should offer a combination of orientation, planning, execution and control. At the same time, the possibilities that VR offers for design should be used (e.g., novel ways of visualization or presentation of assistance) [10].

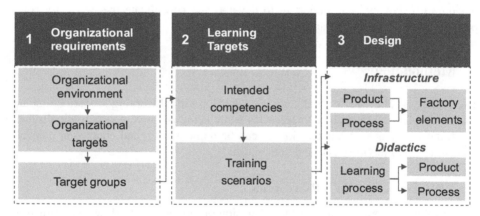

Fig. 1. Approach for competency-oriented design of suitable virtual learning environments [8].

In context of the design, reference is made to the design levels and the design dimensions by Tisch. The design levels are considered vertically at the macro (e.g., learning factory infrastructure), meso (e.g., teaching modules) and micro levels (e.g., specific learning-teaching-situations) [9]. The scenarios are set within a content-related and temporal framework. The individual activities of the scenarios are found on the micro level. Accordingly, it is particularly important to design the micro-level or the individual learning situations. In the context of planning a learning factory, there are five design dimensions that are based on factory-specific or learning-specific problem areas, such as the operator model, the mapped processes or the didactic reference [8]. In addition, a scenario design can follow three different product strategy approaches. A design from product to process, from process to product or a parallel product and process design [8]. In order to exploit the potential of VR, a parallel product and process design can be chosen. Both the processes and the product should in principle be freely selectable. The choice of products for learning factories also offers two paths.

Either industrial products already available on the market are chosen with the aim of completing the learning factory configuration or the development of customized products for fitting into the learning factory concept is aimed at. Mapping in virtual space offers the freedom to adapt processes and products without having physical dependencies [8] (Fig. 2).

	Aspect	Expressions			
V.1	Application time	Before physical training	During physical training	After physical training	
V.2	Duration	< 5 min	5 - 30 min	> 30 min	
V.3	Goal of usage	Personalization	Flexibility	Transfer	
V.4	Hardware connection	Wireless		Wired	
V.5	Output Devices	HMD	Desktop	Mobile Devices	
V.6	Type of use	Completely Virtual		Hybrid	
V.7	Development process	Internal	External	Cooperation	Use of available modules
V.8	Trainer role	Coach	Mentor	Lecturer	Moderator
V.9	Place of implementation	On-Site		Remote	
V.10	Simultaneous physical participants	1	2-5	6-10	>10
V.11	Simultaneous virtual participants	1	2-5	6-10	>10
V.12	Number of production environments	1	2-3	>3	
V.13	Number of processes presented	1	2-3	$\geq$3	
V.14	Number of difficulty levels	1	2	$\geq$3	
V.15	Introduction to the technology	Presentation	Virtual tutorial room	Individual briefing	

Fig. 2. Proposed extension of the leaning factory morphology by Tisch [11] for virtual learning factory trainings (expressions chosen in the research project PortaL underlined)

The potential of being able to change situations in VR through appropriate programming offers many further possibilities for individual adaptation and removes limitations. At the beginning, a selection should be made and can result in a set of difficulty levels from which a participant can choose. By selecting the appropriate difficulty levels, participants should not feel under- or overchallenged. Necessary time differences in the implementation can be put into perspective by the possibility of adaptation. At the same time, learning factory operators and developers are free to use this multitude of degrees of freedom also in the context of the implementation method.

This results in a table with a total of 15 fields of action. This table can be seen as an addition to the learning factory morphology introduced by Tisch [11].

These different expressions allow learning factory developers or learning factory operators to specifically address currently existing weaknesses of the physical learning factory concept. VR can allow remote learning with realistic scenarios as well as the adaptation of difficulty levels and scenarios offered.

4   Results

In the context of the research project PortaL, the approach presented in Sect. 3 was tested with five potential learning environments. In the following, one of these learning environments will be briefly presented on the basis of the three steps *organizational requirements*, *learning targets* and *design* of the approach.

4.1   Organizational Requirements

In the PortaL research project, the contents of the configuration scenarios are to be determined above all, whereby the organizational requirements as well as the learning targets were already available as given. The teaching-learning concept fits into the existing curriculum of the Process Learning Factory CiP at the PTW of the TU Darmstadt and aims in particular at employees of small and medium-sized German companies as a target group. The primary learning content is composed of lean manufacturing topics. In order to further specify the requirements with regard to VR, these were analyzed in a literature-guided manner in the research project and were evaluated in stakeholder interviews [12, 13]. For example, it is required that different learning types are taken into account. This can be done by offering a redundant information output, for example in visual and auditory form.

4.2   Learning Targets

For the application example, the learning targets and intended competencies are organized around the topic of value stream analysis. The learning targets were derived from the already existing physical value stream analysis learning factory workshop, which serves as a basis and is to be enriched by a virtual learning unit as part of the research project. The focus is to provide a complementary virtual scenario for the application of the learned method in a scenario close to the participants own professional environment.

To illustrate the procedure, value stream mapping was chosen as an example application. The primary objective of the value stream analysis is to analyze the actual state of a production value stream. In addition to the recording of process and cycle times, this also includes the recording of process links, inventory types and quantities as well as wastage (e.g., employee waiting times, inventories, etc.) that occur within the value stream. Other aspects, such as drawing the value stream map on the basis of the data obtained, can in principle also be represented in VR. In the context of the teaching-learning scenario presented, however, it was decided to leave this action task in the physical, real space, since a virtual whiteboard application would not offer any advantage over the physical variant with regard to the learning objective.

4.3   Design

With the specifications outlined in the two previous subsections in mind, the learning scenarios were designed. On the one hand, care was taken to ensure that all relevant action tasks could be mapped and, at the same time, that the maximum targeted duration of the virtual scenario does not exceed 30 min. The maximum duration can vary from participant to participant, but the selected value should be possible for the majority of participants without suffering from fatigue or other physiological reactions, such as motion sickness. Based on this restriction of a maximum learning time in VR, three to four virtual production processes should be designed. This amount was also chosen for the later defined scenarios. At the same time, all scenarios were conceptualized in such a way that at least three product variants can be represented in each scenario. Certain other strengths of VR (e.g., different types of visualizations, manipulation of time, etc.) can be considered and implemented later.

For the definition of the used products the development of customized products for fitting into the learning factory concept and a parallel product and process design were chosen. Therefore, on the basis of suitable literature on the topic of competency-oriented learning factories, five first production scenario approaches (for printed circuit boards, sheet metal profiles, design objects, toys and siccors) with a low level of complexity (4–5 production processes for each scenario) were developed and evaluated by experts in the next step. The team of experts consisted of scientific staff of the PTW, learning factory experts, experts from the field of didactics and a developer of VR environments. For the evaluation of the scenarios five criteria which should facilitate a competency-based scenario design were derived from the literature and the requirements of the research project. With the help of a pair comparison, a weighting of the individual criteria was achieved. The criteria and their respective weighting are the following:

- Added value compared to existing value stream (.27)
- Target group fit (.24)
- Low distraction (.20)
- Variant mappability (.20)
- Attractivity (.09)

The five potentially suitable scenarios were evaluated regarding the respective suitability of the scenarios among each other using the criteria named above. A utility analysis was used as an evaluation tool. The fulfilment of the criteria in the respective scenario was evaluated on a six-point scale from *not fulfilled (0)* to *completely fulfilled (5)*. The outcome of this evaluation decided which two approaches would be further developed.

The evaluation of the five scenarios showed (normalized results in brackets) that a production of *printed circuit boards* (1.0) and a production of *sheet metal profiles* (.95) seem to be the most attractive and suitable for the intended purpose of a virtual learning factory workshop with the goal to develop competencies regarding value stream analysis. The other three production scenario options, *design objects* (.90), *toys* (.86), and *scissors* (.84), scored lower in the expert rating and were therefore not further detailed. After the scenarios to be used were determined, they were extensively examined for adaptability and personalization and the possibilities were incorporated in detail into the concepts.

Scenario in Detail: Sheet Metal Profiles
The value stream consists of a total of 4 processes. In addition to a warehouse, the scenario consists of the production processes *cut to length, laser cutting, bending* and *testing*. Three product variants can be manufactured within the value stream. In addition to the execution of all competency-relevant action tasks (recording of relevant times, process links, stock types and waste), the value stream also allows, in particular, the observance of residence times in VR that are suitable from the project point of view. Despite covering all relevant content, the scenario is still kept lean enough to be explored completely within 15–30 min.

5 Discussion and Outlook

In the context of this publication, a procedure for the structured design of competence-oriented teaching-learning environments was presented alongside a morphology for the classification of virtual teaching-learning environments. Both elements emerged as a result of the research project PortaL and were already used there for the development of three teaching-learning scenarios. Although the principal suitability for the creation of teaching-learning scenarios could already be tested by the application for development as well as in smaller application tests with the virtual environments, a detailed evaluation is still pending. This is to take place in the near future to provide results on the suitability of the developed environments for competence development. With confirmation of suitability, the approach can provide a valuable contribution for developers to develop competency-based teaching-learning environments and to classify existing or new environments. Nevertheless, the proposed morphology can be used for the design of future learning scenarios in learning factories.

Acknowledgements. The authors would like to thank the Bundesministerium für Forschung und Bildung (Federal Ministry of Education and Research, BMBF) and the Deutsches Zentrum für Luft- und Raumfahrt (German Aerospace Center, DLR) for financial and administrative support during the project "Virtual Action Tasks for Personalized Adaptive Learning" (01PV18001A).

References

1. North, K., Reinhardt, K., Sieber-Suter, B.: Kompetenzmanagement in der Praxis: Mitarbeiterkompetenzen systematisch identifizieren, nutzen und entwickeln; mit vielen Fallbeispielen, 2nd edn. Gabler Verlag, Wiesbaden (2013). http://site.ebrary.com/lib/alltitles/docDetail.action?docID=10715795

2. Tisch, M., Metternich, J.: Potentials and limits of learning factories in research, innovation transfer, education, and training. Procedia Manuf. **9**, 89–96 (2017). 7th CIRP-sponsored Conference on Learning Factories
3. Abele, E., et al.: Learning factories for research, education, and training. Procedia CIRP **32**, 1–6 (2015). 5th Conference on Learning Factories
4. Weidig, C., Menck, N., Winkes, P.A., Aurich, J.C.: Virtual learning factory on VR-supported factory planning. In: Bayro-Corrochano, E., Hancock, E. (eds.) Progress in Pattern Recognition, Image Analysis, Computer Vision, and Applications: 19th Iberoamerican Congress, CIARP 2014, Puerto Vallarta, Mexico, 2–5 November 2014, vol. 434, pp. 455–462. Proceedings, Springer International Publishing, Cham (2014). https://doi.org/10.1007/978-3-662-44745-1_45
5. Abele, E., et al.: Learning factories for future oriented research and education in manufacturing. CIRP Ann. **66**(2), 803–826 (2017)
6. Thomas, O., Metzger, D., Niegemann, H.: Digitalisierung in der Aus- und Weiterbildung: Virtual und Augmented Reality für Industrie 4.0. Springer, Berlin, Heidelberg (2018). https://books.google.de/books?id=W3VQDwAAQBAJ
7. Heers, R.: "Being There": Untersuchungen zum Wissenserwerb in virtuellen Umgebungen. Dissertation, Tübingen, Eberhard-Karls-Universität Tübingen (2005)
8. Tisch, M., Hertle, C., Abele, E., Metternich, J., Tenberg, R.: Learning factory design: a competency-oriented approach integrating three design levels. Int. J. Comput. Integr. Manuf. **29**(12), 1355–1375 (2016)
9. Tisch, M., Hertle, C., Cachay, J., Abele, E., Metternich, J., Tenberg, R.: A systematic approach on developing action-oriented, competency-based learning factories. Procedia CIRP **7**, 580–585 (2013). 46th CIRP Conference on Manufacturing Systems
10. Riemann, T., et al.: Gestaltung von personalisierten Lernfabrikschulungen in Virtual Reality im Kontext schlanker Produktion, pp. 1–6. GfA-Press, Dortmund (2020). Digitaler Wandel, Digitale Arbeit, Digitaler Mensch?
11. Tisch, M., Fabian, R., Abele, E., Metternich, J., Hummel, V.: Learning factory morphology – study of form and structure of an innovative learning approach in the manufacturing domain. Turkish Online J. Educ. Technol. Special Issue **2**(2015), 356–363 (2015)
12. Riemann, T., Kreß, A., Roth, L., Klipfel, S., Metternich, J., Grell, P.: Agile implementation of virtual reality in learning factories. Procedia Manuf. **45**, 1–6 (2020)
13. Riemann, T., Kreß, A., Roth, L., Klipfel, S., Metternich, J., Grell, P.: Requirements for the Implementation of Virtual Reality in Learning Factories, pp. 1–20. TUPrints, Darmstadt (2020)

Fostering Insights from Real-Time Data

Henrik Saabye[1,2(✉)] and Daryl John Powell[3,4]

[1] VELUX A/S, 8752 Østbirk, Denmark
Henrik.saabye@velux.com
[2] Department of Materials and Production, Aalborg University,
9220 Aalborg, Denmark
[3] SINTEF Manufacturing AS, Horten, Norway
daryl.powell@sintef.no
[4] Norwegian University of Science and Technology, Trondheim, Norway

Abstract. A logical first step for many manufacturers when embarking on a transformation towards digitalization of their production system is to acquire technologies that captures data in real-time to help monitor and improve machines and production performance. Nevertheless, the presence of real-time digital data will not in itself lead to significant improvement in production performance. It also requires manufacturers to enable shop floor workers to generate insights from these real-time captured data to frame and resolve problems. As such, this action research paper presents the outcome of a learning intervention's first action learning cycle at the Danish Roof-top windows manufacturer VELUX. The intervention aimed to institute a lean learning system to enable the further successful digital transformation of the company's production system.

Keywords: Lean · Digitalization · Action learning · Action research

1 Introduction

Today, most manufacturers have initiated a digital transformation to enable their organizations to utilize new digital technologies to cope with changing customer demands [1]. Manufacturers who already have integrated lean practices into their existing production system are therefore exploring how lean and digital manufacturing technologies are supplemental or integrated. Several studies examine the affiliation between lean practices and digital manufacturing technologies, e.g., Bittencourt [2]. However, these studies predominantly address issues related to digitization and not digitalization. Digitization is the technical process of transforming from an analog to a digital form.

In contrast, digitalization is a socio-technical phenomenon about organizations' capability to utilize digital technology and data to transform [3]. Hence, it is the technocentric focus associated with digitization that can prevent manufacturers from capitalizing on their investment in digital technologies and utilizing the increasing amount of digital data available on the shop floor [4]. Therefore, lean manufacturers pursuing a digital transformation must achieve a higher degree of organizational learning capabilities [5] and insights gained through experimentation and reflection [6].

© IFIP International Federation for Information Processing 2021
Published by Springer Nature Switzerland AG 2021
D. J. Powell et al. (Eds.): ELEC 2021, IFIP AICT 610, pp. 68–76, 2021.
https://doi.org/10.1007/978-3-030-92934-3_8

Developing organizational learning capabilities would require manufacturers to supplement their production system with a learning system, e.g., to develop capabilities for utilizing digital data on the shop floor [7, 8]. Therefore, this study seeks to advance our understanding of how manufacturers can institute a lean learning system that simultaneously engages all organizational hierarchy levels in developing shop-floor workers in gaining insight and acting based on digital data. This paper outlines the first action cycle of designing and deploying an action learning intervention as part of an action research project at VELUX to develop a lean learning system. By investigating this first action cycle, the paper seeks to answer the research question: *"How can manufacturers develop and enable shop-floor workers to utilize real-time captured data, gain insights, and initiate actions - based on the scientific method."*

2 Literature Review

2.1 Lean as a Learning System

There is a growing consensus in extant operation management literature that adopting a lean production system requires manufacturers to understand lean as a learning system [9]. When understanding lean as a learning system, the purpose of lean tools and methods is to foster insights among employees and develop an organization of problem solvers instead of a sole objective of improving efficiency [8]. Hence the manufacturers are obliged to ensure that new tools and methods, either analog or digital, are standardized and formalized into the workers' daily work in an enabling way and not perceived as coercive [10]. When shop-floor workers perceive real-time data capturing systems as enabling, they are more likely to be motivated intrinsically to address problems, which is a prerequisite for generating insights, action, and learning without awaiting permission from management or specialists [10].

An underlying element of a lean learning system is action learning, based on the notion that no learning can occur without action and no (sober or deliberate) action without learning [11]. Revans [12] proposes a theory for action learning consisting of the three equally important learning systems of alpha, beta, and gamma. These three systems are not connected linearly or sequentially but are intertwined. System alpha is the task of framing a problem by investigating the external demands and how these correspond with the existing internal capabilities and the problem solvers' values and mental models. System beta is the underlying problem-solving process based on the scientific method [6]. System gamma is the learning process of examining the divergence between expectation and experience by questioning and critically reflecting on the contextual taken-for-granted preconceptions [13]. Moreover, a distinct element of a lean learning system is the presence of a supportive learning environment [4, 14]. A supportive learning environment embodies an atmosphere of psychological safety where workers feel safe expressing their and ideas and spaces to explore each other's ideas, experiment, and reflect together [15].

3   Research Method

To accommodate this study's interventionistic and action-based design, an action research approach is adopted [15] and follows the 5-step action learning cycle process of diagnosing, action planning, action-taking, and evaluation [16]. The research is part of an ongoing industrial Ph.D. project to advance VELUX's ability to utilize existing and future digital technologies as part of a digital transformation. VELUX has developed its production system on lean tools and methods for the past two decades. This paper reports on the first (pilot) of three action learning cycles facilitated by the first author at a VELUX lead factory in Denmark.

4   Results – The Action Research Process

4.1   Diagnosing

As part of the diagnosis phase, the first author followed the implementation of a new real-time digital capturing system on one of the Cladding department's production lines. Six months after commission, production had not improved, nor had the workers used the data to initiate any problem-solving or improvement activities. To understand why the real-time capturing system's implementation was unsuccessful, the first author conducted a series of observations, interviews, and a survey with managers, project managers, specialists, and shop-floor workers [4]. The diagnosis resulted in the following key findings:

- The predominant approach to problem-solving is firefighting [17]. The scientific method is only applied episodically.
- The shop floor workers believe that it is not their job to initiate problem-solving or improvements themselves, but only to notify the maintenance department or manager and perceived the new real-time data capturing system as coercive [10].
- There is no visible presence of learning structures and processes supporting system alpha, beta, and gamma where shop floor workers, specialists, and managers can collaborate on conducting experiments and reflecting [12, 14].
- Leaders did not perceive their job as fostering a supportive learning environment [14]. Instead, they provided the answers and solutions instead of asking questions and developing the scientific method's usage among their direct reports [18].

As part of the action research process, the diagnosis's findings were examined and validated by the research participants. The respondents and the project steering committee agreed to design and test an action learning intervention [15].

4.2   Action Planning

The first author designed an action learning intervention based on Revan's [12] intertwined alpha, beta and gamma systems to resolve the issues identified during the diagnosis phase. The purpose of the action learning intervention is to develop the managers' ability to enable and empower shop-floor workers to apply the scientific method and utilize (digital) production data when solving problems by fostering a supportive learning environment.

The factory leadership team decided to pilot the intervention in two separate departments. The first case took place at the cladding department and involved the department manager and two groups consisting of 3 shop-floor workers. The second case two took place at the flashing department, and besides the department manager, two groups consisting of 6 participating shop-floor workers. In both cases, the aluminum factory manager, the general manager, and the first author participated.

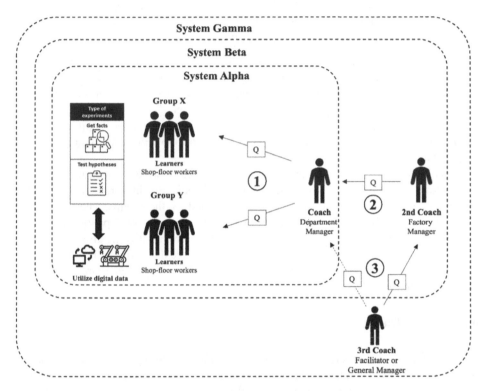

Fig. 1. Action learning intervention for developing a supportive learning environment based on system alpha, beta, and gamma.

System Alpha. As illustrated in Fig. 1, the action learning intervention setup is for the shop-floor workers to find, face and frame a specific operational problem identified with their department manager. To support them in applying the scientific method, the groups meet every morning for 15 min of coaching with the department manager (step 1 in Fig. 1). Grounded on Rother's [19] coaching kata, the department manager firstly asks the groups to visualize the current situation of the performance gap they are working on closing. To foster insights into the group's problem, the department manager facilitates a content reflection [20] on their last experiment as knowledge input to define the next small experiment to be conducted until the next day's meeting. The groups are encouraged to decide between two types of experiments. Either gather facts or test specific hypotheses. For both types of experiments, the groups utilize existing digital production data to solve and validate the experiments. Should the groups, e.g., lack the ability to retrieve or analyze data from the data gathering system, the next-day step is then to learn this ability. To capture the gained learning and insight, the groups use a learning board and an experiment log designed with inspiration from Rother's [19] improvement kata.

Preparation and Problem Identification. In the first case, the department manager decided to invite the shop-floor workers to introduce the learning intervention and discuss the problem the groups should start resolving. In the second case, the department manager decided to start directly with initiated the morning coaching cycles.

Initiating the Daily Learning Cycles. In the first case department, the coaching cycles were conducted as planned, and the groups began to start working on gathering insights for framing the identified performance gap. However, the department manager struggled to help one group frame the problem and identify experiments within their circle of control and not eschewing the problem. Contrary, the other group felt a sense of empowerment and quickly improved their ability to follow the scientific method and utilize data.

Since specific performance gaps for the groups to close were not identified upfront, the second case department struggled with starting their learning process. During the first week, no experiments were defined and initiated by the groups. The department manager realized that just because several performance gaps were apparent to him, it was not apparent for the shop floor workers. He, therefore, invited the groups into a set of meetings where he presented the available performance data and facilitated a process where the groups identified performance gaps to start closing. Afterward, the groups began to identify and execute their first experiments.

4.3 Action Taking

System Beta. System beta concerns forming solutions through cycles of action by having the groups generate insights for closing the concrete performance gaps. During this process, the department manager, as the coach, must focus on the underlying problem-solving process and develop the group's ability to apply the scientific method. The department manager's task is not to propose solutions but to facilitate a cycle of actions and process reflections with the groups of learning how to solve the problem at hand, how the insights of the problem and countermeasures are evolving, and how to conduct and evaluate experiments [20–23].

To develop the department managers' ability to facilitate a cycle of actions and reflections with participating groups, the factory managers observe the conversation between the groups and the department manager. Afterward, the factory managers facilitate a cycle of actions and process reflection [20] with the department manager on developing the groups' scientific method abilities (step 2 in Fig. 1). These conversations are expected to result in gemba-walks [18] to help the groups conduct their experiments.

Conduction Daily Learning Cycles. In both departments, the daily meetings continued for five weeks until solutions were in place to close the identified performance gaps. In the first department, the department manager decided to drop the group that was eschewing the learning process and only helping the other group. In the second department, both groups exceeded the manager's expectations.

4.4   Evaluation

System Gamma. The purpose of system gamma is firstly about reflecting over (and sharing) emergent learning. Secondly, it is about developing the factory managers' ability to learn how to learn and become aware of preconceptions and mental models [13] that hinder the groups from applying the scientific method and for the factory to adopt a lean learning system. The facilitator and the general manager observe both conversations (Steps 1 and 2 in Fig. 1).

Afterward, the facilitator and the general manager engage in premise reflection [20] with the factory manager on developing the department manager's thinking and practice for developing the group's scientific method abilities (step 3 in Fig. 1). This conversation can be described as upstream and downstream learning [24]. 'Upstream' learning can be classified as a critical reflection process of examining the managers underlying mental models, assumptions, and values that govern the 'downstream' behaviors.

After-learning Cycle Evaluation. Derived from observations, performance data, and interviews both during and after the intervention pilot's ending, the participants reported the following learning:

- Both department managers observed that the shop-floor workers took more ownership to improve their work and solve problems. They were proactively experimenting with ideas on improving their production lines. The department managers also observed improved communication among the shop-floor workers where they were helping each other more and openly discuss and distribute tasks in a respectful manner.
- The shop floor workers all agreed that they want to continue working in this new way and worries that it was a one-time activity. They all felt that they obtained a shared understanding of their problems, resulting in more open discussions without co-workers becoming defensive. They also realized that it is much more effective to focus on what they themselves can do to improve performance instead of pointing fingers at other departments and waiting for maintenance.
- The Factory manager observed that the department managers had started to be more critically reflective of their mental models and behavior [13], leading to a more long-term focus beyond resource efficiency.

Operational Outcomes. In the first case, the participating shop floor workers improved OEE (Overall Equipment Efficiency) from 48 to 60. The shop floor workers in the second case improve their daily output by 10%.

5   Discussion

The evaluation of the first action learning cycle indicates:

- Most participating shop-floor workers applied the scientific method by iterating between experimentations and reflections to generate insights from available production data.
- The participating shop-floor workers experienced first-hand how to expand their circle of control by conducting problem-solving or improvement activities themselves without waiting for maintenance or management to take action.
- The participating shop-floor workers experience that data gathering systems can enable them to improve their production lines' performance.
- Instituting a lean learning system with structures and processes build on the intertwined systems of alpha, beta, and gamma and involving the whole organizational hierarchy simultaneously is an effective way to foster a supportive lean learning environment for conducting experiments and reflection [25].

The evaluation also suggests that the department manager identify a potential performance gap with the participating shop-floor workers before initiating the daily morning cycles. The managers must furthermore be patient and expect to set aside time to communicate and coach between the morning coaching cycles in the beginning to avoid groups abandoning the learning process.

6   Conclusion

This paper describes the outcome from the first cycle of an action learning intervention at VELUX. The purpose of the intervention is to build a lean learning system that enables shop-floor workers to gain insight and act based on real-time digital data according to the scientific method. Designing the action learning intervention based on the intertwining systems of alpha, beta, and gamma [12] combined with simultaneously involving all organizational hierarchy levels proved successful. It positively addresses the organizational issues identified in the diagnosis phase and the challenges of developing organizational learning capabilities as part of a digital transformation. However, the study has its limitations. The paper only describes the first learning cycle of the action research project, which only involved 5% of the workforce at the Danish VELUX factory. Moreover, the first learning cycle lasted only five weeks and did not cover the intervention's longitudinal effects. Thirdly, there is a limitation in terms of transferability to other contexts as a single case study.

References

1. Sousa-Zomer, T.T., Neely, A., Martinez, V.: Digital transforming capability and performance: a microfoundational perspective. IJOPM **40**, 1095–1128 (2020). https://doi.org/10.1108/IJOPM-06-2019-0444

2. Bittencourt, V.L., Alves, A.C., Leão, C.P.: Industry 4.0 triggered by lean thinking: insights from a systematic literature review. Int. J. Prod. Res. **59**, 1496–1510 (2021). https://doi.org/10.1080/00207543.2020.1832274

3. Legner, C., et al.: Digitalization: opportunity and challenge for the business and information systems engineering community. Bus. Inf. Syst. Eng. **59**(4), 301–308 (2017). https://doi.org/10.1007/s12599-017-0484-2

4. Saabye, H., Kristensen, T.B., Wæhrens, B.V.: Real-time data utilization barriers to improving production performance: an in-depth case study linking lean management and industry 4.0 from a learning organization perspective. Sustainability **12**, 8757 (2020). https://doi.org/10.3390/su12218757

5. March, J.G.: Exploration and exploitation in organizational learning. Organ. Sci. **2**, 71–87 (1991). http://www.jstor.org/stable/2634940

6. MacDuffie, J.P.: The road to root cause: shop-floor problem-solving at three auto assembly plants. Manag. Sci. **43**, 479–502 (1997). https://dl.acm.org/doi/10.5555/2869050.2869056

7. Fujimoto, T.: The Evolution of a Manufacturing System at Toyota. Oxford University Press, New York (1999)

8. Liker, J.K.: The Toyota Way: 14 Management Principles from the World's Greatest Manufacturer. McGraw-Hill, New York (2004)

9. Powell, D., Reke, E.: No lean without learning: rethinking lean production as a learning system. In: Ameri, F., Stecke, K.E., von Cieminski, G., Kiritsis, D. (eds.) APMS 2019. IAICT, vol. 566, pp. 62–68. Springer, Cham (2019). https://doi.org/10.1007/978-3-030-30000-5_8

10. Adler, P.S., Borys, B.: Two types of bureaucracy: enabling and coercive. Adm. Sci. Q. **41**, 61 (1996). https://doi.org/10.2307/2393986

11. Powell, D.J., Coughlan, P.: Rethinking lean supplier development as a learning system. Int. J. Oper. Prod. Manag. **40**, 921–943 (2020). https://doi.org/10.1108/IJOPM-06-2019-0486

12. Revans, R.W.: Developing Effective Managers. A New Approach to Business Education. Longman, London (1971)

13. Reynolds, M.: Reflection and critical reflection in management learning. Manag. Learn. **29**, 183–200 (1998). https://doi.org/10.1177/1350507698292004

14. Garvin, D.A., Edmonson, A., Gino, F.: Is yours a learning organization? Harv. Bus. Rev. **83**, 109 (2008)

15. Coughlan, P., Coghlan, D.: Action research for operations management. Int. J. Oper. Prod. Manag. **22**, 220–240 (2002). https://doi.org/10.1108/01443570210417515

16. Coghlan, D.: Doing Action Research in Your Own Organisation, 5th edn. SAGE Publications, Thousand Oaks, CA (2019)

17. Tucker, A.L., Edmondson, A.C., Spear, S.: When problem solving prevents organizational learning. J. Organ. Chang. Manag. **15**, 122–137 (2002). https://doi.org/10.1108/09534810210423008

18. Dombrowski, U., Mielke, T.: Lean leadership–15 rules for a sustainable lean implementation. Procedia CIRP **17**, 565–570 (2014). https://doi.org/10.1016/j.procir.2014.01.146

19. Rother, M.: Toyota Kata: Managing People for Improvement, Adaptiveness, and Superior Results. McGraw Hill, New York (2010)

20. Mezirow, J.: Transformative Dimensions of Adult Learning, 1st edn. Jossey-Bass, San Francisco (1991)

21. Coghlan, D., Coughlan, P.: Notes toward a philosophy of action learning research. Action Learn. Res. Pract. **7**, 193–203 (2010). https://doi.org/10.1080/14767333.2010.488330

22. Ballé, M., Jones, D.T., Chaize, J., Fiume, O.: The Lean Strategy: Using Lean to Create Competitive Advantage, Unleash Innovation, and Deliver Sustainable Growth. McGraw-Hill Education, New York (2017)

23. Ballé, M., Chartier, N., Coignet, P., Olivencia, S., Powell, D., Reke, E.: The Lean Sensei. Lean Enterprise Institute Inc., Boston (2019)
24. Reason, P., Torbert, W.: The action turn: toward a transformational social science. Concepts Transform. **6**, 1–37 (2001). https://doi.org/10.1075/cat.6.1.02rea
25. Edmondson, A.C.: The Fearless Organization. John Wiley & Sons, Hoboken (2018)

Shaping Lean Teaching Methods: Recognizing Lean as a Journey vs. A Set of Tools

Khavitha Singh$^{(\boxtimes)}$ and Guy Bowden

Toyota Wessels Institute for Manufacturing Studies, Durban, South Africa
Khavitha.singh@twimsafrica.com

Abstract. The Toyota Wessels Institute for Manufacturing Studies (TWIMS) was founded in 2018 in South Africa. Its mandate is to develop manufacturing executive leadership capabilities in Africa.

Academics that teach and engage on lean transformation journeys recognize that lean management systems are a foundational requirement for sustainable lean capability development. And yet, management practitioners (and consultants) often fail to recognize this. Rather, they see lean as a toolkit applied for quick organizational turnarounds and short-term operational or supply chain management gains.

This paper explores the lean teaching methodology adopted at TWIMS which contextualises the short-term elements of lean within a long-term journey of continuous improvement. The paper explains that such teaching methodologies are vital if students are to implement lean strategies in their organisations that last beyond the initial gains created by short-term lean tools. Finally, the paper finds that TWIMS' teaching methodology is successful in creating a more holistic comprehension of lean among students, which instils a greater appreciation for lean as a long-term strategy.

Keywords: Lean teaching methods · Lean tools · Lean management philosophy

1 Introduction

The effectiveness of lean teaching methods at a post graduate level have been widely documented by various authors over the last few years. This has shaped the teaching methods adopted by practitioners globally in ensuring that these lean methods and concepts, with their associated learnings, secure effective pragmatic comprehension within manufacturing environments. A key consideration is: have these teaching methods been successful in representing lean as a system underpinned by a long-term management philosophy, or has it rather encouraged the view that lean is a set of practices or tools?

This paper sought to shed light on the current lean teaching methods adopted by the academic team at the Toyota Wessels institute for Manufacturing Studies (TWIMS) and understand whether they have been effective in shaping students' interpretation of lean as a management philosophy and part of a long-term journey, rather than a set of discrete tools and methods for realizing short-term benefits. TWIMS conducts two lean

© IFIP International Federation for Information Processing 2021
Published by Springer Nature Switzerland AG 2021
D. J. Powell et al. (Eds.): ELEC 2021, IFIP AICT 610, pp. 77–85, 2021.
https://doi.org/10.1007/978-3-030-92934-3_9

management executive short courses – Lean Supply Chain Management and Lean Operations Management – on an annual basis. While students are taught various lean tools in both courses, these are deeply contextualized in the philosophical and strategic imperatives of implementing lean as a long-term process. Liker's 4P Model, which encompasses Toyota's 14 management principles (Liker 2021), informs a strong foundational ethos for TWIMS' content structure which also emphasizes strategic and journey elements of lean. In addition, the course content shares the history of lean as well as the evolution of the manufacturing automotive sector within this context. The results of TWIMS lean teaching approach are shared and discussed throughout this paper.

2   Literature Review

2.1   Lean as Management Philosophy

Lean manufacturing, over the last few years has been presented by academics, practitioners and consultants as a set of practices and methods, that if integrated correctly into processes and ways of working, will yield positive business results. These results emerge because of reduced waste and improved flow of communication, material, information and movements within a value chain (Womack and Jones 2003).

Authors such as Liker (2021) and Bhasin and Burcher (2006), have postulated that lean goes far beyond those practices and has to be viewed as a philosophy, integrated as a part of broader management culture. Similarly, Ballé, et al. (2015) support this view and encourage leaders to adopt lean as a strategy and support the notion that lean should be seen as journey, rather than a set of tools and practices. Hence practitioners are encouraged to see lean, not as a total production system but rather a total management system (Bhasin and Burcher 2006). Liker (2021), in his book – *The Toyota Way* – where he captures the 14 management principles, also highlights that the success of lean is basing management decisions on a management philosophy.

2.2   Common Lean Teaching Methods and Outcomes

Tortorella and Cauchick-Miguel (2018) share that the effectiveness of learning approaches attached to lean is dependent on several elements, including the teaching methods that are adopted. Successful lean teaching methods have adopted a variety of approaches that have evolved over the last few years - aligned to the spirit of improvement. These varieties have included case studies, guest visits, simulations as well as industry visits.

Bednarek et al. (2020) in their analysis of 39 post graduate programs of lean, identified that one of the effective elements of lean teaching was the inclusion of industry visits to align teachings to application and real business situations. This thought process was supported by Tortorella and Cauchick-Miguel (2018), where they encouraged the teaching efforts to include problem-based teaching methods - aligning real world integration to lean concepts.

In their table, consolidating lean manufacturing teaching methods, objectives largely centered around application and technical skills development. This was confirmed with some of the objectives noted: "Complement the understanding of lean theory so that students become able to apply it outside the classroom", and "Enhance student technical and professional skills and business knowledge", and "Verify how LM techniques and principles can be transferred to companies and students" (Tortorella and Cauchick-Miguel 2018, pp. 305–306).

This supports the view, that whilst teaching methods and designs may be effective, they tend to encourage the view that lean is interpreted as a set of tools or techniques rather than a long-term strategic philosophy and management system. One has to align teaching content and concepts and application to a continuous learning system (Hall and Holloway 2008). Hence one cannot encourage learning with instantaneous once off application.

2.3 Why the Need to Evaluate Lean Teaching Methods?

Only 10% of lean transformation journeys or integration of lean into practices are successful or sustained (Bhasin and Burcher 2006; Poksinska et al. 2013). Similarly, only 10% of organizations have the philosophy element linked to lean embedded as part of the company culture and hence most organizations achieve ephemeral benefits (Bhasin and Burcher 2006; Lodgaard et al. 2016).

Whilst the review of teaching methods has been encouraged by the lean academic community, one must be mindful if the teaching methods are adequate in shifting the wrongful assumptions of lean by those who are expected to implement it in their operations. Ultimately, the teaching methods should reinforce lean as a long-term philosophy which involves the development of leaders, individuals and teams who embody and display the principles of lean in a consistent manner throughout every level of an operation. Therefore, it is essential that this be at the forefront of discussions surrounding lean teaching methods.

3 Research Methodolgy

A total of 38 past participants of TWIMS' Lean Supply Chain Management and Lean Operations Management courses in 2019, 2020 and 2021, where surveyed in order to better understand the effectiveness of TWIMS' teaching methods in shaping lean as a long-term philosophy.

The research methodology consisted of an initial survey of 27 participants which largely consisted of questions that required a qualitative response from participants. Because these answers were in a qualitative form, answers were broken down into key words and phrases. The frequency of keywords and phrases was then measured and tabulated to show the participants' perspectives on TWIMS lean teaching methods.

From the results of the initial survey a second survey was introduced with participants in TWIMS' most recent lean course which ran in August of 2021. The methodology behind this second survey sought to better quantify the change in students' perspectives on lean as a result of TWIMS' teaching method and therefore used several Likert-style and matrix questions that captured students' perspectives on lean before and after the course. Since this survey was only run for one course the sample size for the second survey was only 11. While the sample size for this survey is limiting, it is TWIMS' intention to run this survey after every course in the future, thereby growing the sample size and the reliability of the dataset.

Finally, in cases where both surveys asked the same questions the results of both surveys were combined, and the sample size was indicated as 38 where applicable. Where it was not possible to combine the results of both surveys, the results of one of the surveys was shown and discussed while the key findings of the other were explained alongside this.

4   Analysis of Results and Discussion

Of the 38 participants from both surveys 13 indicated that they had attended a lean management course prior to attending one of TWIMS' lean management courses. Of those students that had already attended a lean course prior to TWIMS' course, both surveys showed a clear predisposition to viewing lean as a means of simply reducing waste. Moreover, none of these students expressed lean through the lens of its philosophical and Kaizen elements as a result of previous courses.

In both surveys students were asked whether they viewed lean as a short-term or long-term process before and after TWIMS' course. Figure 1 showed that TWIMS' course had a clear effect on changing students' views on lean with 100% of students indicating they viewed lean as a long-term practice after the course. This is likely a result of the strong emphasis on management philosophy, strategic intent and journey elements of lean emphasized during the course. While Fig. 1 expresses the results of the second survey since this was more quantifiable it should be noted that a similar trend was observed in survey one. In both surveys, when asked what had informed participants' prior opinions on lean as a short-term practice the majority cited their company's understanding of lean, followed by previous lean courses and interaction with lean consultants.

Figure 2 and Fig. 3 show only results from survey two (i.e. n = 11). Visually the figures show a clear impact of TWIMS teaching methods in changing their interpretations of lean. Terms such as "strategy", "journey", "philosophy", "incremental" and "culture" go from being almost uncited to the most cited. Importantly however, terms such as "just-in-time", "waste reduction", and "cost reduction" stayed significant with only a slight decrease in overall citations. This emphasizes the contextualization of such elements within an overarching long-term philosophy as espoused by TWIMS.

Figure 4 shows how participants interpretations of lean management changed as a result of attending TWIMS' lean courses for both surveys (i.e. n = 38). Encouragingly, an appreciation for "organizational culture and inclusivity" and the "longer-term processes and philosophies of lean" were cited most frequently. Importantly, those elements which could be considered more of the short-term aspects of lean (i.e., lean tools and waste reduction) were still cited among some of the participants. This reflects TWIMS' teaching methodology which contextualizes these tools within a long-term process. The balanced takeaway by students in terms of long-term and short-term

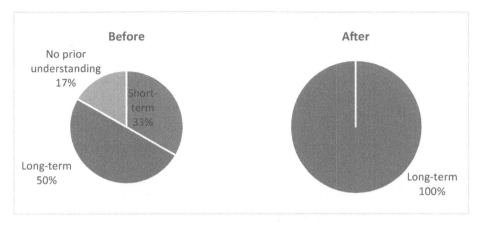

Fig. 1. Views on lean as a short term or long-term process before and after TWIMS' course (n = 11).

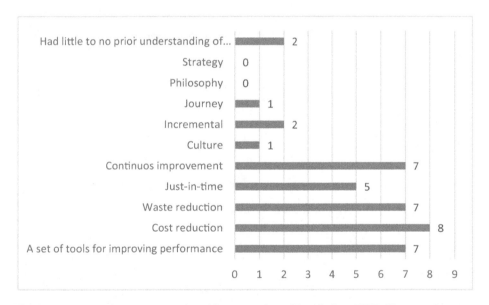

Fig. 2. Key terms that capture students' interpretation of lean **before** TWIMS course (n = 11).

aspects of lean is encouraging and shows a greater appreciation for lean in its totality and not simply a means for short-term benefit.

Figure 5 shows those elements of TWIMS' teaching approach that participants found most influential in changing their interpretation of lean management for both surveys (i.e. n = 38). "Interaction with fellow class participants" was the most frequently cited element that changed students' perspectives on lean. This is likely due to a healthy balance of different manufacturing sectors and backgrounds represented by students within the class. Their different backgrounds allow them to share, engage and

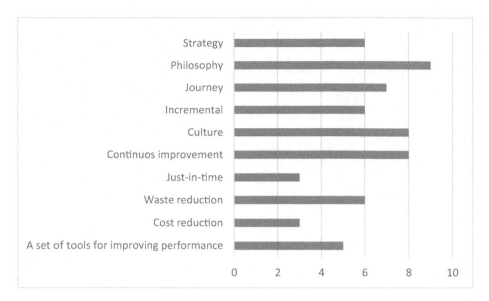

Fig. 3. Key terms that capture students' interpretation of lean **after** TWIMS course (n = 11).

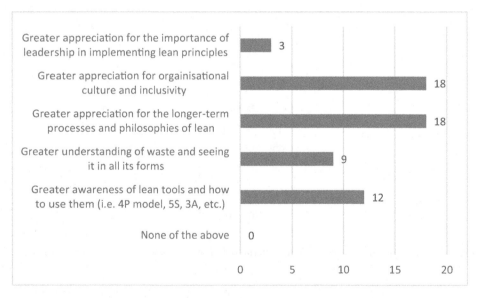

Fig. 4. Participants' interpretation of lean management after attending TWIMS' lean courses (n = 38).

challenge each other through varying perspectives while interacting with the course content and problems faced by each other's firms. Figure 5 also shows the value of practical firm level examples in influencing students' perspectives on lean. Such examples included: visits to automotive OEM suppliers, textile manufacturers and other examples during class – all exposing them to lean best practices and Gemba principles. Students likely attach a high level of value to this as it gives them exposure to different industries and allows them to approach broader business challenges, as well as those associated with the implementation of lean, from a fresh perspective.

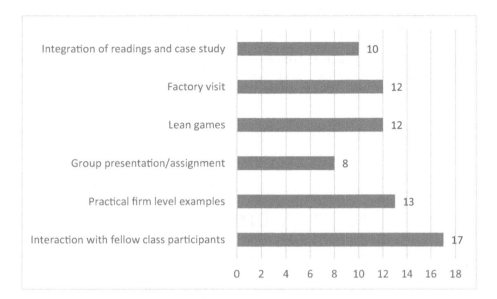

Fig. 5. Elements of teaching approach that changed students' perspective on lean management (n = 38).

Due to the emergence of COVID-19 not all students were able to participate in a factory visit. Thus, "factory visits" would likely have been more frequently cited in Fig. 5. Of those students who did attend factory visits, personal interactions showed this to be a particularly profound learning activity as they were able to see a world class lean factory in operation. In class discussions students explained that this was critical in demonstrating the value of lean in an operation and how it could visibly transform their own operations. Figure 5 also showed "lean games" to be fairly frequently cited. The lean games consisted of a Lego simulation in which participants were part of a value chain. Several iterations of the game were run with each incorporating another layer of lean practices (i.e. going from a batch process to single batch process and then introducing Kanban). The games included a score component and class discussion around these scores allowed students to see the impact of lean practices on even the simplest of tasks. While the games highlighted the value of lean tools, class discussion was used to contextualize this in terms of the long-term philosophical elements of lean. Discussion centered around the importance of leadership, training, and communication within teams so that continuous improvements can be made even once lean tools have been applied.

Figure 6 shows participants' key takeaways from TWIMS' lean course in survey two. The figure shows that TWIMS teaching approach had a clear impact on their understanding of lean as a long-term process. "Greater appreciation for organizational culture and inclusivity" and "greater appreciation for the longer-term processes and philosophies of lean" were by far the most cited takeaways from the course. The importance of lean tools and reducing waste were still frequently cited, but not at the expense of the long-term elements. In circumstances where students do not frequently cite the long-term practices of lean it shows that students do not have a complete understanding of lean. Importantly, participants in survey one showed similar takeaways further supporting the evidence shown in Fig. 6. Therefore, it is encouraging to see students taking away a more holistic view of lean which will have a greater and more sustainable impact on their businesses.

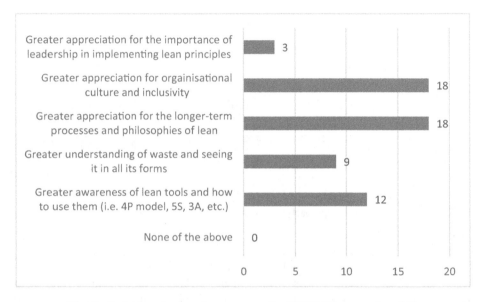

Fig. 6. Insights gained on lean as a result of TWIMS' course (n = 11).

5   Conclusion

Perspectives and teaching methods of lean that reduce it to a set of tools for short term benefits are problematic and in need of revision. This paper showed the experiences of students who participated in lean courses facilitated by TWIMS. The results showed TWIMS' teaching methodology to be effective in contextualizing the short-term elements of lean within in a more holistic understanding of lean as a long-term journey. The importance of attaching the philosophical elements of lean to any organizations lean journey is a key element in the successful implementation of lean. It is also an important factor in ensuring that any short-term benefits realized through the implementation of lean tools can be sustained and built upon into the future.

Factory visits, practical case studies and the contextualization of lean as a long-term management philosophy throughout TWIMS' courses were critical in changing students' misinterpretations of lean. It is important to reiterate that TWIMS teaching approach does not exclude the importance of lean tools and other short-term practices. Instead, such elements that teach these aspects of lean (i.e. lean games, group work and theory) are contextualized, through group discussion, as part of a long-term journey that involves all members of an organization striving for continuous improvement. The result of such teaching methodologies yields stronger business leaders who can better implement long-term lean strategies within their business operations.

References

Ballé, M., Jones, D., Orzen, M.: True lean leadership at all levels. Ind. Manag. **57**(1), 26–30 (2015)

Barnes, J.: Developing manufacturing leadership in South Africa (and regionally): the role of monozukuri. In: South Africa–Japan University Forum Conference (2019)

Bednarek, M., Buczacki, A., Bielakowski, L., Gladysz, B., Bryke, M.: Postgraduate studies on lean management–a review of initiatives. Educ. Sci. **10**(8), 1–21 (2020). https://doi.org/10.3390/educsci10080197

Bhasin, S., Burcher, P.: Lean viewed as a philosophy. J. Manuf. Technol. Manag. **17**(1), 56–72 (2006). https://doi.org/10.1108/17410380610639506

Hall, A., Holloway, L.: Application of lean concepts to the teaching of lean systems. In: ASEE Annual Conference and Exposition, Conference Proceedings (2008). https://doi.org/10.18260/1-2-3836

Liker, J.K.: Introduction and Principle 1 in The Toyota Way: 14 Management Principles from the World's Greatest Manufacturer. McGraw-Hill, New York (2021)

Lodgaard, E., Jonas, A., Ingvaldsena, B., Gammea, I., Aschehouga, S.: Barriers to lean implementation: perceptions of top managers, middle managers and workers. Sci. Direct (2016). https://doi.org/10.1016/j.procir.2016.11.103

Poksinska, B., Swartling, D., Drotz, E.: The daily work of lean leaders–lessons from manufacturing and healthcare. Total Qual. Manag. Bus. Excell. **24**(7–8), 886–898 (2013). https://doi.org/10.1080/14783363.2013.791098

Tortorella, G., Cauchick-Miguel, P.A.: Teaching lean manufacturing at a postgraduate level: integrating traditional teaching methods and problem-based learning approach. Int. J. Lean Six Sigma **9**(3), 301–323 (2018). https://doi.org/10.1108/IJLSS-08-2017-0101

Womack, J., Jones D.: Lean Thinking. 2nd edn. Simon and Schuster, New York (2003)

Learning a Lean Way

Signe Syrrist[(⊠)] [iD]

Western Norway University of Applied Sciences, 5020 Bergen, Norway
signe.syrrist@hvl.no

Abstract. The paper "Learning a Lean Way" reflects on if and how students could improve their learning process by applying a Problem-Solving A3. Three different gains that could be achieved are focused. Applying A3s could heighten students' awareness of their own contribution to their learning processes, the importance of being involved in a thinking process connected to learning, structure knowledge through a common language and sharing results in progress and at the end. Based on three student groups filled-in A3 formats, in addition to students' reflections given in short interviews after course completion, four important results appeared 1) Students expressed that applying the Problem-Solving A3 had, to their surprise, heightened their awareness and understanding of how important own practices were to achieve improved learning. 2) All students seemed to have a quite clear understanding of what "good learning behavior" implies: preparation, actively participation in lectures and afterwork, but only to a minor extent did they usually practice this understanding. 3) In their root-cause analysis students gave many similar explanations as to why they didn't act according to what they knew were the "good learning behaviours". They focused on bad planning, low priority, deficient study techniques and class atmosphere. 4) Most students expressed reluctance to share experiences with their other classmates both during their filling-in sessions and after having completed their A3s. Overall results show that students own participation, reflections and tracking of own learning process improvements, through applying a Lean-thinking Problem-Solving A3, could be a valuable addition to increase students learning and thereby contribute to heightening quality in higher education.

1 Problem Addressed

«We spend 50% of our time at school attending lectures. The learning outcome is not proportionate to the time spend in class."[1] [1] This was the problem addressed in 2013 by a group of five students in chemistry engineering when they, free of topic choice, decided this to be their problem addressed when applying a Problem-Solving A3. Their overall aim was to "Take a lead in your own studies – become a superstudent!" (see Appendix). They focused on "How to maximize learning outcome from lectures". Their systematic approach followed an A3 template handed out to them by the teacher in a TQM-course, developed by Wig (2009, 2014) [2, 3]. Being students in Chemistry,

---

[1] See Appendix for the chemistry students' final A3.

© IFIP International Federation for Information Processing 2021
Published by Springer Nature Switzerland AG 2021
D. J. Powell et al. (Eds.): ELEC 2021, IFIP AICT 610, pp. 86–99, 2021.
https://doi.org/10.1007/978-3-030-92934-3_10

they decided to make improvement in a core chemistry course that semester. As their 5 Key Performance Indicators (KPI's) they listed 1) being present in class, 2) having focus in class, 3) preparations, 4) afterwork and 5) attending class on time. By self-registrations, each group member continuously indicated their performances on a scale from 1–25, monitored every four week. Their summed-up results at the end of the semester revealed that their goal was obtained in the chemistry course. Everyone had reached a minimum score of 20 out of possible 25 points. At their final exams, the students excelled both in their TQM course – and in the chemistry course. Emiliani's words (2003) [4] "Improved process, improved results!" came true for these students.

Finally in 2017, in a Lean-course in Logistics that I was responsible for, I asked these students if they would voluntarily participate in a project similar to the chemistry students. My reluctance not to ask them before was due to a hesitation not to over-burden them. They already had another mandatory assignment and exam to complete in my course in addition to other courses they were expected to finish the same semester. But then, autumn 2017, I asked my students if they recognized and had similar experiences from lectures as the problem the students in chemistry had addressed. All my students nodded and agreed, it was, for them, a well-known problem. They agreed to give the A3 a try on their learning process. The next autumn, in 2018, a new group of students in Logistics agreed to voluntarily apply the same A3 template. To them all, applying an A3, would not be embarking on a new theme, they were already familiar with the method and approaches in Lean-thinking from my course. They chose to address "improved learning", not just from lectures. I made no significant changes in instructions and coaching between the two student groups. Students who sincerely gave the A3 a try received very good grades in their academic essentials. Other explanations could be given but applying the A3 template required students' direct involvement and contribution in their own learning process. Results show that the A3 could benefit students learning. As important, these results could give an additional perspective on how to increase quality in learning in higher educational institutions.

2   Lean in Higher Education

Lean in higher education has received a growing attention. The network Lean Higher Education (LHE) [4], a worldwide (US, Europe, Australia) network, was established back in 2000. Conferences, communication blogs etc. have been initiated to exchange experiences and insights from higher educational institutions. Lean approaches at universities from the three continents USA, Europe and Australia have given remarkable results. Special focus has been on improving administrative processes in higher educational institutions such as the process of finance, library procedures, student accommodation, administering foreign student applications or efforts of cultural change. Many projects have involved students and/or different groups of employees. Studies have shown how time-consuming and inefficient many administrative processes can be (Emiliani 2015) [5]. Great improvements can be achieved by applying Lean principles and focus on waste reduction (ibid), eliminating or reducing non-value-added time-consuming activities.

Few studies have I found investigate how Lean principles could be applied in teaching and learning, with Bob Emiliani [5] as an exception. Attending the LHE-conference in Tromsø in November 2018 [4], it turned out that none of the participants, or in their very widespread networks, were familiar with any Lean-projects to improve students' learning process. In teaching and learning, focus has been on Teaching Lean rather than Lean Teaching and Learning [5]. As an exception, Bob Emiliani [5] has departed from an approach to "identify and eliminate waste, unevenness and unreasonableness in academic processes and strive to achieve flow in all processes", including learning processes [5]. All educational offerings containing Lean concepts, emphasize the value of applying an A3, but applications have been on real or constructed cases presented, rather than asking students to focus on themselves and their own learning process. Applications on real cases have also been extensively focused in courses given to adults and employees as part of further educational programs, then using their own workplace as cases. Not to my knowledge have students been involved in improvement projects departing from their "workplace" and core activity which is their own learning process and participation in it.

3   A Problem-Solving A3

"A3 is a powerful tool. It establishes a concrete structure to implement PDCA management. It helps draw the report author(s) to the problem or opportunity, and it gives insight into how to address that problem" (Shook 2008) [7]. Tracy Richardson gave in 2011 [8] a more detailed description of a variety and purposes of A3s: Companies and organizations, private or public, customize A3s to a variety of situations and purposes: sharing documentations, product-improvements, process-improvements or in strategic planning. In 2008 John Shook presented what he perceived as the essens of A3: "The most fundamental use of the A3 is as a simple Problem-Solving tool. However, the underlying principles and practices can be applied in any organizational setting. Given that the first use of the A3 as a tool is to standardize a methodology to understand and respond to problems, A3s encourage root cause analysis, reveal processes, and represent goals and actions in a format that triggers conversation and learning" [7]. A main point here being that a Problem-Solving A3 focus on understanding problems, plan and implement actions based on a root-cause analysis, through a shared process that inspire discussion and learning.

There are further many different types of Problem-Solving A3s and templates (Richardson 2011, Sobek & Smalley 2008, Shook 2008, Ballé 2011) [8, 9, 10, 11], but these presentations show some shared common characteristic of A3s: They are written on a single sheet of paper, in an A3 size. Problem background and goals to be achieved are essential to be made explicit from the beginning. Then a gap-analysis, between current state and future state, is mapped in a Value Stream Mapping (VSM) chart. Then, departing from a few defined key performance indicators (KIP's) root-causes are analyzed. In analyzing root-causes an Ichikawa diagram (fishbone) or 5 Why could be used. Recognizing alternative root-causes opens up for a variety of countermeasure options and thereby getting a better understanding of the complexity of problem and causes. When detecting countermeasures to the root-causes, Michael Ballé [11] underlines the importance of presenting more than one solution to a root-cause: "Think of two or three distinct solutions to the same problem … Distinct and valid solutions demonstrate that the problem space has been thoroughly understood and explored and make it easier for others to get on board with the proposed countermeasures." [11].

Then a few root-causes and attached countermeasures, are chosen for an implementation plan, outlined in a chart, containing the 5 W + 1H, What to do, Why do it, How to do it, Where to do it, Who will do it and When will it be considered "done" or re-evaluated. Results are registered and followed up, deciding how to proceed by applying a PDCA-circle. A Problem-Solving A3 follow this more general structure and rely on charts and graphics to "tell the story". The overall common characteristics in a Problem-Solving A3 gives a shared "language" for discussions and conducting improvements. All these characteristics were parts in the A3 the students applied.

According to John Shook (2008) [7], a Problem-Solving A3 has, at least, three essential objectives: 1) To heighten awareness of problem, process, solution 2) To structure knowledge and the process of problem solving 3) To make sharing possible through visual presentations. An A3 is a step towards "learning to learn" (ibid). These objectives will be addressed in this paper based both on their filled-in A3s and interviews.

4 Students A3 Focus

«We spend 50% of our time at school attending lectures. The learning outcome is not proportionate to the time spend in class." 1) This was the problem addressed by the students in chemistry engineering when they, free of topic choice, decided this to be their focus for their A3. Their overall aim was to "Take a lead in your own studies – become a superstudent!". They focused on "How to maximize learning outcome from lectures" Their systematic approach followed an A3 template handed out to them by the teacher in the TQM-course (Wig 2009) [2]. Following the template, they described their current state and gave an outline of the future state they intended to reach at the ending of the course in chemistry. Deciding on 5 Key Performance Indicators (KPI's) and indicating their performance on a scale from 1–25 by self-registrations, their progress and goal could be given in numbers, and they continuously monitored their improvements, summing up results every four weeks. Their decided KPI's were 1) being present in class, 2) having focus in class, 3) preparations, 4) afterwork and 5) attending class on time. As their summed-up goal they agreed upon: "Everyone (in their group of four students) were to reach a minimum of a score of 20 out of possible 25 points" by the end of the course in chemistry. At their final exams, the students in chemistry slogan: "Improved process, improved results! (Emiliani 2003) [4].

I had been reluctant to ask students in Logistics, attending my Lean-course, to voluntarily participate in a project similar to the chemistry students'. They all had other mandatory assignments and exams to complete in their studies. My reluctance was due to a hesitation not to overburden them. Overburdening is one of the Mura, Muri and Muda in Lean-thinking which are unwanted and aimed at reducing or eliminating. But autumn 2017 and the following year, I asked my students if they recognized and had similar experiences from lectures as the problem the students in chemistry had addressed. All my students nodded and agreed, it was, for them, a well-known problem. Both years, 2017 and 2018, the students conceded to voluntarily participate, or at least try out an A3 to improve their learning process. To them, applying an A3, would not be embarking on a new theme, they were already familiar with the method and approaches in Lean-thinking from my course. They chose to address own "improved learning", not just learning outcome from lectures. Asing new groups of students to participate were prevented by the Covid-19 pandemic, all lectures had to be given digital. Close coaching was experienced needed by the previous groups of students, not made possible by digital communication.

5   A3 Material from the Students

The research design applied was an explorative design (Grønmo 2017) [12] with the question would and if, how could students be able to benefit from applying a Problem-Solving A3 to improve their learning process? Explorative designs enable investigating an area not previously being the focus for systematic research. An aim in such designs is to get a deeper understanding of a problem-area and the approach allows flexibility and continuously to address unexpected challenges. Lack of previous research and the students' voluntary participation made such a design appropriate to apply.

Proceedings in gathering information from the students' work on their A3s where partly in-progress and partly at the end of the semesters. After an initial plenum session getting started, I met the students every week, in scheduled lecturings, where each group could bring up with me questions or comments and give me a peak into their proceedings. Mid-term they gave me their preliminary A3s, for me to get an impression of how they were moving along and for me to give them additional motivation to continue. At the semester-end they all handed me in their final A3s.

In interviews, after semester-end, with some of the students we had a more informal discussion about their more general experiences, their suggestions of improvements and if would they agree to share their A3s with the other classmates. The questions departed from John Shook's essential objectives: heightened awareness, A3s as a means to structure knowledge and the process of problem solving and making sharing possible through visual presentations. Participants in the interviews were 10 in students from 2017 and 8 students from 2018. In the beginning of every interview, I presented the topics I wanted to discuss with them.

They all allowed me to record the interview sessions. The interviews took place in a small room at the premises and lasted from ¾–1 h. The interviews became a mix of focus groups and dialog between us. To the interview sessions, I had brought with me their own A3s. Many remembered their A3s with excusing smiles, but most importantly, they told me that when they now saw their A3s again they recalled better where they had struggled and where they had been insecure – as well as what they thought had gone well. Seeing their A3s again became to them very helpful to their recollections and aftermath reflections.

6 Students' Experiences

Students struggled with describing their current and future state of learning process, as well as presenting a clearer understanding of what they were aiming at, their goals. Their struggles were partly due to difficulties even defining learning as a process and disagreements in and between groups when to define start and end of their learning processes. Differences spanned from looking at learning process as a process within a lecturing session, within the time-span of a day or even with start and end of process as presented in the education programs for the specified course.

Just as the Chemistry students, all students in Logistics indicated KPI's as what they perceived as "good learning behaviors". Their perceptions coincided with the understandings within higher educational institutions. These were: learning activities before class, participating in class/lectures and learning activities after class. By the students, preparations ahead of lectures were noted as reading relevant syllabus and teaching material, while being active during lectures was perceived as note-taking and participating in discussions. Afterwork contained for them writing assignment, re-read lecturing material and keeping up to date on syllabus, and learning new concepts. In addition to these "good learning behaviors", several students noted some quite different prerequisites for "good learning processes". Among them were: not being tired, hungry or thirsty, and arriving on time/being present in class. Lastly, some also focused on that their learning behavior depended on their motivation and interest. Even though many students had part-time work aside their studies, this was not mentioned as any obstacle to good learning behavior by the students in Logistics. Being well aware of what "good learning behavior" implied, they were now asked to address the question of why they didn't practice such behaviors. Reasons for not practicing them, they discussed as possible root-cause in their Ichikawa diagrams.

Frequently appearing in their root-cause analysis as reasons for not practicing "good learning behavior" were: low priority, bad planning, distractions, class atmosphere and deficient study techniques, in addition to not having necessary learning materiel/equipment available. Being tired, thirsty/hungry and attending part-time jobs were now also given as root-causes for lack of own learning process efforts. The first five reasons appeared in all students' implementation plans, in addition to being tired, as their KPI's.

Addressing low priority, they wrote down reasons of laziness, yesterday's partying, disturbances at night (noise from outside) and their day-and-night rhythm not adapted to lecturing hours. They linked low priority directly to being tired. On being tired, students mentioned they hadn't had enough sleep, they had overslept, had partied too late, slept in "wrong" bed, received visits at night and/or in bed, had poor sleeping pattern, stress, noise, computer gaming and Netflix at night or attending late night concerts. Lack of enough sleep resulted in fatigue. Active and good learning behavior were the "losers" in comparison with many late-night offerings and temptations. Low priority was further associated with being hungry/thirsty. Specifications of being hungry/thirsty as root-cause were given as due to not having enough time for breakfast, time-schedule at school with too early starts and too short brakes. Poor economy was given as reasons for not being able to buy food and drink at the cafeteria. Reasons for bad planning appeared were given as very similar to the reasons for low priority: being tired, hungry/thirsty and some also wrote that bad planning was due to low priority. Priority and planning were linked in these student's A3's. Without priority, lack of planning. Distractions were exemplified by being engage on their i-phones or scrolling on their computers, as well as small talk and "day-dreaming". By one group "wrong seating" in the classroom was given as root-cause for not participating in discussions or being alert in class. This groups preference with seating was up front in the class-room, almost always available. Class atmosphere as root-cause for not being active in class by asking questions or giving comments and feedback to all in plenum as well as participating in discussions, students attributed to a perceived strenuous and silencing class-atmosphere. Nobody wanted to be sticking-out, as if the law of Jante was alive in the class-rooms. They all agreed, in an open lecture-setting, that this was how it had always been in their class, already from the first year and in all courses. They perceived that nothing could be done about that now in their third year of studies. As reasons for not being present in class were mentioned bad weather and transportation problems. Deficient study-techniques were written by students, but it turned out it was more a search for explaining not practicing "good learning behavior", than a definite root-cause. When asked to elaborate on this, they mentioned note-taking and reading techniques, but did not seem familiar with any reflected insights into varieties of study techniques.

Some of the root-causes mentioned by the students lay outside students reach of influence. Deficient study techniques and class atmosphere were also difficult for just a few students to address and change by own actions. Educational schedules were also out of their sphere to influence, while weather and transportation challenges were definitely quite out of their reach to do something about. Students appeared to be conscious about what they could influence even though they would have liked some cultural changes in class-atmosphere.

In their implementation plans they were asked to address three chosen root-causes and suggest countermeasures for them and develop implementation plans. The students' plans were to contain presentations of what they would do (their suggested solutions to the chosen root-causes); why they would do it and saw the action as important (justifications); who were to be responsible for the actions; where the actions would take place and lastly, when they would sum up results for their chosen actions and then decide further proceedings.

To address low priority and bad planning one group wrote "make a plan and stick to it", which this group really did! The group presented a detailed improvement plan showing what to do specified and linked to the next lecture. All group members were to be responsible, and it should be acted upon at school and at home. Detailed registrations were decided made from week to week by self-registrations by each group member. Results were presented in a bar charts.

Many of the students had chosen the same root-cause to address: being tired and distractions. Several groups made plans for improving their sleeping habits with specified actions "going to bed not later than 23.00". Others wrote in more general terms like "getting enough sleep". As justifications of why to remove "distractions" was mentioned "to improve focus in lectures". Others wrote "making time useful in class, be active in class or actively participate in class". Those who chose countermeasures to avoid distractions discovered that they easily could make necessary changes by "removing phones, not taking them up during lectures, putting them back in rucksacks or pockets, or stop scrolling and minimize small-talk in class".

As countermeasures presented to address preparations before or after lectures but students really tried to focus actionable suggestions, but in writing many presented activities more in general. Some wrote just "reading syllabus" or "improve note-taking or participate more in lectures by asking questions", "read ahead, read more, read to-days topic, read lecturing files, be prepared for class". Being more prepared was presented as a means to get a better understanding of learning syllabus content and be more attentive. In interview, students focusing on participating more in class, it turned out that what the really meant was to ask more questions in a face-to-face communication with teacher in breaks or after class. When presenting countermeasures partly in more general terms and partly without based solely on own efforts, it became very difficult for them to register development and, preferably, progress.

Some students had distressing difficulties in giving a specified connection between root-causes and countermeasures. In spite of such difficulties, students reported in interviews that discussing root-causes and countermeasures, and reflect upon them, had contributed to increase their awareness and consciousness of what they themselves could do to improve their learning process. Some reported that they afterwards had felt better prepared for exams, had improved their written hand-ins, had become more focused - all of them means to improve learning process. Many factors could of course have contributed to these perceptions and results, but increased awareness seemed very important to them. Improved learning process, it turned out, was not important to all of them. One student said in interview: "Why should I improve? I am happy with my grades!" In addition, none of the students had chosen motivation or interest as root-causes to discuss countermeasures for.

Improved learning process some also noted and related to their future standings in the labor market; "getting a great job, receive a good salary or be able to implement knowledge and competences in praxis". Students aiming at entering master studies were also more conscientious of improving their learning process. A couple of students also mentioned that enhanced learning process could contribute to their sense of personal achievement.

In all the students' plans, every group member was made a responsible participator to implement their plans both at school, in lecturings, and/or at home. Again alternatives varied and were presented in more general terms. When to register progress, the stop and check as in a PDCA-circle, were presented with even greater vagueness with intervals ranging from "all the time" (give priority), "weekly" (preparations) or on specified weekdays (remove distractions during lectures). The students in chemistry had similar loose statements such as "all the time" (give priority to lectures), "every day", "after agreements" (in the group) or "during lectures".

When choosing a scaling for self-registration of improvements, these varied both in scalings and whether to register for each individual group member or the for the group as a whole. When the students in chemistry had chosen scalings from 1–25 for every group member by self-registrations and then summing up for group every 4 week, other students chose divergent scalings and focus could differ between individual group members and the whole group. Some used scalings from 1–5, others from 1–10 or even

5–20, these scalings were to present goals they wanted to reach in the end. These variation contributed to confusion to them on what the goal could be, what to register in addition to linking estimated effects of their countermeasures to their defined goal. Quite a few presented estimated improvements in percentage, but these percentages seemed more like guesswork and written into their A3s just to fill in something in this section, they later reflected in interviews. As an out-turn of such variations, final results achieved in the end were presented as blanks in their A3s.

Both with sleeping habits and removing distractions, students underlined how they had become aware of how little effort was needed to make improvements in learning process. On an effort-value scale, distraction and bad sleeping habits, little effort was needed, and their actions removing distractions could have significant value to them. In later interviews students told me that reflecting on these root-causes had been an eye-opener for them. Never before had they been aware of how much time they spend on such distractions.

In interviews students told me that the thinking process connected to reasoning about root-causes and implementation plans had been an eye-opener and, sometimes, it was not always very pleasant to acknowledge the extent of their own "weaker learning behavior". Although, they reported gained insights in how to improve their learning process after applying the A3. They now had realized that they could implement actions to make improvements which would have a positive effect on their learning process. They perceived that such action would not necessarily be that demanding.

7   Challenges When Students Were Applying an A3 on Own Learning Process

While the student in chemistry were given the application of the A3 as their mandatory semester assignment, the students in logistics were asked to voluntary participate applying the A3 to improve their learning process. Facing difficulties, these students had the option of quitting the project – as some did after a couple of weeks. These students excused themselves with other requirements in other courses they were following, difficulties meeting up as groups, or other obligations at home, family and children. All students were situated within an educational context, an academic demanding setting with several exams and assignments. Applying the A3 was to be perceived as an add-on to an already tough educational program. In higher educational institutions in Norway many students have less instructional and compulsory educational bindings, they could in many respects dispose more freely their hours. This context gave them opportunities to give priority to other more social activities, if they preferred that, or taking on part-time work or fulfill family obligations.

James Womack argues in "Gemba Walks" (2013) [13] the need for understanding the context for improvement projects. What is needed to grasp is what the context contains of possibilities, boundaries, and characteristics. Higher educational institutions have both an academic and a vocational mission. Students' perceptions of their learning process could and did vary depending on their futures goals either being on outlook for a vocation directly after completing their bachelor studies or seeking- admission to master studies or phd.

Learning represents a core activity in educational institutions, but learning is not a definite knowledge or skill. Learning, either as knowledge, skills or general competences, is an intangible end product of education. Students acquire a competence unique for each student. How and in what degree their competences will be executed, only future circumstances can reveal. Some students presented a more "wait and see" attitude towards their educational outcome and the possible applications of their learning outcomes. To contribute to improve own learning process was for many perceived within this context as not quite their responsibility and problem. When applying the A3 students were challenged with framing learning process as problem – not just for them but improving it by them. To frame problems could be a demanding task. Framing and re-framing problems is discussed by Thomas Wedell-Wedellsborg (2020) [14] in his book "What's Your Problem? To Solve Your Toughest Problems, Change the Problems You Solve". More time spend on framing and reframing learning process as problem, maybe could have given the students a clearer understanding of what they were embarking on. Alternative interpretations of the problem could also have heightened students' comprehensions of own actions when learning.

In Lean-thinking an important perspective is the question of how those on operational level perceive problems, suggest and implement solutions. In short, contribute to create value for the customer. According to Liker & Meier (2006) [15], in Lean-thinking this is a vital concern, but how are students to be perceived when they apply a Problem-Solving A3 on own learning process? Students could hardly be defined solely as customers when they are the learners. Rather they are input, co-producer in collaboration with teachers, but also customers of the educational offerings [16]. As co-producers they are responsible for producing value for themselves, but as customers they are receivers of value created by the institutions intellectual and other resources.

Studies have shown that students more often ascribe responsibility for their learning to teachers. "A majority of students refers to extrinsic motivators as defining factors that make them to invest effort in studies. Consequently, students are more critical regarding teachers' performance than their own and do not always see a direct link between their efforts and study outcome." (Grinfelde 2019) [17]. Teachers are of course invaluable to students in their learning processes. According to the "Respect for people" principle in Lean-thinking from Ballé [18] the reciprocal relationship between students and teachers cannot be underestimated in students learning. Here lies a cultural challenge to be addressed in academia (Fadnavis et al. 2020) [19]. Teachers combine several roles, including as coaches (sensei) not only lecturers. Applying a Problem-Solving A3 can contribute to increase students' consciousness of their own roles in their learning process and their relationships to teachers as well as teachers can become more aware of how students own efforts can contribute to improve their learning process. In Lean-research, surprisingly minor attention has been given to the "Respect for People" (Badurdeen, 2017 [20] pillar compared to studies of applications of tools and methods within Continuous Improvement. (Ballé 2014, Emiliani 2010, Cardon 2015, Netland 2016) [18, 21–23]. Further research may be required. Ballé's [18] eight steps model to achieve results and relationships for sustained performance emphasize many challenges to consider both students and teachers roles. Heightened awareness of own roles students showed a desire for when they were to turn from cases and examples given in lectures to own student experiences. It appeared to me easier for them to conduct an A3 analysis departing from a given case-text.

Several times questions occurred about goals and metrics. Many became very uncertain when asked to attach metrics to goals and performances. In Lean-thinking it is emphasized the importance of being able to register whether or not improvements are achieved, therefor using metrics is recommended. Metrics enables tracking process proceedings and used as checkpoints towards reaching goals. Or in M. Ballé's (2015) [24] words: "what gets measured gets done!" Students expressed that they were very unfamiliar with thinking of improved learning process as goals in metrics, except when they thought of grades – even though these are now in letters. Initially, many therefore wrote «an A» as their goal. To avoid metrics, others wrote «improve learning». Thinking in metrics and measurements required close coaching to all the students. The lack of clearly defined metrics made it difficult for many to connect improvements effects to initial framed goals. Tracy Richarson (2014) [25] recommends to read an A3 first from left to right and then backwards, calling it "Reading in Reverse". This reading enables to discover cause and effect logic in the A3 back and forth. This recommendation could help students understand their A3's as continuously changing working material, which an A3 is when combined with PDCA. It appeared easier for students if their starting point was from the root-cause analysis of problem with chosen KPIs. Flexibility in application of the A3 was required, with continuously moving back and forth in the template, not just forward and backwards. The templates are not intended to be rigorously followed, section by section.

Two distinct problems occurred for students when trying to map their learning process as a Value Stream (Shook 2004) [26]. In their final A3's "start and end point" varied from group to group, ranging from within a day from morning to bed-time to encompassing a whole course-span. Regardless of their choices, none of the students connected any learning process activities to their value streams. In addition, describing a current and future state became for almost all of them such a de-motivating experience that we agreed to leave it unfinished relying on (Entwistle et. al 2015:2) [27, 28]. There has been "… little direct attention to the process of students' learning and the effect of teaching on it.".

Deficient study techniques appeared as a root-cause in a couple of groups A3. These students put increased note-taking during lectures as a countermeasure, but they felt a need for further input on alternatives and increased knowledge. Furthermore, despite efforts to increase their notetaking during lectures and to be more active by asking or answering questions, they did not quite succeed due to what they perceived as a non-inviting and judgmental class atmosphere. Dunlosky (et al. 2013) [29] has reviewed research on different Learning-Techniques and their effectiveness. "Improving educational outcomes will require efforts on many fronts, but … one part of a solution involves helping students to better regulate their learning through the use of effective Learning-Techniques». As a motivational reason for their review they cited McNamara 2010 (Dunlosky (2013:46) [29] "There is an overwhelming assumption in our educational system that the most important thing to deliver to students is content, … teaching students content and critical-thinking skills, whereas less time is spent teaching students to develop effective techniques and strategies to guide learning".

Of the 10 Learning-Techniques they reviewed chose to review, they concluded that two Learning-Techniques could be documented to have high utility: Practice testing (Self-testing or taking practice tests over to-be-learned material) and Distributed Practice (Implementing a schedule of practice that spreads out study activities over time). Utility was defined to what extent the techniques contributed to "Cued recall, Free recall, Recognition, Problem Solving, Argument Development, Essay Writing, Creation of Portfolios, Achievement Tests, Classroom Quizzes" (Dunlosky et al. ibid p. 6 Table 2). Getting a better understanding of learning-techniques with high utility could help students give more detailed suggestions in their accomplishment plans. Both students and teachers could benefit a more thorough understanding of different techniques. Practice Testing could be advantageous for teachers "... teachers could also incorporate some of them into their lesson plans. For instance, when beginning a new section of a unit, a teacher could begin with a practice test (with feedback) on the most important ideas from the previous section" [29].

8 Concluding Remarks

From all three groups of students, it appeared that they knew very well what good learning behavior implied. Initially they were less preponed to act accordingly as students. Compared to the three essential objectives in an A3 implementation, according to Shook (2008) [7], all students confirmed that applying A3 had been an eyeopener realizing how their own conduct influenced their learning process and contribute to improvements. They had gained heightened awareness of problem, process, solution. Structuring knowledge and the process of problem solving, they had experienced as a need of greater flexibility. Only a few groups systematically followed up their plan and finialled their A3. Students who managed to fulfil their projects were quite pleased with themselves and valuated A3 as being both helpful to their learning and an awakening call. In interviews others underlined how surprised they had become of the importance of their own contribution to reach increased learning and how much time they spend on non-educational activities, distractions as phone and computer scrolling. An A3 is both a method and a tool. A most vital point when using an A3 is the thinking process for all those implied (Ballé 2019) [30].

When asked to share their A3s with fellow students, both in progress and at the end, through visual presentations, none of the students in Logistics were willing to share with other classmates outside their group. This opinion, they said, was partly due to the perceived uninviting class atmosphere.

Problem-Solving A3 could benefit students, most of all as a means to reflect on own contribution to improve learning processes. Flexibility, PDCA-focus, learning-techniques could all contribute to improved benefits from applying A3 by students. Full advantage of an A3 requires training and systematic work. A core question remains: students know what good learning behavior implies, but why do they not practice it? Further research is needed to get a deeper understanding of this challenge in higher education. In this explorative research students gave some answers. The A3 approach invited students to reflect and discuss. Giving students this opportunity could represent a contribution to discussions of how to achieve increased quality in higher education, valuable for students and teachers, and everyone involved in quality work.

Appendix

The Chemistry Students A3 (201, filled in. The same template used by the students in Logistics.

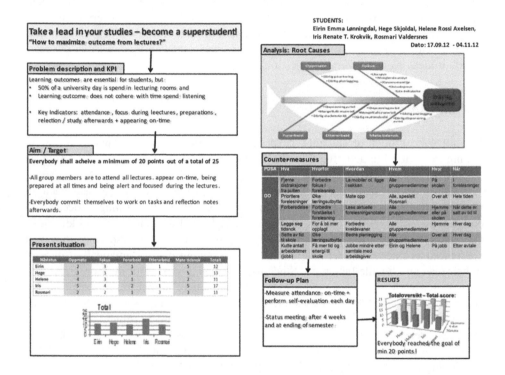

References

1. Appendix: The chemistry students' final A3
2. Wig, B.B.: Det er ledelse: kvalitetsledelse for det 21. århundret. TQM Center Norway AS (2009)
3. Wig, B.B.: Lean. Ledelse for lærende organisasjoner. Gyldendal (2014)
4. Lean in Higher Education https://www.leanhe.org/
5. Emiliani, B., et al.: Better Thinking, Better Results. The CLBM, LLC Wethersfield, Connecticut, USA (2003)
6. Emiliani, B.: Lean Teaching. A Guide to Becoming a Better Teacher. Center for Lean Business Management, LLC (2015)

7. Lean Higher Education Conference, 2018, at the Arctic University Tromsø. https://en.uit.no/tavla/artikkel/562904/2018_lean_in_higher_education_conference
8. Shook, J.: The A3 Process – Discovering at Toyota and what it can do for You (2008). https://www.lean.org/shook/DisplayObject.cfm?o=1445
9. Richardson, T.: What are the different types of A3's? (2011). https://www.lean.org/common/display/?o=1882
10. Sobek, D., Smalley, A.: Understanding A3 Thinking: A Critical Component of Toyota's PDCA Management System. CRC Press, Boca Raton (2008)
11. Shook, J.: Managing to Learn. Lean Enterprise Institute, Cambridge (2008)
12. Ballé, M.: Is there a Right Way to Teach A3? (2011). https://www.lean.org/balle/DisplayObject.cfm?o=1943
13. Grønmo, S.: Samfunnsvitenskapelige metoder. Fagbokforlaget (2016)
14. Womack, J.: Gemba Walks. Lean Enterprise Institute, Cambridge (2011)
15. Wedell-Wedellsborg, T. (ed.): What's Your Problem? To Solve Your Toughest Problems, Change the Problems You Solve. Harvard Business Review Press, Brighton (2020)
16. Liker, Jeffrey K. & Meier, David (2006) The Toyota way. Fieldbook. McGraw-Hill Companies
17. Netland, T., Powell, D.: A lean world (Chapter 40). In: The Routledge Companion to Lean Management. Routledge, New York (2016)
18. Grinfelde, I.: Who is Guilty That I Fail in Classrooms: Students' Perspective on Higher Education (2019). https://doi.org/10.17770/sie2019vol1.3841
19. Ballé, M., Ballé, F.: Lead with Respect. Lean enterprises Institute Corp (2014)
20. Fadnavis, S., Najarzadeh, A., Badurdeen, F.: An assessment of organizational culture traits impacting problem solving for lean transformation. Procedia Manuf. **48**, 31–42 (2020)
21. Badurdeen, F., Gregory, B.: Analyzing the Softer Side of Lean. Industrial Engineer (2017)
22. Emiliani, B.: Real Lean (4th volume). The Centre for Lean Business Management, LLC, USA (2010)
23. Cardon, N.: Respect for people: the forgotten principle in lean manufacturing implementation. Eur. Sci. J. **11**(13) (2015). May 2015 edition
24. Netland, T.: Poor performance? The problem is you! (A review of Lead with Respect) (2015). https://better-operations.com/2015/09/02/book-review-lead-with-respect/
25. Ballé, M.: How do I Measure People Development? (2015). https://www.lean.org/balle/DisplayObject.cfm?o=2943
26. Richardson, T.: Test Your PDCA Thinking by Reading the A3 Backwards (2014). https://www.lean.org/LeanPost/Posting.cfm?LeanPostId=235
27. Shook, J.: Misunderstandings about Value Stream Mapping, Flow Analysis, and Takt Time (2004). https://www.lean.org/common/display/?o=912
28. Womack, J.: Gemba Walks. Lean Enterprises Institute (2013)
29. Entwistle, N., Ramsden, P.: Understanding Students' Learning Process. Routledge (2015). (First published 1983)
30. Dunlosky, J., et al.: Improving Students' Learning process with Effective Learning process Techniques: Promising Directions from Cognitive and Educational Psychology. Sage Publications (2013)
31. Ballé, M., et al.: The Lean Sensei. Go See Challenge. Lean enterprise Institute (2019)

Lean Learning Factories: Concepts from the Past Updated to the Future

Gabriela R. Witeck$^{(\boxtimes)}$ (iD) and Anabela C. Alves$^{(\boxtimes)}$ (iD)

Centro ALGORITMI, Department of Production and Systems,
University of Minho, Guimarães, Portugal
anabela@dps.uminho.pt

Abstract. Lean Production has its roots in the Toyota Production System, introduced before World War II, and is constantly evolving. Its importance as an organizational management model triggers the need to teach it in the academy. Promptly, Lean Education is being taught all over the world. However, teaching Lean using traditional expositive lectures is not effective, and many academics and practitioners are using active learning methodologies. Lean and Learning Factories, which are two concepts that come from the past, are more than alive nowadays. This paper presents a literature review regarding Lean Learning Factories, based on a scientific articles research at Scopus database. The review was conducted for the period from 1990 to 2021 and resulted in a total of 76 papers. Main findings revealed that the first articles within the context of Lean Learning Factories were published in 2006. The learning factories initiatives were developed by universities and the most used learning strategies are simulations and gamification. Also, the latest configurations of these are in Germany, Austria, and Croatia. The results revealed an increase in the number of publications since 2015, reaching 14 publications in 2020.

Keywords: Lean thinking · Learning factories · Gamification · Industry 4.0

1 Introduction

The engineering workforce of the future is being prepared in Higher Education Institutions (HEI). Nevertheless, changes in engineering education are often slow [1], being teacher-centered education, using for example, traditional expositive lectures in a classroom, the most adopted instructional method. Fortunately, new and different instructional methods associated with learning simulation environments, or even, real environments, through partnerships with industry companies and organizations, are also being used in HEI [2–5]. Particularly, for future engineers to visualize and understand their role in the workplace, make sense learn in different environments. That is the reason for "Learning Factories" gaining much attention.

In a "Learning Factory" environment, students/trainees/employees are involved with authentic processes. To promote this involvement, it is established a physical mockup that resembling a production system with a real value chain. Therefore, a physical product is manufactured and a didactic concept that comprises formal, informal, and non-formal learning is learned [6].

© IFIP International Federation for Information Processing 2021
Published by Springer Nature Switzerland AG 2021
D. J. Powell et al. (Eds.): ELEC 2021, IFIP AICT 610, pp. 100–108, 2021.
https://doi.org/10.1007/978-3-030-92934-3_11

The idea of "Learning Factory" was originated in the decade of '90s, when the National Science Foundation (NSF) sponsored a grant to a consortium led by Penn State University to design a learning environment that would promote engineering design projects with industry [7]. These projects integrated students from Industrial, Mechanical, Electrical, Chemical Engineering and Business, also involved 43 faculty members, across five time zones [8]. Later, the National Academy of Engineering reward Learning Factories achievements with the Bernard M. Gordon Prize for Innovation in Engineering and Technology Education [9].

Nevertheless, others authors, namely Foden [10] and Gento et al. [11] pointed out the origin of the concept to the beginning of the last century (1916) by the hands of Herbert Schofield and Loughborough College. At that time, it was named "instructional factory", as the main aim was to instruct workers [10]. "Training on production" was considered an instructional quick and important method for, particularly, emergent situations as was the First World War and Second World War. In the latter, it was adopted on a large scale under the name of "Training Within Industry" [12, 13]. Learning by doing is a key concept under the instructional method. Before and now, it remain preferred method of teaching, mainly when anyone is being prepared for practical and productive work for society [14].

The concept of Learning Factories rises more interesting since the Fourth Industrial Revolution leaps forward and hence labeled Industry 4.0 [15]. As a company converges from traditional automation to a fully connected and flexible system, including information and operations technology. The result is a production system more efficient, with a greater ability to predict and adjust to changes.

However, as all industry processes come from human assets, people are expected to be the key to the process, eliminating wastes (i.e., activities that do not add value to the products in the point of view of the client) that exist in these processes [16]. This concept of "waste", the contrary to the "value", is key to a Lean organization that recognizes people as the most important asset of companies [17–19]. People's behavior can lead to culture organizational and affect, negative or positively, the company's success. More than an organizational model, Lean Thinking [20, 21] is a culture that is not easily understood, even by academics [22]. This contributes to the difficulty to implement Lean [23–26]. The way to success is the symbiosis of people and technology along the entire value creation chain.

In a Lean Learning Factory, people can see with "their own eyes" and make mistakes until they learn, promoting continuous improvement. The learning model from real factories operates next to the industry and provides a dynamic education and assay environment. The combination of Lean learning and other instructional methods such as gamification can provide new and important competencies to professionals once that allow students/trainees/workers to develop such competencies by solving problems and making decisions [27–30]. When it happens, Lean learning becomes effective [31–33] and brings many benefits that impact professionals and personal lives [34–39].

As well, it allows exploring aspects related to the teaching-learning process through interactive and collaborative methods to expand the company's knowledge. It signifies the opportunity to generate greater value within the four walls of the company, and requires suiting the schooling process and evolves education programs within factories.

Interest in Learning Factories is growing. In 2012, Wagner et al. [40] identified more than 25 research and development organizations that have established learning factories. In 2018, almost 30 learning factories were founded in Germany [41]. Abele et al. [30] identified more than 60 learning factories, almost all from the last decade, many related to Lean, others with Industry 4.0, and some related to both concepts.

This paper presents a literature review on Lean Learning Factories to show that these two concepts with different origins in time and fields (industry and academic) are combined to conveniently instruct engineering students in a simulated learning environment. This literature review was based on a scientific articles research at Scopus database. The review was conducted for the period 1990 to 2021. This period was chosen due to this designation is in use for the first time by Penn State University.

The paper structure consists of four sections. The first section is the introduction, where the objectives of this paper are introduced, and a brief contextualization is illustrated. Materials and methods are composed in section two. The third section presents the results of the literature review. Lastly, the fourth section outlines concluding remarks.

2   Materials and Methods

In this research, the authors developed a literature review in the Scopus database in the period 1990 to 2021. The question and sub-questions that guided this research were:

- "Is the Learning Factory used as an instructional method to teach Lean?
  - "Which countries are using this method of teaching Lean?"
  - "Is the Learning Factory promoted/funded by a company?
  - "What learning strategies are being used?"

The string used was "Lean" AND "Learning factory". These were searched in title, abstract, and keywords for the period from 1990 to the present. Figure 1 presents the number of papers obtained.

76 document results

TITLE-ABS-KEY ("learning factory" AND lean) AND PUBYEAR > 1989 AND (LIMIT-TO (LANGUAGE , "English"))

Fig. 1. Number of papers obtained from Scopus search.

3   Results and Discussion

This section presents the results in terms of quantitative statistics and answers to the sub-questions, namely, countries/universities, companies involved and learning strategies used.

3.1   Quantitative Statistics

Seventy-six papers were evidenced with "Lean" and "Learning factory" in the title, abstract and keywords, and 70% of which were published in conferences. Related to the question raised in Sect. 2, it was answered by this result, as the papers, indeed, discussed Learning Factory as an instructional method/learning methodology suitable for teaching Lean.

For the quantitative analysis, the authors used the Scopus functionality "Analyse research results" and the graphics generated by it. Three main graphs were collected: number of documents by year, number of documents by country, and learning strategies used in the learning factories.

Figure 2 shows the number of documents per year. It is possible to realize the growing interest in Lean Learning Factories what corroborates the importance of these two concepts, even currently, as mentioned in the title of this paper. In the last year, 14 papers were indexed in the Scopus database.

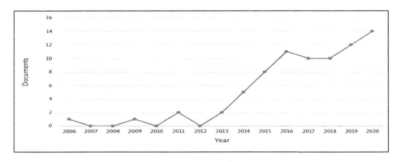

Fig. 2. Number of documents by year.

In total, 30 universities were identified associated with Learning Factories, which aim to teach the fundamentals of concepts related to Industry 4.0 and promote the student's development of Lean competencies and skills.

To answer the first sub-question (country/university), Germany is the country with more papers published, as the Fig. 3 reveals.

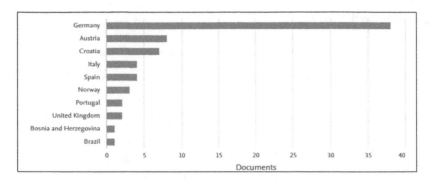

Fig. 3. Number of documents by country.

3.2   Companies Involved and Learning Strategies

To answer the second and third sub-questions raised in Sect. 2, namely, companies involved and learning strategies, the authors analyzed the content of full papers. This analysis allowed to discard eleven papers because nine did not have the full paper and authors could not read their content. Two papers were not related to the Lean organizational model. After applying these exclusion criteria (represented in Fig. 4), it resulted in 65 papers.

Fig. 4. Number of papers obtained after exclusion criteria.

Each paper was carefully read and some information about it was collected. This information was placed in a spreadsheet file. The information was related to the sub-questions: companies associated and learning strategies.

Regarding the sixty-five papers about Learning Factory context to teach Lean Thinking, many and diverse learning strategies are being used for Lean learning. Simulation of the industrial environment is most cited. The detailed analysis of the papers revealed the learning strategies resulted in Fig. 5.

For example, the Simulation learning strategy was used by the authors Crnjac et al. [42] during the development period of a new product, in a Lean Learning Factory. This passive strategy could help to visualize how the new product will "behave" in its environment under the influence of different environmental factors. Then, the students could learn about waste reductions in cost and quality. Moreover, Fu [43] related the logistics simulation teaching remains at the simulation level with a certain gap between real logistics production practice.

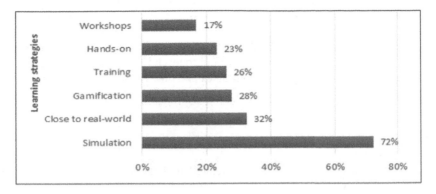

Fig. 5. Percentage of learning strategies used in the learning factories

3.3 Discussion

This discussion intends to highlight some aspects derived from the results. First of all, a growing interest in the Learning Factory as an instructional method, not only for teaching Lean but also other concepts such as the ones related to Industry 4.0. Many authors could say that this is a fact well known but if this is the case why not more universities are investing in it? Probably because of the high investment that is needed. For more details about the investment it could be seen the book from Abele et al. [30]. It seems Germany is the leading country in this investment; at least, more papers were published by this country. The reason for this may lie in the fact that in 2011 the Initiative on Learning Factories was founded in Darmstadt, Germany, as a union of several European Learning Factories. In the same year, the 1st Learning Factories Conference was launched in the same city. Since then, every year this initiative, which has been renamed the International Association of Learning Factories, has organized this conference, now in the 11th edition.

Additionally, in 2014, the International Academy for Production Engineering (CIRP) started a Collaborative Working Group (CWG) on the topic 'Learning Factories'. In this CIRP CWG, the main characteristics of the learning factories are defined for the dimensions of product, process, didactics, setting, and purpose [27].

Therefore, it is not surprising that most papers come from Procedia CIRP and Procedia Manufacturing (87%). Unexpectedly, countries like the US, that which has a long tradition of implementing Learning Factories, did not appear in the result of this literature review.

Regarding the 65 papers, all initiatives are from universities and institutes. Some of these initiatives were supported by national and/or international projects. The research showed few connections to companies/industries (only in 15 papers).

Despite active learning methods (e.g., close to real-world experience; gamification; hands-on or workshops) being recognized as a superior approach of instruction in Lean, a major of articles afford that the learning factories instructors still use passive learning methods (e.g., traditional computer simulations and training). Learning activities, or experiential learning, provide students' response to being actively engaged with the task. But, in the case of courses/programs related to Lean, it is difficult to teach /apply in a classroom environment [31], for this reason, the involvement between academia and industry is so important.

This section discussed and provided the answers to the questions that guided this research. Learning Factories are, in fact, serving to teach Lean concepts. More recently, such contents implied Industry 4.0 technologies and how Lean practices benefited from them.

4 Conclusion

This paper presented a literature review related to Lean Learning Factories. These concepts are not new, but this fact does not inhibit the growing interest in them. Lean Thinking is, in fact, a difficult content to teach because it is more than a content [32, 41], it is a competency needed to be developed. The concept of Learning factories has a long history, keeping an appropriate learning environment that explores the approaches and everything that instructors need, as this research reveals.

This research illustrated 76 papers from the Scopus database, which Learning Factories settled all over the world and associated with universities and/or companies that funded it. As findings of this research could be identified that many learning strategies are being applied in the context of the Learning Factory to teach Lean Thinking.

To effectively achieve the Lean learning strategies such simulation and gamification are the preferred approaches reported in the literature as effective teaching approaches. Furthermore, the research showed few connections to companies/industries, this fact may be associated with the systemic innovation learning is still in the process of being present in companies.

Limitations of this study are associated to the unique database research and the keywords. It is preliminary research, and much more could be done in future work, namely, obtaining the outcomes achieved by these Lean Learning Factories.

Acknowledgements. This work has been supported by FCT – Fundação para a Ciência e Tecnologia within the R&D Units Project Scope: UIDB/00319/2020.

References

1. Kolmos, A.: Engineering Education for the Future. In Engineering for Sustainable Development. UNESCO (2021)
2. Alves, A.C., et al.: Final year Lean projects: advantages for companies, students and academia. In: Project Approaches in Engineering Education, pp. 1–10 (2014)
3. Municio, A.G., Pimentel, C., Ruano, J.P.: Lean School: an example of industry-university collaboration. In: Proceedings of the Fifth European Lean Educator Conference (ELEC2018) "Lean Educator's Role in Lean Development", p. 10 (2018)
4. Dinis-Carvalho, J., Fernandes, S., Lima, R.M., Mesquita, D., Costa-Lobo, C.: Active Learning in Higher Education: developing projects in partnership with industry. In: Proceedings of INTED2017 Conference, pp. 1695–1704 (2017)
5. Lima, R.M., Dinis-carvalho, J., De Campos, L.C., Mesquita, D., Sousa, R.M., Alves, A.: Projects with the Industry for the Development of Professional Competences in Industrial Engineering and Management (2014)

6. Abele, E.: Learning factory. In: Laperrière, L., Reinhart, G. (eds.) CIRP Encyclopedia of Production Engineering, pp. 1–5. Springer, Heidelberg (2016). https://doi.org/10.1007/978-3-642-35950-7_16828-1 The International Academy for Production Engineering

7. Abele, E., et al.: Learning factories for research, education, and training. Procedia CIRP **32**, 1–6 (2015). https://doi.org/10.1016/j.procir.2015.02.187

8. Lamancusa, J.S., Jorgensen, J.E., Zayas-Castro, J.L.: The learning factory-A new approach to integrating design and manufacturing into the engineering curriculum. J. Eng. Educ. **86** (2), 103–112 (1997). https://doi.org/10.1002/j.2168-9830.1997.tb00272.x

9. Lamancusa, J.S., Zayas, J.L., Soyster, A.L., Morell, L., Jorgensen, J.: 2006 Bernard M. Gordon Prize Lecture*: the learning factory: industry-partnered active learning. J. Eng. Educ. **97**(1), 5–11 (2008). https://doi.org/10.1002/j.2168-9830.2008.tb00949.x

10. Foden, F.E.: Herbert Schofield and Loughborough College. Vocat. Asp. Educ. **15**(32), 231–246 (1963). https://doi.org/10.1080/03057876380000271

11. Gento, A.M., Pimentel, C., Pascual, J.A.: Lean school: an example of industry-university collaboration. Prod. Plan. Control, 1–16 (2020). https://doi.org/10.1080/09537287.2020.1742373

12. Dietz, W., Bevens, B.W.: Learn By Doing: The story of Training Within Industry (1970)

13. Huntzinger, J.: The roots of lean: training within industry, the origin of Kaizen. Target **18**(9–22) (2002)

14. Dewey, J.: Democracy and Education An Introduction to the Philosophy of Education. Free Press, New York (1916)

15. Kagermann, H., Wahlster, W., Helbig, J.: Recommendations for Implementing the Strategic Initiative INDUSTRIE 4.0, München (2013)

16. Bittencourt, V.L., Alves, A.C., Leão, C.P.: Industry 4.0 triggered by Lean Thinking: insights from a systematic literature review. Int. J. Prod. Res. **59**(5), 1496–1510 (2021). https://doi.org/10.1080/00207543.2020.1832274

17. Sugimori, Y., Kusunoki, K., Cho, F., Uchikawa, S.: Toyota production system and Kanban system Materialization of just-in-time and respect-for-human system. Int. J. Prod. Res. **15**(6), 553–564 (1977). https://doi.org/10.1080/00207547708943149

18. Takeuchi, H., Osono, E., Shimizu, N.: The contradictions that drive toyota's success. Harv. Bus. Rev. (2008)

19. Ohno, T.: Toyota Production System: Beyond Large-Scale Production, 3ª Edição. New York (1988)

20. Womack, J.P., Jones, D.T.: Lean Thinking: Banish Waste and Create Wealth in your Corporation. Free Press, New York (1996)

21. Amaro, P., Alves, A.C., Sousa, R.M.: Lean thinking: from the shop floor to an organizational culture. In: Lalic, B., Majstorovic, V., Marjanovic, U., von Cieminski, G., Romero, D. (eds.) APMS 2020. IAICT, vol. 592, pp. 406–414. Springer, Cham (2020). https://doi.org/10.1007/978-3-030-57997-5_47

22. Alves, A.C., Leão, C.P., Uebe-Mansur, A.F., Kury, M.I.R.A.: The knowledge and importance of Lean Education based on academics' perspectives: an exploratory study. Prod. Plan. Control **32**(6), 497–510 (2021). https://doi.org/10.1080/09537287.2020.1742371

23. Amaro, A.P., Alves, A.C., Sousa, R.M.: Context-dependent factors of lean production implementations: 'Two sides of the same coin.' J. Mechatron. Autom. Ident. Technol. **5**(3), 17–22 (2020)

24. Schonberger, R.J.: The disintegration of lean manufacturing and lean management. Bus. Horiz. **62**(3) (2019). https://doi.org/10.1016/j.bushor.2019.01.004

25. Spear, S., Bowen, H.K.: Decoding the DNA of the Toyota production system. Harv. Bus. Rev. **77**(5), 96–106 (1999).http://search.ebscohost.com/login.aspx?direct=true&db=buh&AN=2216294&site=ehost-live

26. Womack, J.P., Jones, D.T.: From lean production to the lean enterprise. Harv. Bus. Rev. **72** (2), 93–103 (1994)
27. Tisch, M., Hertle, C., Abele, E., Metternich, J., Tenberg, R.: Learning factory design: a competency-oriented approach integrating three design levels. Int. J. Comput. Integr. Manuf. **29**(12), 1355–1375 (2016). https://doi.org/10.1080/0951192X.2015.1033017
28. Enke, J., Glass, R., Kreß, A., Hambach, J., Tisch, M., Metternich, J.: Industrie 4.0 – competencies for a modern production system. Procedia Manuf. **23**, 267–272 (2018). https://doi.org/10.1016/j.promfg.2018.04.028
29. Abele, E., et al.: Learning factories for future oriented research and education in manufacturing. CIRP Ann. - Manuf. Technol. **66**(2), 803–826 (2017). https://doi.org/10.1016/j.cirp.2017.05.005
30. Abele, E., Metternich, J., Tisch, M.: Learning Factories. Springer, Cham (2019). https://doi.org/10.1007/978-3-319-92261-4
31. Alves, A.C., Flumerfelt, S., Moreira, F., Leão, C.P.: Effective tools to learn lean thinking and gather together academic and practice communities. In: Volume 5: Education and Globalization, vol. 5, pp. 1–10, November 2017. https://doi.org/10.1115/IMECE2017-71339
32. Carvalho Alves, A., Flumerfelt, S., Kahlen, F.-J. (eds.): Lean Education: An Overview of Current Issues. Springer, Cham (2017). https://doi.org/10.1007/978-3-319-45830-4
33. Adam, M., Hofbauer, M., Stehling, M.: Effectiveness of a lean simulation training: challenges, measures and recommendations. Prod. Plan. Control, 1–11 (2020). https://doi.org/10.1080/09537287.2020.1742375
34. Alves, A.C., Leão, C.P., Maia, L.C., Amaro, A.P.: Lean education impact in professional life of engineers. In: Volume 5: Education and Globalization, vol. 5, p. V005T06A044, November 2016. https://doi.org/10.1115/IMECE2016-67034
35. Kahlen, F.-J., Flumerfelt, S., Sinban-Manalang, A.B., Alves, A.: Benefits of lean teaching. In: ASME 2011 International Mechanical Engineering Congress and Exposition, IMECE 2011, vol. 5, pp. 351–358 (2011)
36. Black, J.T., Phillips, D.T.: Lean Engineering The Future Has Arrived. Virtualbookworm.com Publishing, College Station (2013)
37. Dombrowski, U., Wullbrandt, J., Fochler, S.: Center of excellence for lean enterprise 4.0. Procedia Manuf. **31** (2019). https://doi.org/10.1016/j.promfg.2019.03.011
38. Alves, A.C., Dinis-Carvalho, J., Sousa, R.M.: Lean production as promoter of thinkers to achieve companies' agility. Learn. Organ. **19**(3), 219–237 (2012). https://doi.org/10.1108/09696471211219930
39. Alves, A.C.: Competencies driven by Lean Education: system-thinking, sustainability and ethics. In: International Conference on Active Learning in Engineering Education (PAEE_ALE2019), vol. 9, pp. 710–713 (2019)
40. Wagner, U., AlGeddawy, T., ElMaraghy, H., MŸller, E.: The state-of-the-art and prospects of learning factories. Procedia CIRP **3**, 109–114 (2012). https://doi.org/10.1016/j.procir.2012.07.020
41. Sudhoff, M., Prinz, C., Kuhlenkötter, B.: A systematic analysis of learning factories in Germany - concepts, production processes, didactics. Procedia Manuf. **45**, 114–120 (2020). https://doi.org/10.1016/j.promfg.2020.04.081
42. Crnjac, M., Aljinovic, A., Gjeldum, N., Mladineo, M.: Two-stage product design selection by using PROMETHEE and Taguchi method: a case study. Adv. Prod. Eng. Manag. **14**(1), 39–50 (2019). https://doi.org/10.14743/apem2019.1.310
43. Fu, H.: Integration of logistics simulation technology and logistics learning factory in a two-stage teaching method for logistics management courses. iJET **12**(9), 62–72 (2017)

The Impact of Different Training Approaches on Learning Lean: A Comparative Study on Value Stream Mapping

Matteo Zanchi[1]([⊠]) [iD], Paolo Gaiardelli[1] [iD], and Daryl John Powell[2] [iD]

[1] Department of Management, Information and Production Engineering,
University of Bergamo, Viale Marconi 5, Dalmine, BG, Italy
{matteo.zanchi,paolo.gaiardelli}@unibg.it
[2] Norwegian University of Science and Technology, Trondheim, Norway

Abstract. More and more often, companies that follow a lean implementation path realize that the results deriving just from the application of the most basic tools often guarantee temporary results that are only satisfactory in the short term. An aspect that is often overlooked, but essential for the effective achievement of a lean structure, consists in the training of operators and, more generally, of all actors in a given company. The aim of this paper is to understand how different training methods, such as the instructor-led classroom training, on-the-job training, vestibule training, and coaching lead to different learning results in the context of a lean implementation. The results are contextualized within the process mapping phase, specifically in the adoption of Value Stream Mapping at three different companies, for each of which a different training program was designed.

Keywords: Training · Lean management · Value stream mapping

1 Introduction

Since its popularization in the 1990s, Lean Manufacturing (LM) has proven its effectiveness in increasing and maximizing productivity and efficiency [1, 2]. However, though this improvement paradigm has been embraced by countless manufacturing companies, very few have been able to fully exploit and make LM methodologies their own. Indeed, as pointed out by many studies, 70–90% of companies adopting LM achieve short-term relevant results that rapidly vanish over the course of a little amount of time [3]. The main reason for this unsatisfactory outcome lies in the adoption of "ready-to-use" practices, which grant immediate satisfying results but without developing a mindset of continuous improvement and learning [4]. This weakness underlines the strategic importance of properly transferring lean thinking through appropriate and efficient approaches. In this respect, "teaching lean" literature describes the way workers and employees should be trained on LM principles, concepts and tools, and then transfer them to their day-by-day activities [5]. Nevertheless, although multiple

© IFIP International Federation for Information Processing 2021
Published by Springer Nature Switzerland AG 2021
D. J. Powell et al. (Eds.): ELEC 2021, IFIP AICT 610, pp. 109–117, 2021.
https://doi.org/10.1007/978-3-030-92934-3_12

training approaches with unique features in the learning process have been elaborated over the years by scholars and practitioners, they are often proved to be inadequate or at least characterized by inefficient training methods, thus limiting the development of successful LM programs [6]. Such a limitation has led to an urgent need to identify the most suitable training approaches enhancing learning of LM principles, not only in terms of acquired knowledge or skills, but also in relation to the attitude and confidence users place in the adoption of LM tools and techniques.

On these premises, this paper proposes case-based research aiming to understand the effects of the application of different training methods in learning LM principles. In particular, the study, conducted in three companies that have recently launched a lean manufacturing program through the introduction of Value Stream Mapping (VSM), focuses on a set of training methods identified in literature to assess their influence on participants' learning. The paper is structured as follows: the next section provides a theoretical background on training theories and learning process; Sect. 3 illustrates the methodology through which the different case studies were analyzed, while the case study and the achieved results are presented in Sect. 4; the conclusions, limitations and further developments are finally reported in the last section.

2 Theoretical Background

Training plays a fundamental role for the development of any organization [7]. In particular, proper training of human resources is considered necessary for the development of successful change management programs. According to Chatzimouratidis et al. [8] the main training approaches currently adopted in industry can be categorized into nine groups, as described in Tables 1 and 2.

Table 1. Main training approaches [8]

Training approach	Description
On-the-job training	Employees are directly trained on a one-to-one basis while they perform their job. Thanks to its effectiveness, on-the-job training is considered very popular
Coaching	A person with advanced experience, called mentor or sensei, personally follows the development of a worker with very little experience, named mentee [9]
Apprenticeship	Skills are acquired in the workplace while getting paid, through a combination of on-the-job training and classroom instruction [9]
Vestibule training	Employees perform the job in a simulated environment, similar to the real workplace [10]. This prevents any training-related incident

Although many training approaches are available, none can be considered suitable for any training program in isolation. Indeed, since training is a situational process, each specific context requires adopting an appropriate training strategy [12], often

Table 2. Main training approaches [8]

Training approach	Description
Web-based learning	Self-paced instructions are provided through interactive multimedia, training software and teleconferencing programs [11]
Instructor-led classroom training	Face-to-face training is provided by an instructor in a classroom. Trainees can ask questions and complete tasks and exercises under the strict guidance of the instructor
Programmed self-instruction	The employee studies training material on his/her own, without the guidance of any teacher or mentor, in a self-paced way
Role playing	Interactive approach that allows employees to act out situations that may happen in their workplace
Systematic job rotations	Employees are transferred to different jobs or rotated among different workstations in order to acquire a wider knowledge and increase motivation

combining different approaches into an integrated and distinctive training plan. In this regard, Chatzimouratidis [8] proposes an evaluation criterion based on cost, time, applicability, efficiency, and employee motivation through which companies can identify the training approach best suited to the context in which it is implemented.

Regardless of the implemented training strategy, any training plan must necessarily be evaluated to measure its effectiveness. In this regard, Kirkpatrick's model represents the most common and versatile framework for the evaluation of training [13], as it assesses both formal and informal training methods and rates them considering four levels of criteria [14]:

1. **Reaction:** measuring the initial response of participants allows the trainer to understand whether the content is aligned with expectations of attendees and participants perceive what he/she desires to communicate.
2. **Learning:** measuring how much information has effectively been absorbed during the training program and how much skills have changed from before to after the training, provides the trainee with a clear indication of the effectiveness of the training program.
3. **Behavior:** assessing how participants apply achieved skills on their day-by-day activities allows the trainer to understand whether what it was delivered has become part of the participant's behavior.
4. **Results:** measuring and analyzing the impact of training on productivity, quality, efficiency, and customer satisfaction, allows the trainer to understand if what has been learned has translated into a tangible benefit for the company.

3   Research Design and Methodology

As the aim of this research is to understand how different training approaches convey LM principles, four training methods were selected for the study: 1) instructor-led classroom training; 2) vestibule training; 3) on-the-job training and 4) coaching. The choice was made in accordance with the criteria of applicability and motivation of employees indicated in Chatzimouratidis' framework. Indeed, when training goal concerns transferring knowledge about tools that must be implemented on a daily basis, such as a VSM, full involvement of employees becomes crucial, as employee's training fulfilment increases post-training organizational commitment, self-efficacy and motivation [15, 16].

The selected training methods were then applied to three SMEs that have recently launched a lean optimization project characterized by a similar organizational structure and based on the implementation of a VSM. For each company a basic instructor-led classroom training was first carried out to develop a common knowledge on LM principles and VSM tools. Subsequently, a second training session was conducted using a distinctive training approach, namely on-the-job training for Company A, coaching for Company B and vestibule training for Company C.

As the main goal is to evaluate the influence of each training method in terms of acquired skills, competencies and knowledge, as well as attitude, confidence and commitment gained during the training process, the evaluation was carried out mainly referring to the 'Learning' phase of Kirkpatrick's model. Consistently, the Italian Ministry of Education's learning evaluation model [17] was selected to assess participants learning according to its dimensions, namely autonomy (the ability to use what has been learnt without the help of the trainer), continuity (the ability to continuously demonstrate what has been learnt), context (the ability to apply what has been learnt in different contexts and situations), resource (the ability to use the right tools and competences to solve a problem). Table 3 provides a brief description of 4 different evaluation levels.

Table 3. Learning levels

Level 4 Advanced	The trainee completes his/her tasks autonomously and with continuity, adapting the available resources to the context in which he/she operates
Level 3 Intermediate	The trainee completes tasks in known situations independently and continuously. He/she faces unfamiliar situations using multiple resources discontinuously and not autonomously
Level 2 Pre-intermediate	The trainee completes his/her tasks in known situations and uses the resources provided autonomously but discontinuously, or rather with continuity but with the support of the trainer
Level 1 Basic	The trainee completes his/her tasks only in known situations and with the full support of the trainer, using specific tools and resources provided

To avoid any interference or bias in the evaluation, assessment of each participant was carried out by an independent university researcher with long term experience in LM, different from the (external) training provider, who evaluated the progress of the mentees in two different moments: right after the instructor-led classroom training session, by testing participants ability to properly adopt VSM tool to solve a simple case exercise of which main issues, improvement suggestions and most appropriate solutions were already provided by the text [18], and at the end of the second part of the training program, by evaluating through direct observation how each participant applied the learnt notions on a real case study carried out in their company. Finally, achieved results were gathered and analyzed, to identify the best training method as well as to understand how each training method influences personal learning.

4 Case Studies

Company A (an SME) is a producer of meal distribution systems that deployed a VSM program with the support of an external consultant. The latter also acted as responsible for the training plan of the implementation team, made up of 6 members, namely the operations manager, the foreman of production and 4 workers. To properly train the participants about the VSM tool, an instructor-led classroom training followed by an 'on-the-job' training session approach was adopted, where the mentor assisted the operators in the process of observation, identification of different activities and issues within the process.

Company B is an Italian SME specialized in the production of lift components. In 2020 it launched a VSM program training to enhance the productivity of its operations. The group of participants in this project included the production manager, 2 foreman and 3 workers. The training plan consisted of instructor-led classroom training followed by a 3-session coaching in which, upon a simple tour of the production department, the mentor assigned the participants tasks to be accomplished, concerning the mapping process or possible ideas for improvement, consistent with the lean methodology. Progresses were then monitored in the following sessions and updated according to individual pathways.

Company C is a company involved in the extraction and production of inert materials, concrete and asphalt. Similar to Company A, training was held by an external consultant, specialized in optimization topics, hired by the company itself. Participants to the project were 6: the chief of the production department, the foreman, the maintenance worker of the factory and 3 factory workers. For this third case, an instructor-led classroom training followed by a vestibule training session was undertaken.

5 Results and Discussion

As shown in Fig. 1, which displays the average training evaluation, according to the Italian Ministry of Education's model, inclusive of all the 6 operators who were involved in the lean program for each company, the instructor-led classroom training approach has led pretty much to the same results across all companies, with slight changes due to different contexts and everyone's predisposition towards learning. One third of the participants have maintained a basic level, while two thirds have reached a pre-intermediate level. Conversely, what has changed significantly is the final outcome following the second phase of training provided with the on-the-job training (company A), coaching (company B), and vestibule training (company C) approaches, respectively. In particular, collected data indicate that coaching is the best training method when it comes to training people on the adoption of VSM (3,17/4,00). In fact, all operators who have experienced this type of training have reached an intermediate level except for one operator who has achieved an advanced level of learning. The on-the-job training also brings appreciable results (2,83/4,00), while the vestibule training with a final evaluation of 2,33/4,00 seems the less appropriate approach to adopt. In this case only 2 out of 6 participants have reached an intermediate level of learning while the level of the other participants remains pre-intermediate. The rationale mainly depends on the nature of vestibule training, which takes place in the laboratory without providing any support on the understanding of the working environment. This results in a very limited approach in terms of potential, usefulness and applicability of a tool such the VSM.

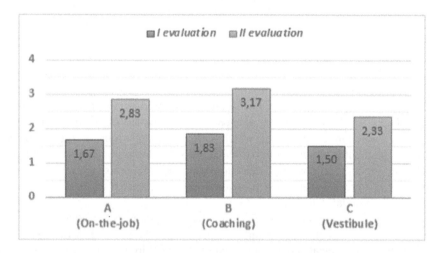

Fig. 1. Intermediate and final results for each case study

Despite the strategic relevance played by coaching activities, each training method provides marked outcomes for some specific dimensions of learning, as shown in Fig. 2. For instance, vestibule training, although being the less effective method overall, provides the best results in terms of resources. Results of the analysis show that approximately 83% of participants in this type of training were able, at the end of the training program, to use all the resources made available. This peculiarity is probably due to the nature of this training approach, that lets mentees deepen and consolidate the acquired knowledge, as learners can receive immediate feedback and ask questions more easily [19]. On the other hand, on-the-job training and coaching approaches show similar results, with a slight preference for coaching approach. This outcome is probably referable to the duration of the training process: while on-the-job training took place only in a one-day session, coaching sessions were organized on three occasions, given the more 'long-term' nature of this method.

In conclusion, the achieved results suggest that training approaches are not alternative, but rather complementary to each other, and may be all considered whenever

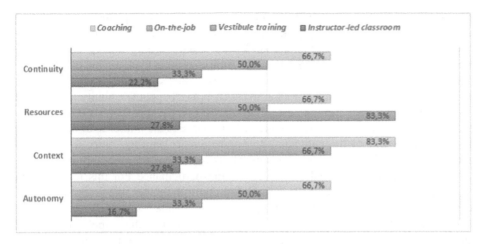

Fig. 2. Training outcomes according to the four training evaluation dimensions

designing a training program, especially if the training plan is developed around the Shu-Ha-Ri cycle [20]. In this perspective, instructor-led classroom and vestibule training emerge as fundamental to provide the basic knowledge required in the Shu (obey the rules) phase. Conversely, on-the-job training and coaching can prove useful as they put the mentee in front of the real challenges on the shopfloor, prompting him/her to acquire a complete awareness of the tool and adjust its usage, in relation to the context in which he/she operates.

It's then up to each company to decide whether adopting all the mentioned training approaches rather than, depending on available resources in terms of costs and time, focus more on specific methods. Typically, a firm may start with a fist instructor-led classroom training followed by an on-the-job training session which may eventually evolve to a coaching training, if the training time frame goes for the long-term.

6   Conclusion

The aim of this research was to find out how different training methods have a specific impact on the learning of LM practices and in particular the implementation of VSM. Results show that, following a first instructor-led classroom training session, which levelled out the level of lean knowledge among the different companies, vestibule training, on-the-job training and coaching respectively have an increasing effectiveness on learning results. Moreover, all the adopted training approaches emerge as valid in some way, as they provide significant outcomes in specific dimensions of learning and, so, can be integrated to accomplish a valid and comprehensive LM training program.

Nevertheless, the considerations emerging from this study refer to the learning phase only and therefore cannot be considered exhaustive since, according to the theory, a successful training project presumes that the acquired competences should be applied to the work context to generate tangible results for the company. In addition, consistently with the lean thinking paradigm a successful implementation of LM practices involves transferring of skills, competences, attitudes, and commitment to other workers to establish a virtuous circle of mutual learning among people. This consideration suggests the need to expand the future research by assessing the effects of different training methods on the other dimensions of the Kirkpatrick's model, namely reaction, behavior, and results, as well as to assess not only people's ability to take hold what has been learnt but also to transfer it to others. Finally, this analysis does not take into account any variations in the learning outcomes due to the influence of the working environment/context or the individual predisposition to learning. Also, the personal predisposition of the external mentor towards the different training approaches may have played a role in the final assessment of the participants. For this reason, even though quantitative results have been provided along the paper, the research may be intended more as led with a qualitative approach, to orientate research into a defined set of training methods. Therefore, further research regarding this aspect may be of interest, as well as testing other training approaches, of the nine listed in the second section, to check if they can eventually lead to better results.

Acknowledgements. The authors would like to acknowledge support from the research council of Norway for the project Lean Digital.

Bibliography

1. Kovács, G.: Productivity improvement by lean manufacturing philosophy. Adv. Logist. Syst: Theory Pract. **6**(1), 9–16 (2012)
2. Netland, T., Powell, D.J.: The Routledge Companion to Lean Management. Routledge, New York (2016)
3. Bhasin, S.: An appropriate change strategy for Lean success. Manag. Decis. **50**(3), 439–458 (2012)
4. Powell, D.J., Coughlan, P.: Rethinking lean supplier development as a learning system. Int. J. Oper. Prod. Manag. **40**(7/8), 921–943 (2020)
5. Dinis-Carvalho, J.: The role of lean training in lean implementation. Prod. Plann. Control **32** (6), 441–442 (2021)
6. Shrimali, A.K., Soni, V.K.: A review on issues of lean manufacturing implementation by small and medium enterprises. Int. J. Mech. Prod. Eng. Res. Dev. **7**(3), 283–300 (2017)
7. Pennathur, A., Mital, A.: Worker mobility and training in advanced manufacturing. Int. J. Ind. Ergon. **32**(6), 363–388 (2003)
8. Chatzimouratidis, A., Theotokas, I., Lagoudis, I.N.: Decision support systems for human resource training and development. Int. J. Hum. Resour. Manag. **23**(4), 662–693 (2012)
9. Mihiotis, A.: Human Resource Management. Hellenic Open University, Patras (2005)
10. Business Jargons. Vestibule Training (2021). https://businessjargons.com/vestibule-training. html. Accessed 04 July 2021
11. Lee, H.-M., Davis, R.A., Chi, Y.-L.: Integrating XML technologies and open source software for personalization in E-learning. Int. J. Web-Based Learn. Teach. Technol. **4**(3), 39–54 (2009)
12. Tüzün, İ.K.: General overview of training effectiveness and measurement models. Gazi Üniversitesi Ticaret ve Turizm Eğitim Fakültesi Dergisi (1), 144 (2005)

13. Kirkpatrick, D., Kirkpatrick, J.: Evaluating Training Programs: The Four Levels. Berrett-Koehler Publishers (2006)
14. Andriotis, N.: How to evaluate a training program: The definitive guide to techniques & tools (2019). https://www.talentlms.com/blog/evaluate-employee-training-program/. Accessed 30 June 2021
15. Tannenbaum, S.I., Mathieu, J.E., Salas, E., Cannon-Bowers, J.A.: 'Meeting trainees' expectations: the influence of training fulfillment on the development of commitment selfefficacy, and motivation. J. Appl. Psychol. **76**(6), 759–769 (1991)
16. Bartlett, K.R.: The relationship between training and organizational commitment: a study in the health care field. Hum. Resour. Dev. Q. **12**(4), 335–352 (2001)
17. Il Ministro dell'Istruzione. LINEE GUIDA - La formulazione dei giudizi descrittivi nella valutazione periodica e finale della scuola primaria (2021). https://www.istruzione.it/valutazione-scuola-primaria/allegati/Linee%20Guida.pdf. Accessed 10 July 2021
18. Rother, M., Shook, J.: Learning to see: value stream mapping to add value and eliminate muda. Lean Enterprise Institute (2003)
19. Clark, D.R.: Design methodologies, instructional, thinking, agile system (2014). [from OLABIYI, S. O. (2020). ENHANCING JOB SATISFACTION THROUGH VESTIBULE TRAINING IN CONSTRUCTION INDUSTRIES IN LAGOS STATE. Nigerian Online Journal of Educational Sciences and Technology, 1(2), 35–48.]
20. Capobianco, D.A.: Shu-ha-ri how to break the rules and still be agile. In: Ciancarini, P., Sillitti, A., Succi, G., Messina, A. (eds.) Proceedings of 4th International Conference in Software Engineering for Defence Applications, vol. 422, pp. 231–238. Springer, Cham (2016). https://doi.org/10.1007/978-3-319-27896-4_19

Teaching Lean in the Digital Era

Lean Courses in Process Form - Do as We Learn, Success or Not?

Eivind Arne Fauskanger[(✉)] and Roland Hellberg

University of South-Eastern Norway, Hasbergsvei 36, 3616 Kongsberg, Norway
{eivind.fauskanger, roland.hellberg}@usn.no

Abstract. Experience of successful improvements in organizations based on lean methods shows that participation and involvement are the keys. Based on this, a university-level continuing course has been created that focuses on quality improvements. Students gain knowledge of various lean and quality tools, as well as of implementation processes. The students learn essential tools and implementation processes gradually, while at the same time using this knowledge.

The course is structured in modules with intermediate work steps, where the intention for students is to use their knowledge in quality improvement projects at their workplace. The underlying idea is that the course modules correspond to the Plan, Do, Check, Act (PDCA) methodology. The exam consists of submitting a folder with reports from the intermediate work steps and a final report.

This course has been offered and completed twice, with the third in progress as of this date. We describe the background for our choice in quality and process tools as well as the examination form. In addition, results from a survey among the participating students on their opinions of the course content, structure, and examination form, are presented.

We conclude that students who take this further education in parallel with their regular work are of great benefit to their employers, as the course is module-based, and the participants work on a project at their workplace. In addition, the students appreciate interweaving theory and knowledge training. The course grades are determined on a final report based on sub-assignments where the students do an academic reflection on their improvement project.

Keywords: Folder examination form · Lean education · Process-based training

1 Background

Many universities and colleges, for example NTNU, Chalmers, DTU, Aachen, provide courses in lean and quality improvement. However, curricula from quality courses show that many of these courses are traditional, where students learn more about tools and methods, and less about improving processes and implementation. When this course was developed, it focused on providing knowledge about basic quality tools/methods and lean methodology for improvements and on employees' commitment and organizational development to achieve improvement objectives. The course is structured in modules where students gradually learn different quality tools, methods,

© IFIP International Federation for Information Processing 2021
Published by Springer Nature Switzerland AG 2021
D. J. Powell et al. (Eds.): ELEC 2021, IFIP AICT 610, pp. 121–131, 2021.
https://doi.org/10.1007/978-3-030-92934-3_13

and different aspects of improvement processes. Between modules, students practice what they have learned in improvement projects in the workplace.

The course is titled "Management of quality enhancement and continuous improvement." It is focused on modern management tools and the main elements of Total Quality Management (TQM) and lean. In addition, the course focuses on continuous improvements as a method and how organizational culture affects changes. In this context, we also look at which organizational changes are required and how they can be implemented to achieve the objectives for the desired improvement.

The target group for this professional development course is employees working with quality and improvement work. Prerequisites are general study competence (in the Norway education system, or the equivalent) and two years' work experience.

The course has been offered twice (2019, 2020), and a third-round has started in the spring of 2021. There have been approximately 20 students in each course. The study questionnaire indicated that most students have leading positions in public administration, health or work in the private sector.

2   Purpose

We wanted to explore participating student's opinions about the course content, module structure, examination form, and experiences with working on a concrete improvement project in the workplace. Based on students' evaluation of the course, the idea was to develop the course further.

The purpose was to obtain students' views from the two completed courses in retrospect and determine to what extent they anticipate being able to apply knowledge from the course at their workplace. Furthermore, we wanted to learn whether course participants can contribute to improvement work in their organization. In other words, we wanted to find out if our way of implementing the course and testing knowledge through applied improvement work was successful.

3   Course Structure and Content

The course covers 15 ETC based on the European credit system. The course consists of four sessions of two full days. Between sessions, participants work on a current quality/improvement project at their workplace. The assignment is based on that work. The assignments form the basis of a final report, which is also the exam for grading.

The idea is to gradually give the participants methods and tools used in TQM and lean and use them in their in-house project. The PDCA methodology permeates the structure of the course and the associated assignments. The PDCA cycle is also known by two other names, the Shewhart cycle and the Deming cycle (e.g., Johnson 2016). There is very little research on the success of concept-based education (such as PDCA, the main tool of the course), hence the value of finding out more about it (Laverentz and Kumm 2017).

Successful implementation of improvement projects is based on strong involvement of people who work close to the operations in organizations or other workplaces.

Therefore, the course also provides knowledge of staff commitment, change management, and organizational development.

Our view of the concept of quality is characterized by diversity. Quality is a much more complicated term than it may appear. It seems that every quality expert defines quality in a somewhat different way. Various perspectives can be taken in defining quality (e.g., customer's perspective, specification-based perspective). Garvin's five different perspectives on quality (Garvin 1988) particularly resonates with us.

A contemporary definition of quality derives from Juran's "fitness for intended use" (e.g., Juran 2014), meaning that quality is meeting or exceeding customer expectations. According to Deming (1986), the customer's definition of quality is the only one that matters.

Based on this reasoning, it is essential to understand the customer, the customer's needs, and the environment where the customer will use the product/service. In this context, we have included tools such as the Kano model and Quality function deployment (QFD) or House of Quality. The Kano model offers insight into how product attributes are perceived by customers (Kano et al. 1984) and into how customer's needs can be met or even surpassed (e.g., Oakland 2000). House of Quality is a method to transform qualitative user demands into quantitative parameters (Akao 1994).

The course also provides insight into traditional quality management tools such as Ishikawa's seven basic tools of quality (e.g., Tague 2005) and Failure Modes and Effects Analysis (FMEA). The syllabus also includes a briefing on statistical process control and Six Sigma as a quality management method.

Unlike the seven basic QC tools, which measure quality problems that already exist, the seven new QC tools make it possible for managers to plan wide-ranging and detailed TQC objectives throughout the entire organization. These tools, some borrowed from other disciplines and others explicitly developed for quality management, include relations diagrams, affinity diagrams, systematic diagrams, matrix diagrams, matrix data analysis, process decision program charts (PDPC), and arrow diagrams (Mizuno 1988).

Working towards improvements in quality is important, beneficial, and rewarding. We have included several different angles on quality improvement work, for example the fact that accidents are often due to human error and poor construction. Therefore, elements are also included on how various human error causes can be reduced and engineered and planning designs improved for reliability and redundancy.

In addition to quality tools and methods, the course provides insight into TQM as an overall concept and International Organization for Standardization (ISO) systematics, including certification. TQM in this context can be a system used by customer-centric organizations that involve all its employees in the process of continuous improvement. TQM is essentially a management practice that focuses on meeting or exceeding customer expectations. A TQM-centric organization focuses on process measurements and controls to achieve continuous improvements in the business process. Thus, it is an integrated approach to improve productivity by using both qualitative and quantitative concepts.

With this integrated approach, there are parallels between TQM and lean, although these two concepts differ significantly in other areas. The philosophy behind continuous improvement based on lean tools is called Kaizen. It involves identifying benchmarks of excellent practices and instilling a sense of employee ownership of the process. Some other lean tools are 5S, seven waste, suggestions for improvement, stand-up meeting, visualization, A3, Kanban, extended value stream mapping, and Andon, Jidoka, Gemba walks.

The basis of the continuous improvement philosophy is the belief that virtually any aspect of an operation can be improved. The people most closely associated with an operation are in the best position to identify the changes that should be made towards improvement. Consequently, employee involvement plays a significant role in continuous improvement programs. Workplace cleanliness, visualization (panel meetings), and employee involvement (proposal activities) are essential components of continuous improvement work. Involving all staff in the lean operation creates participation.

Our starting point in implementing the TQM improvement philosophy is not only that quality and quality improvement is about the customer's perceived quality and how to manage quality with different tools and models, but it is equally an improvement focus on the processes involved in creating the organization's products and services.

Deming's theory of profound knowledge (Braughton 1999) is a management philosophy grounded in systems theory. It is based on the principle that each organization comprises a system of interrelated processes and people who make up the system's components (e.g., Braughton 1999). We believe that Deming's idea that the parts of a system are interconnected is central to implementing changes. Not least, cognitive insight is essential.

To stabilize and streamline processes, it is in some cases necessary to create new processes (BPR) (Andersen 2007). We lean towards the thinking based on continuous improvements as part of a lean approach where we push hard for employee participation (Rolfsen 2014).

Change of attitude and behaviour is required to achieve quality improvement in business. Basic knowledge of realizing cultural and organizational changes in a quality context is also required. It is mainly based on Kurt Lewin's theories of groups and group dynamics/ways of change and links to systems theory and organizational culture theory (e.g., Huarng and Mas-Tur 2016; Hussein et al. 2018).

4 Pedagogic Approach

The pedagogical approach has been to adapt the course to students who work full time. Therefore, we chose a session-based structure with four sessions of two days (9–16). Furthermore, in line with Bloom's taxonomy (Anderson and Krathwohl 2001), we wanted the participants to practice what they learn.

Problem-based learning is the pedagogical approach for the course. This is self-driven learning, problem-solving, and peer collaboration skills (Pettersen 2005). In Norway, this method is used in medical education (Lycke et al. 2006).

The objective was that the students work on a project in their workplace between sessions. The assignments are handed out after each session and submitted before the next session with students receiving feedback at the next session. During the sessions, students present what they have done, and everyone must present at some point during the course. The assignments are based on the PDCA methodology. Therefore, the PDCA approach permeates the structure of the course. Another important aspect is that participants learn from each other, and time is allocated during the course for socializing and exchange.

4.1 Evaluation Form – 3 Phases

There are three compulsory assignments, one oral presentation and one final report. The assignments are assessed as approved/not approved. All three assignments must be approved before the student can submit the final report, which is graded A-F. The content of the submission assignments should be based on issues from the student's workplace, if possible, with companies' current improvement projects or process-oriented change tasks.

First submission; a presentation of the company's management system and the description of an improvement project that will be completed during the course.

Second submission: a written plan for the improvement project based on the PDCA methodology.

Third submission: describe what has been done in the project. If the project has not started - build further on the implementation description.

Final report (exam): Based on the three assignments, the student writes an analytical and reflective report. In the text, course literature is references.

5 Method

We worked with participants in the 2019 and 2020 courses. We investigated the students' exchange of knowledge from this course format and examination form. We also explored how students experienced working with an actual improvement project in the company where they work while they were in the process of completing the course. Finally, we asked whether the student project has, or will have, a (lasting) effect on the company. This was done through a web-based survey of all course participants who completed the course.

As the target group for our survey were students who had already completed the course and left the university, it was easy to contact them by e-mail with the questionnaire. There were 31 students from 2019 and 2020. As this group was small, we wanted to reach them all. Questionnaire surveys are a systematic method of obtaining data (Groves et al. 2004). Questionnaire surveys are a structured form of standardized questioning: all respondents are asked the same questions in the same way (Ringdal 2013). We chose to use "Nettskjema" as a tool, where the respondents are anonymous. The online tool has been developed and operated by the University of Oslo.

There were four areas we wanted to investigate. It was our objective to determine what the participating students thought about the course content, module structure, examination form, and how they experienced working with an improvement project in the workplace.

As it was not initially a question of testing existing theories, the questions were not linked to a theoretical frame of reference but directly to the four areas of investigation.

The questionnaire was semi-structured, which is a combination of pre-coded, graded, and open-ended questions. Pre-coded questions are questions with several stated answer alternatives (e.g., Johannessen et al. 2016). Some of our questions were designed so that it was possible to choose one or more answers, in addition to free-text answers. Several questions were statements, where the respondents scaled the extent to which they agreed or disagreed with the statement. We used a 7-point scale for the degree of agreement. There are differing opinions among researchers as to whether a neutral middle category should be included. Some believe that such a category is an invitation to the respondent to not really think through difficult questions (Jacobsen 2005). On the other hand, others have good experiences with neutral survey response options (Johannessen et al. 2016). The 7-point scale used in the study included neutral response questions.

The questions in the form were divided into four parts. Part 1 covers the respondent's background. These questions were simple and neutral, which helps increase the respondent's motivation to complete the survey (Haraldsen 1999). Parts 2–4 included questions related to our four areas of interest. For brevity, the survey questions are not included here.

When the questionnaire was completed, we asked three people with relevant backgrounds, from outside the study group, to review the form and offer constructive feedback. This resulted in adjustments according to the scope and understanding of specific questions and word choices. We wanted the survey to take around 10 min to complete.

The questionnaire surveys were sent in May 2021 as a link by e-mail to all 31 participants. The e-mail addresses provided by students during the course were used. We had an initial deadline of 10 days. However, it became necessary to extend the deadline when we received few responses, and we further encouraged participation. This resulted in several responses, and we ended up with 14 usable surveys. This is a response rate of 45%, which is considered very good (e.g., Nulty 2008). As the answers are anonymous, we have not been able to carry out any deviation analysis of the 14 who responded. The respondents spent between seven and 14 min, in line with our goal.

6 Findings

Of those who responded, 10 were women and four men. Five respondents were aged 50–59, eight were aged 40–49, and one was aged 30–39. Three of the four men had high school qualifications, and the fourth had a bachelor's degree. Four works in the private sector, others in the public sector, of which five in the health sector. Six works in a position as top manager/management team, one as a middle manager. Four works to a large extent with quality-related work, nine work to a lesser extent with quality and one did not work. Four worked for more than five years with quality-related work. Eight have worked 1–5 years with quality.

Regarding our first area of interest, what the students thought about the course content, we found that the course content is relevant concerning both expectations (see Fig. 1) and that it enabled students to complete an improvement project (methods, tools, skills) at their job, see Fig. 2. When asked whether the student can apply what they learned as soon as they returned to work, 12 out of 14 answered yes, and one does not know. This indicates a relevant selection of principles and methods within quality management.

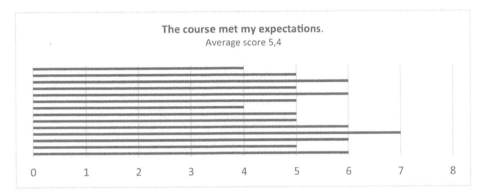

Fig. 1. How the course met the students' expectations. (1 indicates strongly disagree and 7 indicates strongly agree.)

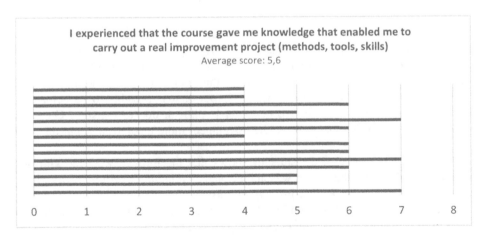

Fig. 2. The extent to which the course enabled students to carry out a real improvement project (methods, tools, skills). (1 indicates strongly disagree and 7 indicates strongly agree.)

In the second area of interest, what the students thought about the module structure, the responses were a bit more fragmented. However, most considered that both the workload and the time between sessions were good with an average of 4.8 on the 7-point scale (where value 1 indicates that the workload/time was too small/short and value 7 indicates that the workload/time was too large/long.) We interpret this to mean that the current module structure is a good model for students who are working.

Regarding our third area of interest, what the students thought about the examination form, 12 students preferred an assessment form with work requirements (assignments) and a home exam (final report), for two it did not matter (see Fig. 3). For most students, the form of assessment contributed to improvement work in the company, which we interpret as a positive effect.

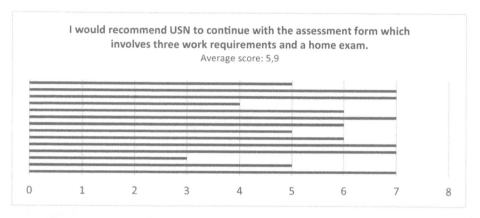

Fig. 3. The extent to which students would recommend continuing with the assessment form of three work requirements and a home exam. (One strongly disagrees and seven strongly agree.)

Regarding the fourth area of interest, the students' experiences with working with an improvement project in the workplace to apply acquired knowledge, 10 out of 14 students had found it easy to identify a suitable project to work with at the workplace. Four had found it difficult. It is problematic for this course if students do not have the opportunity to work on an actual project. Of course, it is possible to participate in another student's project, but in practice, this is difficult due to logistics. On these occasions, students were able to solve the assignment more theoretically, but this was not ideal.

A comment from one of the students shows that the form of working with one project in parallel with the course was appreciated "*I think it was a great way to put theory and practice together in a gradual process.*"

Twelve out of 14 students continued with other improvement tasks according to lean and PDCA thinking after completed the course. The students' projects influenced their workplace (see Fig. 4), which we interpret as the course creating benefit for both the individual and employer.

Some comments that show how the students experienced the course and what the course contributed.

"*I became more aware and also got tools for how not only I should carry the load but involve others and get this way of working into our [company] culture.*"

"The course helped the improvement project that was planned. It became more concrete and contributed to employees and managers showing interest. Also contributed to facts being obtained, this created the basis for the work further. Facts and involvement contributed to the project continuing."

"I would have worked on the improvement tasks anyway, but maybe in a slightly different way. I gained new knowledge and competence in the course, which meant that I adjusted for example my approach."

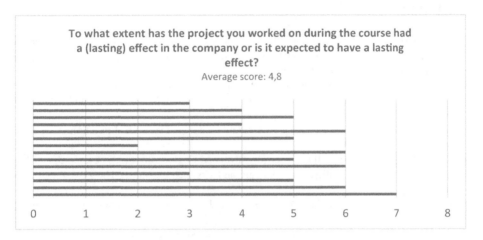

Fig. 4. The extent to which the project worked on for this course had a (lasting) effect on the company. (One reflects a small effect and seven a large effect.)

7 Conclusion

We wanted to explore what students thought about the course content, module structure, examination form, and how they felt about working with an improvement project.

The answers we received clearly show that our selection of methods, techniques, and knowledge of implementation processes has been useful for the course participants. There may be methods and techniques unknown to us, and therefore not used during the course, which could enhance the value of the course. However, what we have chosen has been useful.

A modular structure with full-day collections and intermediate work steps applied to improvement projects in the workplace has worked well for participants who work in parallel with their course implementation.

The examination approach with three assignments and a grade-based final report with reflections and theoretical connections is an appreciated examination form.

Applying knowledge to an improvement project in the workplace has been appreciated. However, some participants were unable to gain access to projects in the workplace, which meant that they did theoretical assignments instead, which is not ideal.

It is our hope that knowledge from this study can be used by others who want to develop courses with an interactive structure and examination form.

To answer the heading question "Do as we learn, success or not?", we would like to say that it has been a success.

Acknowledgements. We greatly appreciate the help to prereview the questions in our survey by Ass. Prof. Jon Hovland Honerud, University of South-Eastern Norway.

References

Akao, Y.: Development History of Quality Function Deployment. The Customer Driven Approach to Quality Planning and Deployment. Asian Productivity Organization, Minato, Tokyo (1994). ISBN 92-833-1121-3

Andersen, B.: Business Process Improvement Toolbox. Quality Press (2007)

Anderson, L.W., Krathwohl, D.R.: A Taxonomy for Learning, Teaching, and Assessing, Abridged edn. Allyn and Bacon, Boston (2001)

Braughton, W.D.: Edwards Deming's profound knowledge and individual psychology. Individ. Psychol. **55**(4), 449 (1999)

Deming, W.E.: Out of Crisis. MIT Center for Advanced Engineering Study, Cambridge (1986)

Garvin, D.A.: Managing Quality. The Strategic and Competitive Edge. The Free Press, New York (1988)

Groves, R., Fowler, F., Couper, M., Lepkowski, J., Singer, E., Tourangeau, R.: Survey Methodology. Wiley, Hoboken (2004)

Haraldsen, G.: Spørreskjemametodikk etter kokebokmetoden. Ad Notam Gyldendal AS, Oslo (1999)

Huarng, K.-H., Mas-Tur, A.: Turning Kurt Lewin on his head: nothing is so theoretical as a good practice. J. Bus. Res. **69**(11), 4725–4731 (2016). https://doi.org/10.1016/j.jbusres.2016.04.022, ISSN 0148-2963

Hussein, S.T., Lei, S., Akram, T., Haider, M.J., Hussain, S.H., Ali, M.: Kurt Lewin's change model: a critical review of the role of leadership and employee involvement in organizational change. J. Innov. Knowl. **3**(3), 123–127 (2018). https://doi.org/10.1016/j.jik.2016.07.002, ISSN 2444-569X

Jacobsen, D.I.: Hvordan gjennomføre undersøkelser? Innføring i samfunnsvitenskapelig metode (2. utg.). Høyskoleforlaget AS - Norwegian Academic Press, Kristiansand (2005)

Johannessen, A., Tufte, P., Christoffersen, L.: Introduksjon til samfunnsvitenskapelig metode (5. utg.). Abstrakt forlag, Oslo (2016)

Johnson, C.N.: Best of back to basics: the benefits of PDCA use this cycle for continual process improvement. Qual. Prog. **49**(1), 45 (2016)

Juran, J.M.: Architect of Quality: The Autobiography of Dr. Joseph M. Juran, 1 edn. McGraw-Hill, New York City (2004)

Kano, N., Nobuhiku, S., Fumio, T., Shinichi, T.: Attractive quality and must-be quality. J. Jpn. Soc. Qual. Control (in Japanese) **14**(2), 39–48 (1984). ISSN 0386-8230. Archived from the original

Laverentz, D.M., Kumm, S.: Concept evaluation using the PDSA cycle for continuous quality improvement. Nurs. Educ. Perspect. **38**(5), 288–290 (2017). https://doi.org/10.1097/01.NEP.0000000000000161

Lycke, K.H., Grøttum, P., Strømsø, H.I.: Student learning strategies, mental models and learning outcomes in problem-based and traditional curricula in medicine. Med. Teach. **28**(8), 717–722 (2006)

Mizuno, S.: Management for Quality Improvement, The 7 New QC Tools. Productivity Press, New york (1988). ISBN 9780915299294

Nettskjema. https://www.uio.no/tjenester/it/adm-app/nettskjema/mer-om/

Nulty, D.D.: The adequacy of response rates to online and paper surveys: what can be done? Assess. Eval. High. Educ. **33**(3), 301–314 (2008). ISSN 0260-2938 print/ISSN 1469-297X online

Oakland, J.S.: Total Quality Management: Text with Cases. Butterworth-Heinemann, Oxford (2000)

Pettersen, R.C.: Kvalitetslæring I høgere utdanning. Innføring i problem- og praksisbasert didaktikk. Universitetsforlaget, Oslo (2005)

Ringdal, K.: Enhet og Mangfold; Samfunnsvitenskapelig forskning og kvantitativ metode (3. utg.). Fagbokforlaget, Bergen (2013)

Rolfsen, M.: Lean blir norsk: Lean i den norske samarbeidsmodellen. Fagbokforlaget, Bergen, Norway (2014)

Tague, N.R.: The Quality Toolbox, 2nd edn. ASQ Quality Press, Milwaukee (2005). ISBN 978-1-62198-045-2

The Digitalisation and Virtual Delivery of Lean Six Sigma Teaching in an Irish University During COVID-19

Olivia McDermott(✉)

National University of Ireland, Galway 91TK33, Ireland
Olivia.McDermott@nuigalway.ie

Abstract. This research discusses how lecturers in an Irish university transferred their classroom-based blended learning Lean Six Sigma modules to online delivery. The transfer from a practical classroom environment to an online classroom needed to be seamless in the students Lean active learning experiences. The output of the paper is to discuss the designing of appropriate delivery methods and practical examples, games, scenarios, exercises in a flipped online classroom. Problem-based learning is ideal for teaching lean manufacturing, driven by a problem-solving culture that values learning as a critical output. The design of a "practical problem based" online Kaizen utilising the virtual classroom as an obeya room enabled students to learn Lean Six Sigma tools and practically deploy the tools. Qualitative and quantitative measures were deployed to assess the success of the transition.

Keywords: Digitalisation · Lean · Online delivery · University teaching · Obeya · Virtual kaizen · Lean Education · Flipped classroom

1 Introduction

The digital era encourages the use of Information Technology (IT) in the education sector [1]. However, due to the COVID-19 pandemic, most educational institutions across the world have moved their teaching online and put their efforts into preparing online distance education to ensure learning and teaching continued uninterrupted [2]. The COVID-19 pandemic brought significant disruption to classroom-based learning and activities. Before COVID-19, Lean was taught to university students via classroom games and activities and practical exercises. Large classrooms with plenty of wall space for "paper" based exercises and containing several whiteboards meant Lean training could be delivered to up to 30 students at a time in a practical and blended manner. Problem-based learning focuses on small groups using authentic problems to help participants obtain knowledge and problem- solving skills. This approach makes problem-based learning ideal for teaching lean manufacturing, driven by a culture of problem-solving that values learning as one key output of manufacturing production[3].

This blended delivery mode helped replicate core concepts of LSS training the application of brainstorming, aided teamwork and aided replication of a fundamental

D. J. Powell et al. (Eds.): ELEC 2021, IFIP AICT 610, pp. 132–143, 2021.
https://doi.org/10.1007/978-3-030-92934-3_14

organisational problem-solving environment [4]. In addition, blended learning can help students assimilate more quickly to online environments [5].

Transferring Lean education online or digitalising is not straightforward. The highly interactive nature of LSS education within the student peer group and with the lecturer or facilitator needed to be replicated. This replication needed to ensure that the quality of learning, the qualitative student learning experience and even the academic quantitative results obtained were not adversely affected. In ramping up the university capacity to teach remotely, schools and colleges took advantage of asynchronous learning, which works best in digital formats. Online teaching should include varied assignments and design student assessments at first to help teachers focus [6].

The goal of this research is a case study on developing a user friendly, virtual learning environment wherein the students studying Lean would be able to apply lean tools in a case study game format in a hypothetical manufacturing facility. The purpose is to educate and acquaint students with real-life problems in an organisation based on real-life scenarios. Furthermore, the students would implement lean tools in a virtual setting, thus fostering the students' development through active learning and improving students' learning, motivation, and retention [7]. This paper explains the main challenges, assignment design, and the integration of various lean tools incorporated in the virtual classroom.

Thus the summary of the research questions are:

- How can Lean Six Sigma education be transferred to the online classroom to emulate the physical classroom?
- How can the quality of learning, understanding and student experience of the methods be assured, applied and measured?
- What were the challenges and the pros and cons of virtual online classroom delivery versus a physical classroom?

2 Literature Review

Lean today has changed from its origin as a manufacturing methodology to an ideology that ties in all aspects of the organisation and can be deployed in services, healthcare, financial organisations. This demands engineers and practitioners with lean, solid basics. Therefore, it is essential to know about the lean tools, but it is even more critical to understand how to apply these tools most effectively [7].

Today's engineering education requires a curriculum that allows students to utilise and learn of the latest technologies [8]. Irrespective of the pros and cons of virtual online delivery, which have been discussed by many authors [9, 10], the COVID -19 pandemic meant the only option available to deliver modules was online.

The online delivery format's perspective does not affect student learning outcomes, dubbed the "no-significant-difference" perspective [11]. However, Gillespie (1998)

[12] put forward that online learning tasks should be designed to help learners develop higher-level thinking skills and evaluate their understanding, mediated by sharing ideas and problems with the content using interactive or collaborative online formats.

Research has shown that "flipped" classroom scores higher than a conventional, lecture-oriented set up on the following criteria: student involvement, task orientation, and innovation and promoting collaborative learning [4]. Thus there is a need to structure the learning tasks to fruition of a flipped classroom exercise, albeit if a virtual one. This approach can be taken with Lean teaching as the methodology requires practical tools and skillset application [3].

Problem-based learning (PBL), active learning, blended learning, flipped learning, and Simulation & Gaming are experiential learning formats. These approaches are all conducive to teaching Lean. Moreover, a PBL approach is more involving and enjoyable than more traditional approaches as learning is active [2, 13].

Literature on teaching Lean virtually or in a flipped classroom is not as prevalent as the giant body of research and journals related to online teaching in general. However, the learnings around online teaching methods can most certainly be leveraged somewhat and applied to Lean teaching in a virtual environment. However, there are still many related studies of online Lean education and virtual industry-based Kaizen events [7, 14–17], with more studies published on lean teaching and virtual kaizen e vents since the COVID-19 pandemic. In addition, a sense of community is central to student engagement and satisfaction in a virtual classroom, and breakout rooms help develop a sense of community [18].

Simulation has been very much presented as a best practice for online Lean virtual events and teaching [3, 19–21]. However, the simulation software available may not always be relevant and does not allow tailoring to lecturer designed case studies and applications. In addition, the de sign of the simulation exercise and practical implementation and learning may take time that is not available. It is also expensive to purchase and develop.

Lean Six Sigma techniques and tools are considered the cornerstones for eliminating waste are thus referred to as "Kaizen building blocks" [22]. A Le an training, approach, deployment, education can begin by implementing basic Lean and Six Sigma techniques and tools such as 5S, Kaizen teams, standardisation and elimination of waste (Muda), unevenness (muri) and overburdeness (Mura) in working processes [15, 23]. Lean Six Sigma thinking evolves towards more complex techniques and tools that are considered to be part of Lean thinking, such as just-in-time (JIT) manufacturing), Kanban setup, poka-yoke (error-proofing), single minute exchange of dies (SMED), and Hejunka (levelling production) [7]. Give n this, the research suggests that learning about Lean within a virtual classroom can aid this learning, application and understanding about LSS.

3 Methodology

3.1 Development of Online Lean Learning Module

This section of the paper describes the main case study Kaizen assignment, which was developed as a series of Lean and Six Sigma exercises that could be carried out in the virtual classroom, which became the online "kaizen" room or "obeya" (Table 1).

The Kaizen room or obeya room, a potent tool for facilitating teamwork and better managing projects, was considered the "control centre" to deploy the Lean training and learning [24].

The students in the university have not worked in a manufacturing environment whatsoever and were not familiar with Lean or operations. However, as postgraduate students, online delivery is more amenable to the se learners as they generally have greater self-regulation and acquiring learning strategies and can adjust to online environments relatively quickly [5, 25].

This assignment, or "kaizen" as it was framed, is based on a theoretical company called "ABC Manufacturing" who produce and deliver sandwiches and are arguably a "service" industry also. "Sandwiches" were picked as a product as opposed to "widgets" or other products as they are uncomplicated and straightforward to make, and students are familiar with them and their "components". The Kaizen case study was designed to present the worst- case scenario or demonstrate an ineffective, poorly managed organisation with poor productivity, high defect rates, late deliveries, extensive customer complaints, poor communication, poor leadership and other inefficiencies. The online Kaizen needed to emulate the active and blended classroom learning environment and an organisational environment of brainstorming, huddles, teamwork and practical completion of process maps, value stream maps, cause and effects via collaboration. Lean Engineering Education calls for both content and competency mastery, and this assignment was designed to provide opportunities to demonstrate these competencies. This combination is necessary for professional engineering career success [17].

The case study game given to the students contained information about the company performance and some background concerning key performance indicators (KPI's). In addition, an explanation is given of how orders are received, processed and downloaded onto the manufacturing floor, details of the supply chain procurement process, incoming receiving, warehousing, production, shipping, and delivery. The students carry out the following activities are outlined in Table 1.

Table 1. Lean Principles and Concepts explored and applied within the virtual Kaizen online obeya classroom.

Lean/Six Sigma Tools Utilised and taught and applied in Online Kaizen Classroom
SMART problem statement The students were given enough information to develop a problem statement and set goals and objectives for the kaizen activity
Key Performance Indicator (KPI) Scorecard Students were given a suite of data and performance measures related to Productivity, Delivery, Quality and Cost
Non-Value Add wastes Several Lean wastes are presented within the case study for the student to identify (more than 30 examples of the 8 Lean Wastes were contained therein
Value Stream Mapping and Takt time Based on the case study information, they were asked to identify non-value wastes and potential areas which could be causing problems. Finally, takt time was to be established based on the VSM and data provided
Pull and Flow principles Students were asked to look at the process and ascertain where pull and flow were lacking and where they could be improved
Check sheets, Histograms, Pareto, Control Charts Data was provided to enable students to utilise and learn about essential quality management tools
Poke Yoke an example of process errors in the case study was presented, and students asked to error-proof the process
5S students were given samples of untidiness with the organisation in the warehouse, production floor, and offices; they were then asked to state how they would carry out a 5S based on their information. They were asked to develop a 5S audit template
Hejunka students were given examples of inadequate flow and unevenness within the order scheduling process and within the outgoing shipping process

(continued)

Table 1. (*continued*)

Lean/Six Sigma Tools Utilised and taught and applied in Online Kaizen Classroom
and asked to brainstorm how they would improve it using Hejunka to implement evenness
Cause & Effect diagrams   To root cause issues identified throughout the Kaizen, students were encouraged to apply the C&E diagram to 2 problems; 1) Reasons for deliveries taking up to 5 h and 2) reasons for high defect rates and complaints
5 Whys   Students were asked to utilise and apply the 5 Whys tools to develop a root cause further and identify corrective actions for identified issues within the C&E process
Future State Value Stream Map   Students were asked to brainstorm and design a new future VSM with improvements in flow, pull and waste reduction and new Takt times
New KPI Scorecard   Students estimated how changes and actions implemented had affected the original KPI metrics
Reflection & Kaizen Closeout   Students asked to reflect on Lean and methods and how tools helped

3.2   The Virtual Obeya Kaizen Room

The virtual learning environment (VLE) platform utilised was Blackboard which is the university VLE of choice as shown in Fig. 1. Within the virtual classroom, students were divided into breakout rooms. The breakroom exercises followed a DMAIC problem-solving approach so that each breakout room exercise built on to the previous activity and task. Within the breakout rooms, the lecturer could recreate the teamwork and brainstorming aspects of Lean in the workplace and physical classroom. To evaluate and ensure learning, the lecturer moved between breakout rooms to chat with and advise the students. Each breakout room team had control of the screen and whiteboard, so students could brainstorm and apply Lean tools to aid problem- solving. After each activity, the lecturer would bring the teams back into the virtual classroom, and each group would present their progress. The presentation of progress was essential to ensure that the exercise was understood and provide feedback to the students and share ide as within the class. The lecturer presented some theory and background to each Lean tool or practice and various Lean principles before commencing with the next breakout room exercise.

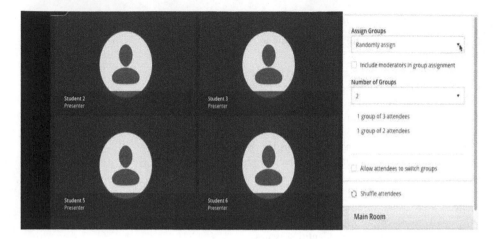

Fig. 1. Screenshot of basic online breakout rooms

4   Results

In this research, postgraduate students applied Lean Six Sigma tools such as Value Stream Mapping (VSM), 5S, Visual Management, Single Minute Exchange of Dies (SMED), Kanban, Plan-Do-Check-Act (PDCA), DMAIC, process and effect, 5Whys, pores mapping, poke yoke, JIT and Kanban amongst others (Table 2). They developed the project or case study virtual kaizen integrated into teams, as per a problem-based learning, active learning and obeya system situation as if they had been in a flipped physical university classroom. In each instance, the students submitted with their groups (alternating in presenting). This helps provide instruction from the lecturer, validated the learnings, and gives both positive and constructive feedback.

Results from the blended learning classroom Lean teaching in previous years were compared qualitatively and quantitatively with the virtual classroom Lean cohorts. Students always performed very highly in the blended flipped classroom kaizen assignments, and the virtual classroom results emulated previous cohorts. Qualitative feedback forms completed by the students measured on a Likert scale were very favorable compared to prior years, which were university campus classroom-based. A satisfaction mean average of 4 out of 5 was achieved in comparing both cohorts. This comparison was similar and didn't demonstrate that virtual learning was better or worse than classroom learning. Students have completed some university classroom-based continuous improvements workshops before the COVID-19 lockdown so that we're able to reach the flipped classroom experiences.

Some of the commentary selected from the students was as follows:

" *I enjoyed how some of us came up with so many different solutions*", "*I enjoyed the scenario -it felt real*", "and "*I didn't feel as if I was in a lecture -as I was busy, active and learning*".

Students also highlighted critical thinking, enjoyability and teamwork interaction. "*Confidence*" was a repeated theme, as was "*I can use these tools*", "*I understand the tools* ", and "*Lean is not hard*".

Table 2. How Lean Principles and Concept Learning was applied, demonstrated, shared and reviewed within the virtual Kaizen online obeya classroom

(Note: All screenshot imagesare taken from examples of online screenshots and uploaded online submissions of student work shared during the online Kaizen)

Lean/Six Sigma Tools Utilised and taught in Online Kaizen Classroom	Learning demonstrated and deployed in Virtual Classroom BREAKOUT ROOMS	Review, Sharing & Feedback in Virtual Classrooms
SMART problem statement The students were given enough information to develop a problem statement and set goals and objectives for the kaizen activity.	Students collaborate in the online classroom to develop a problem statement and set goals and objectives for the kaizen activity.	From a manufacturing point of view, ABC Sandwich Company's operations team has the capacity to produce 3 units/hour, with a TAKT time of 4 units per hour meaning that they are not meeting customer demand. This LSS project will aim to meet customer demand by reducing process cycle time in manufacturing by 50% from 20 minutes to 10 minutes per sandwich within three months/ by the end of quarter 3. Lecturer & Peer Reviewin the online classroom
Key Performance Indicator (KPI) Scorecard Students were given a suite of data and performance measures related to Productivity, Delivery, Quality and Cost.	Students collaborated and developed a KPI scorecard based on the information given.	 Lecturer & Peer Reviewin the online classroom
Process Mapping Students were asked to draw a process map based on the steps outlined in the case study.	Students collaborated and designed a "Current" Process Flow.	 Lecturer & Peer Reviewin the online classroom
Non-Value Add wastes Several Lean wastes are presented within the case study for the student to identify (more than 30 examples of the 8 Lean Wastes were contained therein.	Students brainstormed and presented all of the 8th waste types observed in the case study.	 Lecturer & Peer Reviewing the online classroom
Value Stream Mapping and Takt time Based on the case study information, the students were asked to identify non-value wastes and potential areas which could be causing problems. Finally, takt time was to be established based on the VSM and data provided.	The students worked on creating a VSM within the classroom breakout rooms utilising a virtual whiteboard. Students presented a virtual VSM and takttime calculations.	 Lecturer & Peer Reviewin the online classroom

(continued)

Table 2. (*continued*)

Pull and Flow principles. Students were asked to look at the process and ascertain where pull and flow were lacking and where they could be improved.	Students brainstormed idea on pull and flow improvement and presented in the online classroom.	 Lecturer & Peer Reviewin the online classroom
Check sheets, Histograms, Pareto, Control Charts. Data was provided to enable students to utilise and learn about essential quality management tools.	Students presented examples of tool applications and learnings.	
JIT & Kanban Just in Time and Kanban explained, opportunities were presented within the case study to explain the theory.	Students brainstormed where Kanban and JIT may be utilised in the case study.	 Lecturer & Peer Reviewin the online classroom
Poke Yoke an example of process errors in the case study was presented, and students asked to error-proof the process	Students gave examples of error proofing about the issuespresented	One-Owner Elimination of manual order entry – real-time E-orders Lecturer & Peer Reviewin the online classroom
5S Students were given examples of untidiness with the organisation in the warehouse, production floor and offices; they were then asked to state how they would carry out a 5S based on the information they had. Finally, they were asked to develop a 5S audit template.	Students presented a 5S program and 5S audit.	 Lecturer & Peer Reviewin the online classroom
Hejunka Students were given examples of inadequate flow and unevenness within the order scheduling process and within the outgoing shipping process and asked to brainstorm how they would improve it using Hejunka to implement evenness.	Students presented where Hejunka was required and how it could be utilised.	 Lecturer & Peer Reviewin the online classroom

(*continued*)

Table 2. (*continued*)

SMED Students give examples of slow turnarounds and, in line with 5S examples, brainstormed where SMED could be applied.	Students presented SMED opportunities in the customer ordering processes and within the production line.	 Lecturer & Peer Reviewin the online classroom
Cause & Effect diagrams To root cause issues identified throughout the Kaizen, students were encouraged to apply the C&E diagram to 2 problems; 1) Reasons for deliveries taking up to 5 hours and 2) reasons for high defect rates and complaints.	Students presented C&E diagrams and how they applied cause screening to the issues identified in the cause and effect and prioritised the issues based on a high, medium, and low potential for causing problems and fixing.	 Lecturer & Peer Reviewin the online classroom
5 Whys Students were asked to utilise and apply the 5 Whys tools to develop a root cause further and identify corrective actions for identified issues within the C&E process.	5 Whys scenarios presented for various root causes	**1. Why are the box labels being put on incorrectly?** Operators placing incorrect label on product boxes i.e. mixing up label for secondary and tertiary packaging **2. Why are operators mislabeling boxes?** They are mixing up the labels that are required for each stage of the packaging process **3. Why are they mixing up labels?** Labels are not always legible and can be difficult to read. **4. Why?** Issues with printing equipment and staff skills using the equipment **5. Why?** Staff have not received adequate training and machines are not being maintained adequately **Solution:** Staff training and starting a routing maintenance schedule for equipment including label printer Lecturer & Peer Reviewin the online classroom
Future State Value Stream Map Students were asked to brainstorm and design a new future VSM with improvements in flow, pull and waste reduction and new Takt times.	Future VSM with improvements presented	 Lecturer & Peer Reviewin the online classroom
New KPI Scorecard Students estimated how changes and actions implemented had affected the original KPI metrics.	New KPI scorecard presented with justification for reducing costs, quality defects, improved delivery, etc.	 Lecturer & Peer Reviewin the online classroom
Reflection & Kaizen Closeout Students asked to reflect on Lean and methods and how tools helped.	Reflection discussion and Kaizen close out held. Congratulations to Team.	 Lecturer & Peer Reviewin the online classroom

5 Discussion and Conclusion

The effectiveness of the virtual delivery training was confirmed through feedback from over 100 learners. The research results aim to demonstrate the learning to practical examples and scenario learning due to the relatively simple design of the problem under study. The course evaluations of the learners rated the course 4.5 out of 5 stars. While these results were very similar to the previous classroom blended learning carried out, it demonstrates the quality and standard of the delivery ensured effective learning could take place online. 98% of learners reported that they will apply the new skills learned and felt competent in using Lean Six Sigma tools and applications.

In any form of education, it is important to use a suitable learning environment for the intended purpose of the training, the education, and for the participant group and online to provide that environment if designed correctly. Particularly with Lean education, where the is a suite of tools and techniques associated with the methodology, the above statement is true. The fact that the learner satisfaction ratings and average grades achieved in the virtual classroom didn't differ from the physical classroom demonstrates blended learning can be effective in both environments.

Lean students are the future professionals of organisations, and their learning must be aligned with industry and society needs. Being taught in Lean education, in the form proposed by this paper, students will develop competencies and will have the ability to meet problem-solving and tool applications to a high degree of complexity and application. While simulation type software and Virtual Reality (VR) would be helpful and enhance future online lean, learning was a limitation in terms of time to include in this case study [21]. The author would like to expand the usage of these technologies for future online lean virtual education. An ultimate aim would be to simulate a complete virtual manufacturing process. This would give the students even more insight into the dependence of lean (especially JIT) on the supply chain and logistics.

References

1. Junus, K., Santoso, H.B., Putra, P.O.H., Gandhi, A., Siswantining, T.: Lecturer Readiness for online classes during the pandemic: a survey research. Educ. Sci. **11**, 139 (2021). https://doi.org/10.3390/educsci11030139
2. Tortorella, G.L., Narayanamurthy, G., Sunder M.V., Cauchick-Miguel, P.A.: Operations Management teaching practices and information technologies adoption in emerging economies during COVID-19 outbreak. Technol. Forecast. Soc. Change **171**, 120996 (2021). https://doi.org/10.1016/j.techfore.2021.120996
3. Badurdeen, F., Marksberry, P., Hall, A., Gregory, B.: Teaching lean manufacturing with simulations and games: a survey and future directions. Simul. Gaming **41**, 465–486 (2010). https://doi.org/10.1177/1046878109334331
4. Prashar, A.: Assessing the flipped classroom in operations management: a pilot study. J. Educ. Bus. **90**, 126–138 (2015). https://doi.org/10.1080/08832323.2015.1007904
5. Arbaugh, J.B.: What might online delivery teach us about blended management education? Prior perspectives and future directions. J. Manag. Educ. **38**, 784–817 (2014). https://doi.org/10.1177/1052562914534244
6. Daniel, S.J.: Education and the COVID-19 pandemic. Prospects **49**(1–2), 91–96 (2020). https://doi.org/10.1007/s11125-020-09464-3

7. Gadre, A., Cudney, E., Corns, S.: Model development of a virtual learning environment to enhance lean education. Procedia Comput. Sci. **6**, 100–105 (2011). https://doi.org/10.1016/j.procs.2011.08.020
8. The National Academic Press: Front Matter—Educating the Engineer of 2020: Adapting Engineering Education to the New Century—The National Academies Press. The National Academies Press, Washington DC (2005)
9. Daft, R.L., Lengel, R.H.: Organizational information requirements, media richness and structural design. Manage. Sci. (1986). https://doi.org/10.1287/mnsc.32.5.554
10. Kock, N., Verville, J., Garza, V.: Media naturalness and online learning: findings supporting both the significant- and no-significant-difference perspectives. Decis. Sci. J. Innov. Educ. **5**, 333–355 (2007). https://doi.org/10.1111/j.1540-4609.2007.00144.x
11. Summers, J.J., Waigandt, A., Whittaker, T.A.: A comparison of student achievement and satisfaction in an online versus a traditional face-to-face statistics class. Innov. High. Educ. **29**, 233–250 (2005). https://doi.org/10.1007/s10755-005-1938-x
12. Gillespie, F.: Instructional design for the new technologies. New Dir. Teach. Learn. **76**, 39–52 (1998)
13. Tortorella, G, Cauchick-Miguel, P.A.: Combining traditional teaching methods and PBL for teaching and learning of lean manufacturing. IFAC-PapersOnLine **51**, 915–920 (2018). https://doi.org/10.1016/j.ifacol.2018.08.465
14. Venegas, C.: How to have a successful office kaizen: co-located, distributed, or virtual (Presentation) - ProQuest. In: IIE Annual Conference Proceedings (2007)
15. Suárez-Barraza, M.F., Ramis-Pujol, J.: Implementation of lean-kaizen in the human resource service process: a case study in a Mexican public service organisation. J. Manuf. Technol. Manag. **21**, 388–410 (2010). https://doi.org/10.1108/17410381011024359
16. Alves, A.C., Flumerfelt, S., Moreira, F., Leão, C.P.: Effective tools to learn lean thinking and gather together academic and practice communities. Presented at the ASME 2017 International Mechanical Engineering Congress and Exposition, 10 January (2018). https://doi.org/10.1115/IMECE2017-71339
17. Flumerfelt, S., Kahlen, F.J., Alves, A., Siriban-Manalang, A.: The future state of content & competency-based engineering education: Lean Engineering Education. In: Lean Engineering Education. ASME Press (2015)
18. Berry, S.: Teaching to connect: community-building strategies for the virtual classroom. Online Learn. **23**, 164–183 (2019). ERIC - EJ1210946
19. Kuriger, G.W., Wan, H., Mirehei, S.M., Tamma, S., Chen, F.F.: A web-based lean simulation game for office operations: training the other side of a lean enterprise. Simul. Gaming **41**, 487–510 (2010). https://doi.org/10.1177/1046878109334945
20. Verma - Deliverable for Phase-0 NSRP- ASE – 0301-1.pdf, https://nsrp.org/wp-content/uploads/2015/10/Deliverable-2004-323-Lean_Enterprise_Simulation_Final_Report-Old_Dominion_University.pdf
21. Wan, H., Chen, F.F., Saygin, C.: Simulation and training for lean implementation using web-based technology. Int. J. Serv. Oper. Inf. **3**, 1–14 (2008). https://doi.org/10.1504/IJSOI.2008.017701
22. Imai, M.: Kaizen - the Key to Japanese Competitive Success. Random House Business Division, New York (1989)
23. Byrne, B., McDermott, O., Noonan, J.: Applying lean six sigma methodology to a pharmaceutical manufacturing facility: a case study. Processes **9** (2021). https://doi.org/10.3390/pr9030550
24. Priolo, R: What is an Obeya? https://planet-lean.com/what-is-obeya/. Accessed 07 July 2021
25. Arbaugh, R.B., Hwang, A.: Does "teaching presence" exist in online MBA courses? Internet High. Educ. **9**, 9–21 (2006). Learning & Technology Library (LearnTechLib)

A Lean Six Sigma Training Providers Transition to a 100% Online Delivery Model

Olivia McDermott[1(✉)], Patrick Walsh[2], and Lorraine Halpin[3]

[1] National University of Ireland, Galway H91TK33, Ireland
Olivia.McDermott@nuigalway.ie
[2] Limerick Institute of Technology, Limerick V94 EC5T, Ireland
Patrick.Walsh@lit.ie
[3] SQT Training, Limerick V94 Y8P, Ireland
lhalpin@sqt-training.com

Abstract. This research is a case study on SQT a leading Irish Lean Six Sigma training provider and their transition to online training and the digitalisation of their Lean Six Sigma training programs and other associated programs during the COVID-19 pandemic. The changes and challenges in transitioning from the existing classroom-based training model are discussed. A quantitative survey and qualitative interviews were carried out with the customers (trainee's and sponsoring employer organisations/clients) of the Lean Six Sigma trainer provider for 9–12 months. The results of the survey on the customers learning experiences with online Lean training is analysed. The results will demonstrate that the move to online Lean training was positive for both the customers and the training provider in terms of quality of delivery, cost minimisation, elimination of non-value-add travel and classroom time, improved online teamwork, program structure and engagement and enhanced benefits of the application of the learning in the workplace.

Keywords: Digitalisation · Lean · Online delivery · Training · Virtual learning

1 Introduction

The digital era encourages Information Technology (I.T.) in the education sector [1]. Since Coronavirus Disease 19 (COVID-19) outbreak, strict rules of social distancing have been applied worldwide [11], leading to a substantial negative impact on any types of classroom training with interruption of in-house training activities. The COVID-19 pandemic brought significant disruption to Lean training providers. This case study involved SQT - a leading Irish based Lean Six Sigma (LSS) training provider. Business stalled and stopped with the advent of COVID-19, and many H.R. departments and training departments deprioritised internal and external training agendas and initiatives. Uniquely within this training providers supply chain, the customers (clients) or companies who utilise the provider's services and put forward employees for training stayed open during the pandemic. Many of these were deemed essential by the Irish government, e.g. medical devices, food processing, pharmaceuticals etc. [2]. As it was business as usual for these companies, they still had a training

D. J. Powell et al. (Eds.): ELEC 2021, IFIP AICT 610, pp. 144–154, 2021.
https://doi.org/10.1007/978-3-030-92934-3_15

need, but one that they nor the training provider could meet in a non- socially distanced classroom. There was also a demand from students who had completed specific Lean Six Sigma belt training and certification levels, e.g. Yellow Belt and Green belt, to receive training to progress to the next level, e.g. Black belt. In order to remain in business, maintain training pipelines required by customers, deliver training, and keep tutors and admin staff in employment, the decision was made to transfer training online. As much of the LSS training is blended or classroom- based, involved team-based activities, practical exercises, brainstorming and working on a company project (from the trainee's workplace or organisation), this transition was not straightforward [3]. The highly interactive nature of LSS training with the trainer and trainees was something that the provider did not want to compromise as it would affect training quality, training experience and results. As trainees work on an ongoing work-based problem or projects utilising Lean Six Sigma tools, the mentorship and interaction that happens in the training classroom needed to be replicated online [3].

The research questions are:

1. How can classroom training be transferred to a virtual online environment?
2. What were the advantages, challenges and learnings of the virtual training deployment?

2   Literature Review

LSS as a continuous improvement methodology is utilised in organisations and can be deployed in services, healthcare, financial organisations. However, this demands that engineers and practitioners have solid basics in LSS. Therefore, it is essential to know about the tools, but it is even more critical to understand how to apply these tools most effectively [4].

2.1   Training Design for Online Training Deployment

Online learning tasks should be designed to help learners develop higher-level thinking skills, measure their understanding, and encourage and facilitate sharing ideas and problems within the training content using interactive or collaborative online formats [4]. There are essential criteria within an online classroom: student involvement, task orientation, and innovation and promoting collaborative learning[5, 6]. Thus, there is a need to structure the learning tasks in classroom exercises, albeit virtual. This approach can be taken with LSS teaching as the methodology requires practical tools and skillset application [5]

Problem-based learning (PBL), active learning, blended learning, flipped learning, and Simulation & Gaming are experiential learning formats. These approaches are all conducive to teaching Lean. Moreover, a PBL approach is more involving and enjoyable than more traditional approaches as learning is active [2, 6, 13].

Literature on teaching Lean virtually or in a flipped classroom discusses active learning, problem-based learning, and simulation and games in particular as a means of ensuring experiential learning [6–11]. A sense of community is also central to student engagement and satisfaction in a virtual classroom, and breakout rooms help develop a sense of community [12].

LSS techniques and tools are considered the cornerstones for eliminating waste. Therefore, a Lean training, approach, deployment, education can begin by implementing basic Lean and Six Sigma techniques and tools [10, 13]. Then, LSS thinking evolves towards more complex techniques and tools that are considered to be part of Lean thinking, such as just-in-time (JIT) Kanban setup, poka-yoke (error- proofing), single minute exchange of dies (SMED), and Hejunka (levelling production) [6]. Given this, the research suggests that learning about Lean within a virtual classroom can aid this learning, application and understanding about LSS.

2.2 Advantages of Online and Virtual Training Delivery

Many factors affect an organisations decision to transition to online and virtual training delivery. The advantages include cost savings, shorter training delivery times, flexibility and convenience of training delivery and accessibility, training accessibility, consistency of content and training delivery, enabling and facilitating knowledge management and no need for travel [14]. The disadvantages include lack of human contact, ability to read and respond to body language, resistance to change, confusion about technology, broadband reliability and lack of organisational resources [14].

Selecting the proper infrastructure and content for e-learning is not always the easiest thing [15]. Companies can be confused by many vendors, content providers, and tools available in the market that promise to deliver a complete e-learning solution [16].

3   The Research Project

3.1   The Research Company

The training provider SQT in this study is one of the largest training providers in Ireland, having been established over 30 years ago. The provider employs over 39 tutors and 13 administrative staff. Before COVID-19, training was delivered in public locations and in-company training classrooms. Depending on the type and level of LSS training being delivered, training courses could last from 0.5 up to 20 days with small classes of approximately 8–12 learners. The providers' typical customers of their services are multinational corporations, Irish indigenous industries, public sector organisations, and employed adult learners interested in professional development and training.

3.2   Research Methodology

The research aimed to identify the effects of moving to an online delivery model and the perceived advantages and disadvantages of virtual delivery training implementation through a case study implementation with a mixed-method qualitative and quantitative analysis.

The case study method is used to facilitate the researcher by focusing on a specific case, learning more about the subject in question, and providing an inductive approach to the relationship between theory and practice [17, 18].

The case study presented enhances the understanding of the adoption of virtual learning and training in an online classroom. For this specific research, the authors have utilised an "intrinsic case study" [19, 20] of a specific company picked up because of its size and reputation and because of the challenges presented to transform their training delivery model completely. The case study research builds an in-depth, contextual understanding of the case, relying on multiple data sources [17] rather than on individual stories as in narrative research. In addition, mixed-method data via quantitative survey data and qualitative interviews were also collected. Attendees were asked to complete the survey questionnaire for each training course. The questionnaires listed a series of questions about the online training delivery mainly measured on a Likert scale such as 5 = Excellent, 4 = Very Good, 3 = Good, 2 = Adequate, 1 = Poor or other relevant choices as demonstrated in Table 1.

Qualitative semi-structured interviews were carried out with trainers and tutors and admin staff within the training provider to assess the challenges of transitioning to an online LSS virtual delivery module. Finally, survey records, e-learning training materials, virtual learning environments, project outputs, and assessment outputs were also reviewed.

Table 1. List of questions in quantitative survey

#	Question
1	Considering the general objectives of the course, what was your overall rating? 5 = Excellent, 4 = Very Good, 3 = Good, 2 = Adequate, 1 = Poor
2	How well did the course deliver to the "Learning Outcomes"? 5 = Excellent, 4 = Very Good, 3 = Good, 2 = Adequate, 1 = Poor
3	Will you apply the new skills learned? Yes or No
4	Tech Check in advance of course commencement 5 = Excellent, 4 = Very Good, 3 = Good, 2 = Adequate, 1 = Poor
5	If you contacted the training provider, how did you find Customer Support? (Enquiry response, booking confirmation etc.)? 5 = Excellent, 4 = Very Good, 3 = Good, 2 = Adequate, 1 = Poor
6	Tutor's presentation skills 5 = Excellent, 4 = Very Good, 3 = Good, 2 = Adequate, 1 = Poor
7	Use of technology to aid learning (e.g. Zoom) 5 = Excellent, 4 = Very Good, 3 = Good, 2 = Adequate, 1 = Poor
8	Tutor's ability to answer questions 5 = Excellent, 4 = Very Good, 3 = Good, 2 = Adequate, 1 = Poor
9	Encouragement to participate 5 = Excellent, 4 = Very Good, 3 = Good, 2 = Adequate, 1 = Poor
10	Pace of course delivery 5 = Excellent, 4 = Very Good, 3 = Good, 2 = Adequate, 1 = Poor
11	Ease of access to the virtual classroom (e.g. Zoom) 5 = Excellent, 4 = Very Good, 3 = Good, 2 = Adequate, 1 = Poor
12	How would you rate the clarity of assessment requirements? 5 = Excellent, 4 = Very Good, 3 = Good, 2 = Adequate, 1 = Poor
13	Would you recommend this course to a colleague? Yes or No

4   Results

4.1   Virtual Classroom Design

Technology utilised for delivery is vital to successful online delivery. Having trialled and researched many video conferencing platforms, Zoom was selected as the preferred software. Some of the critical reasons for the training provider choosing Zoom as its delivery platform were as follows: 1) It is lightweight, 2) It is dependable, 3) It is extremely high-quality, 4) It is easy to use, and 5) it can be accessed without downloading additional software. As Zoom is a web-based video conferencing tool with a local, desktop client and a mobile app that allows users to meet online, with or without video, it was considered the most applicable. A key advantage of Zoom is its ability to securely record and store sessions without recourse to third-party software. Other critical security features include user-specific authentication, real-time encryption of meetings, and the ability to backup recordings to online remote server networks ("the cloud") or local drives, which can then be shared securely for collaboration [21, 22]. Where a company does not allow Zoom, the training provider decided to use alternative platforms such as M.S. Teams and WebEx as the alternative options. The Virtual Learning Platform (VLE) utilised by the training provider before the transition was Moodle, and that VLE was maintained. The training provider implemented practices and guidelines to ensure Zoom meetings and activities were as safe as possible. These measures include the following seven (7) characteristics: 1) using Zoom V5.0, which includes the latest security enhancements, 2) not sharing web links through Zoom during the session, 3) working with small groups and only those registered will be provided with the link to join, 4) sessions are only routed through the U.S. & European Data centres, 5) Join before host option is disabled, 6) A random meeting I.D. is associated with the meeting rather than a personal meeting ID and 7) waiting room functionality has been enabled on all meetings.

A series of technical supports needed to be developed to ensure that both trainers and trainees could access the VLE and video conferencing platforms for the training to run effectively. The following three (3) technical supports were put in place for all virtual programmes to solve and diagnose any potential I.T. issues 1) one week before the course, the training provider schedules a 'Tech Check' with all delegates, 2) on the morning of the training course a member of the training providers support staff logs into the Zoom course ensure that all delegates can successfully log in and all equipment is working correctly, and 3) during the sessions dedicated I.T. support staff are available to deal with any Zoom issues. In addition, a dedicated support email address is used for queries in relation to accessing the VLE prior to, during and post-training sessions.

4.2   Virtual Classroom Delivery

Within 3 weeks or so of the 1st Lockdown all tutors and trainers attended virtual training Design and Delivery courses as part of an immediate plan to transition to an online model. This training was delivered by online educational consultants and was virtual. Within the virtual classroom, students were divided into breakout rooms. The lecturer could recreate the teamwork and brainstorming aspects of LSS in the work-place and physical classroom within the breakout rooms. To evaluate and ensure learning, the lecturer moved between breakout rooms to mentor and advise the students. After each activity, the lecturer would bring the teams back into the virtual classroom, and each group would present their progress. The progress presentation was essential to ensure that the exercise was understood and provide feedback to the students and share ideas within the class. The class sizes remained at 8–12 participants to optimise the trainee experience and ensure that the tutor could give individual feedback, mentoring and support.

Several changes were made to the traditional classroom delivery and blended delivery offered by the training provided to transition to virtual delivery, as outlined in Table 2. Virtual delivery took place over 4–5 months instead of blended delivery and classroom delivery, which took place over 3–4 months and over 1–2 months. The duration of the entire program from training to project submission extended slightly to 10–11 months, but that was not deemed negative by trainees or the organisational stakeholders involved in the design based on feedback discussed in later sections.

Twenty-five online training hours were delivered over sessions consisting of 2.5 hours in duration either on zoom or the client organisation specified platform. The training was delivered via shorter sessions due to feedback that online training required more concentration and was more intense. The course learning and decision to award the appropriate LSS belt level was assessed by submitting an organisational LSS project- based on a problem statement or project proposal and a report demonstrating LSS tool application and usage and a final project poster or storyboard.

Table 2. Virtual delivery characteristics versus classroom and blended delivery

Course Format	Classroom Delivery	Blended Delivery	Virtual Delivery
Duration of Training	1-2 months	3-4 months	4-5 months
Programme Duration	7-8 months from training commencement to project submission	9-10 months from training commencement to project submission	10-11 months from training commencement to project submission
Structure	2 days and 3 days (public - hotel) 2 days, 1 day, 2 days (inhouse – on site or hotel)	1 day classroom 2 x 2.5 hour virtual sessions (could have 4x2.5) 2 days classroom (could have 1 day) 2 x 2.5 hour virtual sessions	10 x 2.5 hour sessions via Zoom or equivalent company specified digital platform
Training Hours	40 hours	31-34 hours (depending on whether 2 or 3 days classroom)	25 hours
Self Directed Learner Hours (non project related)	10 hours (study/research – includes familiarisation with statistical software)	16-19 hours (as per Classroom + additional pre-reading & course work)	25 hours (as per Classroom + additional pre-reading & course work)
Assessment	100% Project based - Proposal 20% - Report 70% - Storyboard 10%	100% Project based - Proposal 20% - Report 70% - Storyboard 10%	100% Project based - Proposal 20% - Report 70% - Storyboard 10%
Support Platform	Moodle, includes messenger feature	Moodle, includes messenger feature + Zoom (or equivalent)	Moodle, includes messenger feature + Zoom (or equivalent)
Training Aids	Templates, videos, guideline documents, moodle library	As per Classroom + additional pre-read information, video aids and course work instructions	As per Classroom + additional pre-read information, video aids and course work instructions

4.3   Quantitative Survey Results

Over 19 LSS courses were delivered to 160 trainees. In addition, the stakeholder feedback from survey data (see Table 3) collated from March to December 2020 (with a 65% response rate) from participants was positive. Therefore, it can be concluded that the quality of the online delivery providers LSS courses has not been compromised by this new model of provision.

Table 3. Quantitative survey results

#	Question	Overall Response
1	Considering the general objectives of the course, what was your overall rating? 5 = Excellent, 4 = Very Good, 3 = Good, 2 = Adequate, 1 = Poor	93% gave a rating of "4" or "5" or "Excellent"
2	How well did the course deliver to the "Learning Outcomes"? 5 = Excellent, 4 = Very Good, 3 = Good, 2 = Adequate, 1 = Poor	99.5% gave a rating of "5" or "Excellent"
3	Will you apply the new skills learned? Yes or No	98% Replied 'Yes'
4	Tech Check in advance of course commencement 5 = Excellent, 4 = Very Good, 3 = Good, 2 = Adequate, 1 = Poor	87% gave a rating of "5" or "Excellent"
5	If you contacted the training provider, how did you find Customer Support? (Enquiry response, booking confirmation etc.)? 5 = Excellent, 4 = Very Good, 3 = Good, 2 = Adequate, 1 = Poor	91% gave a rating of "5" or "Excellent"
6	Tutor's presentation skills 5 = Excellent, 4 = Very Good, 3 = Good, 2 = Adequate, 1 = Poor	92% gave a rating of "5" or "Excellent"
7	Use of technology to aid learning (e.g. Zoom) 5 = Excellent, 4 = Very Good, 3 = Good, 2 = Adequate, 1 = Poor	88% gave a rating of "5" or "Excellent"
8	Tutor's ability to answer questions 5 = Excellent, 4 = Very Good, 3 = Good, 2 = Adequate, 1 = Poor	94% gave a rating of "5" or "Excellent"
9	Encouragement to participate 5 = Excellent, 4 = Very Good, 3 = Good, 2 = Adequate, 1 = Poor	91% gave a rating of "5" or "Excellent"
10	Pace of course delivery 5 = Excellent, 4 = Very Good, 3 = Good, 2 = Adequate, 1 = Poor	85% gave a rating of "5" or "Excellent"
11	Ease of access to the virtual classroom (e.g. Zoom) 5 = Excellent, 4 = Very Good, 3 = Good, 2 = Adequate, 1 = Poor	91% gave a rating of "5" or "Excellent"

(continued)

Table 3. (*continued*)

#	Question	Overall Response
12	How would you rate the clarity of assessment requirements? 5 = Excellent, 4 = Very Good, 3 = Good, 2 = Adequate, 1 = Poor	87% gave a rating of "5" or "Excellent"
13	Would you recommend this course to a colleague? Yes or No	96% Replied 'Yes'

96% of trainees stated that they would recommend the LSS training courses to a colleague, with 93% giving the course a rating of excellent and 99.5% responding that the learning and training met the learning objectives. The survey and course feedback was compared with data from classroom-based training courses over the previous four years delivered by the training provider. There was no negligible difference between virtual delivery and classroom-based delivery on comparison of the satisfaction ratings. An average of 4.5 out of 5 was consistently achieved for some based LSS training, and the virtual training satisfaction rating average was consistent at 4.5 out of 5 in the sample selected.

4.4 Qualitative Interview Results

A series of semi-structured interviews were carried out with the training providers, management team, tutors, administration staff, and client organisational management teams. A sample size of 12 was deemed appropriate as it provided a good mix of and representative of the mix of stakeholders under this single case study [23, 24]. The interview questions aimed to ascertain the benefits, challenges, and opportunities with moving to its virtual LSS online delivery. The training providers management team highlighted and reiterated the financial benefits more than once in not conducting "inhouse" or "public" training. Before the virtual delivery transition, courses were held in-house at the training providers larger training facility or were held in various locations around Ireland in hotel conference rooms. There was substantial infrastructure investment costs, but these were mainly upfront once off investments that will pay off over time.

In some cases prior to COVID-19, training delivery may have taken place on-site within the clients own organisation, but the majority of training was carried out either in the training providers own venue or in hotels around Ireland. The training provider had zero costs in relation to hiring venues or paying tutor travel expenses and accommodation to and from venues with the virtual transition. Client organisations discussed the benefits of "*not having to send 12–13 people offsite for a day or more at a time*". Having spaced out smaller online virtual training slots meant better utilisation and flexibility with employee time. The training providers management has pointed out that more significant virtual interaction has led to "*further engagement with many stakeholders. This engagement has led to many opportunities for new and innovative suites of programmes across several sectors*".

From an administration point of view, it was commented by a member of the training providers support staff that, "Since the introduction of virtual delivery, handwritten feedback forms have been replaced by Survey Monkey Evaluation forms, which are integrated into our Management Information System. This is a significant quality enhancement as it allows for immediate feedback, timely analysis and reporting on from both a qualitative and quantitative data perspective".

Tutors commented on the benefits in terms of "no travel", better work-life balance due to less travel", "less administration collecting feedback forms, attendee lists, no submitting of expenses and keeping receipts".

The challenges were met by the tutors as they had to innovate and work harder - tutors noted that "*they had to work harder to verify learner engagement in the virtual environment*". While Zoom was proven to be a very effective platform, the tutors must be "*very active and engaging*" and "*constantly eliciting learners to contribute comments or feedback*" instead of waiting for them to come involuntarily. In order to enhance the delivery experience, tutors have implemented several strategies, such as using a printed list of attendees to rotate questions through them during the class. This helps to check for a better understanding and confirm clarity. Other challenges were "*ensuring participation and active listening*" -this was overcome by requesting that cameras remain turned on at all times (where possible).

The use of breakout rooms and class polls "have been critically important to assist with learner interaction and engagement". While "sharing the screen and document function has been extremely effective" for integrating feedback from breakout rooms and exercises.

Many trainees "brought" a problem or project from their workplace to the training in order to work on the project and apply Lean tools as they were learning them in the virtual classroom.

One employer stated, "the benefits to the organisation have been fantastic, we have had several projects completed and more trainees are getting involved in new projects upon completing their current projects".

The trainees stated that "I applied my learning and training to working on our productivity issues and we utilised the Lean tools to help root cause and fix our problems -yielding a 30% improvement". Also "I have used the training in my job to gain a Green Belt and I would like to progress to a black Belt".

On the experience of learning online the trainees stated, "I had never attended online training before but I was surprised at how much I learned and was able to use in work".

5   Discussion and Conclusion

There were some challenges to achieving an online virtual training delivery. However, the advantages have outweighed the disadvantages in terms of business results and trainer and trainee experiences. Challenges raised were actioned and continue to be reviewed and assessed to improve performance.

The effectiveness of the virtual delivery training was confirmed through feedback from over 160 learners on 19 courses. The course evaluations of the learners rated the courses an average of 4.5 stars out of 5 stars. The quality and standard of the delivery ensured the learning was applicable -with the learnings applied in the LSS project completed by the trainees. 99% of learners reporting that they will apply the new skills learned and 98% reporting that they would recommend their course to a colleague.

Based on qualitative feedback the learning was applied successfully in the work-place of the participants and utilised in projects.

Many customers have confirmed a preference for virtual training (organisation specific) from a future business perspective, and this is a growth area not realised before COVID-19. However many local Irish business LSS networks have given feedback to the provider to express a preference for a blended delivery model of LSS courses once COVID-19 restrictions are lifted. The training provider is confident that virtual delivery will continue to expand and broaden its target market in the long term The training provider will continue to offer a classroom-based model of delivery. A limitation of this study is that the research could not be carried out over a longer timeframe and evaluate the lessons learned and learners' skills acquisition over a more extended period. Future opportunities are to study how effective the application of the training is in the workplaces of the learners.

In LSS training, it is essential to use a suitable training environment for the intended purpose of the training and the participant group, and online training can provide that environment if designed correctly. Therefore, conducting LSS training online and virtually, when designed correctly, can benefit both trainer providers and training participants.

References

1. Junus, K., Santoso, H.B., Putra, P.O.H., Gandhi, A., Siswantining, T.: Lecturer readiness for online classes during the pandemic: a survey research. Educ. Sci. **11**, 139 (2021). https://doi.org/10.3390/educsci11030139
2. Carswell, S.: Medtech sector seeks special status to keep Covid supplies flowing, https://www.irishtimes.com/business/medtech-sector-seeks-special-status-to-keep-covid-suppies-flowing-1.4208262, (2020)
3. Homitz, D.J., Berge, Z.L.: Using e-mentoring to sustain distance training and education. Learn. Organ. **15**, 326–335 (2008). https://doi.org/10.1108/09696470810879574
4. Prashar, A.: Assessing the flipped classroom in operations management: a pilot study. J. Educ. Bus. **90**, 126–138 (2015). https://doi.org/10.1080/08832323.2015.1007904
5. Badurdeen, F., Marksberry, P., Hall, A., Gregory, B.: Teaching lean manufacturing with simulations and games: a survey and future directions. Simul. Gaming **41**, 465–486 (2010). https://doi.org/10.1177/1046878109334331
6. Gadre, A., Cudney, E., Corns, S.: Model development of a virtual learning environment to enhance lean education. Procedia Comput. Sci. **6**, 100–105 (2011). https://doi.org/10.1016/j.procs.2011.08.020
7. Tortorella, G., Cauchick-Miguel, P.A.: Combining traditional teaching methods and PBL for teaching and learning of lean manufacturing. IFAC-PapersOnLine. **51**, 915–920 (2018). https://doi.org/10.1016/j.ifacol.2018.08.465

8. Tortorella, G.L., Narayanamurthy, G., Sunder, M.V., Cauchick-Miguel, P.A.: Operations management teaching practices and information technologies adoption in emerging economies during COVID-19 outbreak. Technol. Forecast. Soc. Change **171**, 120996 (2021). https://doi.org/10.1016/j.techfore.2021.120996

9. Venagas, C.: How to have a successful office kaizen: co-located, distributed, or virtual (Presentation) - ProQuest. In: IIE Annual Conference proceedings (2007)

10. Suárez-Barraza, M.F., Ramis-Pujol, J.: Implementation of Lean-Kaizen in the human resource service process: a case study in a Mexican public service organisation. J. Manuf. Technol. Manag. **21**, 388–410 (2010). https://doi.org/10.1108/17410381011024359

11. Alves, A.C., Flumerfelt, S., Moreira, F., Leão, C.P.: Effective tools to learn lean thinking and gather together academic and practice communities. Presented at the ASME 2017 International Mechanical Engineering Congress and Exposition, 10 January 2018 (2018). https://doi.org/10.1115/IMECE2017-71339

12. Berry, S.: Teaching to connect: community-building strategies for the virtual classroom. Online Learn. **23**, 164–183 (2019). ERIC - EJ1210946

13. Byrne, B., McDermott, O., Noonan, J.: Applying lean six sigma methodology to a pharmaceutical manufacturing facility: a case study. Processes **9** (2021). https://doi.org/10.3390/pr9030550

14. Stamatiadis, F., Petropoulou, M.: The decision to learn on-line: a case study 10 (2006)

15. Eseryel, D.: A framework for evaluation & selection of e-learning solutions. Presented at the E- Learn: World Conference on E-Learning in Corporate, Government, Healthcare, and Higher Education (2002)

16. Fry, K.: E-Learning Markets and Providers: Some Issues and Prospects **43**, 233 (2001). Learning & Technology Library (LearnTechLib). learntechlib.org

17. Yin, R.K.: Qualitative Research from Start to Finish. The Guilford Press, New York (2011)

18. Creswell, J.W.: Mixed-method research: Introduction and application. In: Handbook of educational policy, pp. 455–472. Elsevier (1999)

19. Leggette, H.R., Black, C., McKim, B.R., Prince, D., Lawrence, S.: An intrinsic case study of a post- secondary high-impact field experience. NACTA J. **57**, 129–138 (2013)

20. Bustamante, C., Moeller, A.J.: The convergence of content, pedagogy, and technology in online professional development for teachers of german: an intrinsic case study on JSTOR **30**, 82–84 (2013)

21. Archibald, M.M., Ambagtsheer, R.C., Casey, M.G., Lawless, M.: Using zoom videoconferencing for qualitative data collection: perceptions and experiences of researchers and participants. Int. J. Qual. Methods **18**, 1609406919874596 (2019). https://doi.org/10.1177/1609406919874596

22. Archibald, M.M., Ambagtsheer, R.C., Casey, M.G., Lawless, M.: Using Zoom Videoconferencing for Qualitative Data Collection: Perceptions and Experiences of Researchers and Participants (2019). https://journals-sagepub-com.libgate.library.nuigalway.ie/doi/full/10.1177/1609406919874596. Accessed 20 July 2021

23. Antony, J., Setijono, D., Dahlgaard, J.J.: Lean Six Sigma and Innovation – an exploratory study among UK organisations. Total Qual. Manag. Bus. Excell. **27**, 124–140 (2016). https://doi.org/10.1080/14783363.2014.959255

24. Guest, G., Bunce, A., Johnson, L.: How many interviews are enough? An experiment with data saturation and variability. Field Methods **18**, 59–82 (2006)

Teaching in Virtual Reality: Experiences from a Lean Masterclass

Torbjørn Netland[1]([⊠]) [iD] and Peter Hines[2] [iD]

[1] ETH Zurich, Zürich, Switzerland
tnetland@ethz.ch
[2] Waterford Institute of Technology, Waterford, Ireland
peter@enterpriseexcellencenetwork.com

Abstract. Virtual reality offers an immersive, remote alternative to in-person teaching. We teach a Lean masterclass entirely in real-time virtual reality. By summer 2021, we had taught five masterclasses for 117 senior-level managers from more than 50 leading organisations. In these courses, participants located all over the world can interact with each other almost as in a physical seminar setting. Drawing on formal course evaluation surveys and personal experience, we discuss the appropriateness of real-time virtual reality as a teaching platform and the benefits and challenges of this approach. Based on our experiences, and taking into account the rapid ongoing technological development, we imagine that this form of teaching and learning will accelerate in importance and application.

Keywords: Immersive teaching · Immersive learning · Virtual reality · Metaverse · Oculus · Lean

1 Introduction

The COVID-19 pandemic made in-person teaching difficult, and, for courses with international participants, impossible. Since March 2020, there has been a surge in the quest for remote access technologies for both everyday work and education. Virtual reality (VR) is one such technology that has been proposed due to its ability to connect people without the need for travelling and because it can provide an enhanced learning experience [1]. This paper is the first to report experiences of teaching a business concept such as Lean in a real-time, immersive VR platform. Rather than studying how to teach Lean *with* VR [2, 3], we study how to teach Lean *in* VR, which is totally different.

We have worked with The Leadership Network (TLN) in teaching a masterclass called 'Intelligent Lean' in their state-of-the-art, real-time VR platform called 'Gemba' [featured in 4–6]. In Gemba, participants meet in real time as avatars in an immersive virtual environment and interact much as they would in a classroom or seminar. Based on the first five teaching events with participants from all over the world, we discuss pros and cons and how we see the future of this type of VR for business education.

To inform our discussion, we draw on our own experiences, feedback sessions during the masterclasses, and feedback surveys from the masterclass participants. After

© IFIP International Federation for Information Processing 2021
Published by Springer Nature Switzerland AG 2021
D. J. Powell et al. (Eds.): ELEC 2021, IFIP AICT 610, pp. 155–162, 2021.
https://doi.org/10.1007/978-3-030-92934-3_16

each masterclass event, the teachers and TLN representatives organised debriefing sessions and reflected on how to improve the technical VR environment, the content taught and the teaching process. To rule out self-bias and provide a robustness check of the findings presented in this paper, it has been reviewed and commented on by other masterclass teachers and previous masterclass participants.

Judging from the self-reported learning experience of participants and our own experience as teachers, we conclude that real-time, immersive VR is well suited to teaching Lean and other business curricula. The technology is available and effective, but the technical and administrative efforts required to implement it should not be underestimated. VR education has fundamental advantages, such as cutting travel time, budgets and CO_2 emissions to the minimum. We predict that future education will very soon see increasing use of real-time, immersive VR platforms. We have barely experienced the emergence of this technology.

2 Immersive Teaching with Virtual Reality

According to a definition by NASA, 'Virtual reality is the use of computer technology to create the effect of an interactive three-dimensional world in which the objects have a sense of spatial presence' [7]. VR is a computer-generated 3D environment that users experience as close to real. While other studies use a more relaxed definition of 'VR', including virtual computer games (in 2D or 3D) [e.g. 8, 9], we refer in this paper to technologies that are 'fully' immersive in the sense that they require VR headsets and give the user an impression of being somewhere else. The virtual environment can be virtual copies of a real-life context (captured via 360° footage or photos) [e.g. 3, 10, 11], a fully virtual, computer-generated environment (resembling or not resembling a real-life context) [e.g., 12] or a mix of the two. The VR environment can be 'offline'—without real-time interaction with other users—or 'online' with real-time interaction.

VR has previously been explored as an element of Lean teaching in a small number of cases, including offline applications at ETH Zurich lead by this paper's first author [2, 3, 13]. The ETH Zurich applications involve immersive experiences in which students are provided a recorded VR environment of real factories as part of their course. The students actively explore the virtual content guided by assignment questions. This was initially done through open-access, 360° VR archive material from ABB factories [3], then 360° VR material from YouTube including a Toyota factory [2] and most recently with 360° VR material developed in partnership with the Hilti Corporation [5]. These applications are examples of teaching Lean *with* VR.

VR technology has recently had technological breakthroughs. For example, the first versions of the VR technologies Oculus Rift and HTC Vive, shipped in 2016, and Oculus Quest shipped in 2019. In August 2021, Facebook launched 'Horizon Workrooms', a VR workspace accessible with Oculus and part of the company's strategy to develop a 'metaverse' where people can be 'inside the internet' [14]. Due to such recent advancements, and accelerated by the COID-19 pandemic, VR now offers opportunities for immersive real-time, human interactions in a virtual environment for conferencing, meeting, lecturing and gaming. Yet, due to the novelty, the literature offers no reports on teaching Lean or other business concepts *in* VR. This paper addresses this gap.

3 Teaching Intelligent Lean in TLN's Gemba Platform

We use a platform called Gemba designed, managed and marketed by TLN. This platform has received considerable press coverage, including features in *The Financial Times* [5] and *Forbes* [e.g. 4, 6] among other high-profile publications. Gemba won the 'Best Use of VR/AR' in the 2020/21 Go:Tech Awards.

The authors lead a masterclass entitled 'Leading Intelligent Lean'. The origin of the course was a three-day physical course that was planned for delivery at a factory site in Germany in April 2020. However, it was not possible to run this course due to the COVID-19 outbreak. The course was switched to the VR platform that had already been in development by TLN as an offering separate from their physical courses. By summer 2021, the VR Lean masterclass had been organised five times. It is an executive masterclass delivered over five half-day sessions almost exclusively through VR.

To give an impression of how Gemba looks and works, Fig. 1a–d shows four different VR environments in Gemba (as of spring 2021).

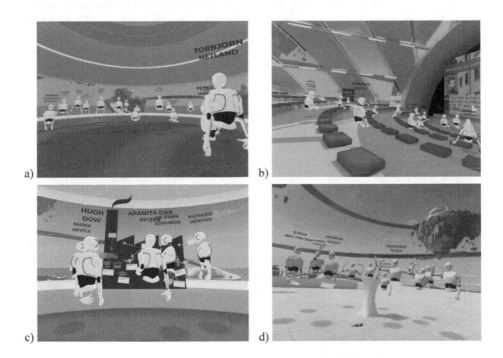

Fig. 1. Snapshots from the Lean Masterclass in the Gemba app

Figure 1a shows the avatars of the teachers and participants in a breakout room. Figure 1b shows the lecture of an external industry speaker in an auditorium. Figure 1c shows group work in a breakout room using virtual yellow notes and icon stickers on a factory-shaped black board. Figure 1d shows a 'sky bar' used for networking and quiz games. The participants—all wearing Oculus headsets at their locations around the world—act as their own avatars and can hear, see and interact with the other participants in real time. A typical masterclass has around 25 participants.

The masterclass 'Leading Intelligent Lean' is split into five half-days of a week. This schedule is easier on the eyes than spending full days in VR and allows the busy participants to handle other responsibilities during the week. We teach and discuss how Lean concepts evolve with the advent of the fourth industrial revolution and how to lead transformational change. Typical participants are C-level and senior-level managers and transformation leaders. The course is organised in the afternoon on European time to also allow participants from the Americas and the Far East to attend (albeit at partly inconvenient times). The entire course is taught in VR, with some extracurricular actives during regular video conferences (e.g., participant introductions, continuous helpdesk and discussions on topics of special interest).

In this course, we define Intelligent Lean as 'making the most of human intelligence and digitalisation in support of sustaining the Lean end-to-end transformation'. The course is carefully designed as a mix of frontal lectures from the leaders and invited guest speakers, group breakout sessions, ideation workshops, reflection sessions and networking (e.g., quizzes). Overall, the course design has similarities to that of a physical classroom setting, but the administration, delivery, facilitation and content are tailored to the opportunities and challenges of current state-of-the-art VR technology.

4   Feedback and Experiences

After each course, a detailed feedback survey was distributed to the participants. The survey asked participants to score the masterclass, leaders, guest lecturers and the administration and organisation. It also requested open feedback on challenges and opportunities. Overall, 81 of the 117 participants completed the feedback form (a 69% response rate). Together with TLN, we actively used this feedback to improve the course incrementally for every instalment.

Due to the novelty of the topic and the fast-moving nature of the technology, we present here primarily the reactions of the participants (i.e., reports of the learning experience). The overall feedback and our experiences suggest that participants were very satisfied with the course content and delivery. The median course rating—as well as the scores for most of the evaluated categories (presentations, content, technology, etc.) —is 8/10, which is in the range of high-scoring physical TLN masterclasses before COVID-19.

Mirroring the positive feedback relating to learning experiences, VR also works surprisingly well for teaching. Teaching can be done from any silent room with wireless internet access. However, special preparation was required to adapt presentation slides to a VR-friendly format before uploading them to Gemba. Because current VR technology makes quick personal coordination between event organisers and lecturers difficult (everyone in the same place would hear it), we communicated via a messenger service on our phones. This allowed us to make quick decisions and adjust the program as needed. After each masterclass, we reflected on and discussed personal observations and notes that we had recorded during the class. We used this to improve the masterclass step-by-step by adapting the program and platform accordingly.

5 Discussion and Outlook

Is VR an appropriate platform for teaching Lean? We conclude that it is. Obviously, it is not sufficient on its own. Learning Lean requires actual observation and interventions in real-life processes. However, as a replacement for classroom teaching, VR now works very well. We could run virtual exercises and breakout rooms, use post-it notes and hold 'classroom' lectures with Q&A sessions without problems. Table 1 summarises the main advantages and challenges regarding teaching in real-time VR we have discovered.

Table 1. Advantages and challenges of teaching in real-time VR

Advantages	Challenges
No travel time or budget	Requires a VR platform and headsets
Minimal CO_2 emissions	Technological limitations
Pandemic-safe	Less 'watercooler moments'
Full immersive learning, no distractions	Very limited human body language
Inspiring VR environments	Required preparations of leaders and material
More equality (all look the same in Gemba)	Less personality (all look the same in Gemba)
Use of new technology gives 'wow effect'	New technology for participants to learn
Ease of use (Oculus and Gemba)	Weight and heat of VR headset
Affordable (Oculus Quest 2 costs USD 300)	Difficult to take notes while in VR
Enable blended information and 3D models	Potential cybersickness/tiredness
Enable computer-based gamification	Limited pedagogy available for VR teaching

5.1 Advantages of Teaching in Real-Time VR

As listed in Table 1, one set of advantages relates to the time and space dimension. VR has obvious advantages when it comes to bringing people from all over the world together quickly without the need for travel or the risk of spreading infectious diseases. This reduces travel time, travel costs and environmental footprints – all very important metrics. Considering the participant locations of the first four courses, we calculated the average CO_2 savings per participant to be about 1,000 kg from flights not flown [based on 15]. This corresponds to approximately 10% of the annual CO_2 footprint of an average UK citizen (lower percentage for a US citizen, higher percentage for a Swiss citizen) [cf. 16]. In the five first courses, we had participants from more than 20 countries (including, Brazil, South Korea, the USA, the Philippines, India, the UAE, and many European countries).

A second set of advantages relates to the immersive aspects of VR technology. Participants appreciate that the fully immersive experience allows them to be fully focused on the content and exercises (in contrast to checking emails or phones during an in-class lecture or video conference). They enjoy the immersive learning in the VR platform and how 'extraordinary' or 'remarkable' the experience was. Half of the participants explicitly noted this in the open area of the feedback survey. Many also emphasised that it was 'fun' and 'exciting'. A legitimate question is whether this form of positive technological enthusiasm will remain, increase or diminish as VR and 'virtual fatigue' become more commonplace. For most of our participants, this masterclass is their first experience in real-time VR.

Thirdly, many advantages stem from the technological opportunities (and limitations). For example, in this course, we use neutral, identical avatars. This makes everyone equal and removes distractions and personal biases associated with physical appearance (which are not to be underestimated in physical lectures or – perhaps even more so – in online conference platforms where one constantly sees oneself). We also noted that listening rates tend to be higher in VR than in real life, simply because there is not much more to do than listen (e.g., no emails, no smartphones). We also note that the technology is affordable (Oculus headsets are selling at USD 300) and easy to use and learn with some assistance. VR enables blended information and 3D models to be shown anywhere, which is an advantage over classroom teaching. VR allows building imaginative worlds that can inspire participants to think beyond the restrictions of the real world. This year, TLN has also integrated 360° factory visits in their VR platform by using recorded VR footage from a Toyota factory in the UK.

5.2 Challenges of Teaching in Real-Time VR

VR also has several limitations. Obvious limitations relate to the technology. Teaching in VR first requires an online platform (e.g., Gemba) and VR headsets present at all participant locations. All participants must have access to high-speed wireless internet. Because installing VR software on the headsets is still cumbersome, this was done by TLN in our case. TLN shipped and collected VR headsets to participants around the world using a third partly logistics provider. This is a logistical and administrative challenge. In addition, all material must be in a VR-friendly format (e.g., large font and toned-down colours).

From a pedagogical point of view, a real challenge is the lack of human body language and few opportunities for serendipitous 'watercooler' discussions. This limits both direct feedback from the students and the overall experience of the participants. The neutral look of the avatars also reduces the potentially positive influence of a charismatic or lively individual, although some of these characteristics will be transferred via voice. It is also a limitation that participants cannot currently take notes in TLN's VR platform, a problem Oculus recently developed a fix for.

Many participants reported tiredness, and some reported cybersickness, especially during the first days in VR. Our own experience suggests that cybersickness diminishes with practice. All these technical challenges are likely to be reduced with future developments in VR technology.

Virtual fatigue is perhaps the greatest challenge for teaching in VR. As VR becomes more commonplace, participants will sharply raise their expectations. Therefore, we expect that a blended teaching approach will be most effective. Figuring out where VR has its definitive edge over classroom and online teaching, is still subject to experimentation and research.

6 Conclusions

This paper discussed the opportunities and challenges related to teaching a course entirely in real-time, immersive VR. We showed that this form of teaching generally works well and is developing rapidly. Taken together, we summarise the following four propositions:

P1. Real-time, immersive VR is a viable and available alternative to classroom and online teaching.

P2. Real-time, immersive VR will accelerate as a future teaching mode.

P3. Real-time, immersive VR is best used as part of a blended teaching approach.

P4. Real-time, immersive VR will require and foster new teaching pedagogy.

Although we have attempted to be comprehensive in our reflections, we see this paper as solely exploratory. We have only reviewed one course; we have only explored one VR platform (TLN's Gemba); and we have only used one type of headset (Oculus Quest). In addition, we have not collected longitudinal information from participants in the months after the course to test their learning outcomes. These shortcomings leave ample opportunity for future research into what is certainly a disruptive technology that educators must consider for future use.

References

1. Dede, C.: Immersive interfaces for engagement and learning. Science **323**(5910), 66–69 (2009)
2. Netland, T., Lorenz, R., Senoner, J.: Teaching lean with virtual reality: Gemba VR. In: Rossi, M., Rossini, M., Terzi, S. (eds.) ELEC. LNNS, vol. 122, pp. 29–37. Springer, Cham (2020). https://doi.org/10.1007/978-3-030-41429-0_4
3. Netland, T.H., Flaeschner, O., Maghazei, O., Brown, K.: Teaching operations management with virtual reality: bringing the factory to the students. J. Manag. Educ. **44**(3), 313–341 (2020)
4. Koetsier, J.: Are shared virtual experiences the future of meetings and work? Forbes.com. Forbes (2021)
5. Conboye, J.: Gaming inspires new worlds of virtual work, p. 19. Financial Times, New York (2021)
6. Kite-Powell, J.: Terraforming the enterprise with virtual reality. Forbes.com. Forbes (2021)
7. NASA: Virtual reality: Definition and requirements. https://www.nas.nasa.gov/Software/VWT/vr.html. Accessed 22 Aug 2021
8. Gamlin, A., Breedon, P., Medjdoub, B.: Immersive virtual reality deployment in a lean manufacturing environment. In: 2014 International Conference on Interactive Technologies and Games (iTAG). IEEE (2014)
9. De Vin, L.J., Jacobsson, L., Odhe, J.: Simulator-assisted lean production training. Prod. Manuf. Res. **7**(1), 433–447 (2019)

10. Belkin, D.: At Penn State, Field Trips Go Virtual, in The Wall Street Journal. Dow Jones, New York (2018)
11. Snelson, C., Hsu, Y.-C.: Educational 360-degree videos in virtual reality: a scoping review of the emerging research. TechTrends **64**, 404–412 (2019)
12. Krajčovič, M., Gabajová, G., Furmannová, B., Vavrík, V., Gašo, M., Matys, M.: A case study of educational games in virtual reality as a teaching method of lean management. Electronics **10**(7), 838 (2021)
13. Gottini, G., Solari Bozzi, L., Kunde, M., Lorenz, R., Netland, T.: Creating VR content for teaching operations management. Whitepaper, ETH Zurich (2021). www.pom.ethz.ch
14. Klar, R.: Facebook launches virtual reality office space as part of 'metaverse' plan. The Hill (2021)
15. BP: Drive down your carbon footprint. https://www.bp.com/en_gb/target-neutral/home/calculate-and-offset-your-emissions/travel.html/. Accessed 13 July 2021
16. Ritchie, H., Roser, M.: Co_2 and greenhouse gas emissions. https://ourworldindata.org/co2-and-other-greenhouse-gas-emissions#co2-and-greenhouse-gas-emissions-country-profiles. Accessed 13 July 2021

Lean and Digital

When Digital Lean Tools Need Continuous Improvement

Marte D.-Q. Holmemo$^{(\boxtimes)}$ ⓘ, Jonas A. Ingvaldsen ⓘ,
and Eirik B. H. Korsen ⓘ

Department of Industrial Economics and Technology Management,
Norwegian University of Science and Technology, Trondheim/Gjøvik, Norway
marte.holmemo@ntnu.no

Abstract. Lean practitioners have traditionally been reluctant to automate and digitalize production. Over the last years the combination of lean and digitalization has been actualized in academic publications, but still there are unanswered questions. In this paper, we address the relationship in a qualitative case study of a performance management system in a lean production context in a department of a Norwegian light-metal production company. We find synergies between lean and digitalization, as the digital system supports visualization of performance, problem analysis and continuous improvement through employee involvement. Nonetheless, we also find that digital solutions might be a barrier for motivation and further production improvement when IT systems are not developed aligned with the continuous improvement on the shop floor. We encourage organizations to find alternative ways of organizing the relationship between distributed continuous improvement and centralized IT development to strengthen the synergy of digitalization and lean.

Keywords: Digital lean · Continuous improvement · Organization design

1 Introduction

It is wrong to automate something just because we can – Taiichi Ohno [1], p. 93.

The wise sayings of Taiichi Ohno and other Toyota senseis have led many lean enthusiasts to be reluctant to automate and digitalize production [2]. Improving work processes prior to automation, and keeping things simple, "low-tech" and flexible have been, and still remain, the ideal for many lean practitioners.

In recent years however, both academics and practitioner are pointing to the attractiveness of combining lean management with information technology, digitalization and automation, often referred to as Industry 4.0. Pinho and Mendes [3] reviewed publications, showing evidence that lean and IT were both complementary and mediators, but also rivals in practice, calling for further empirical studies. In a later review, Lorenz et al. [4] argued for synergy: 1) digital solutions support lean and 2) lean supports digitalization. As examples of the first point, IT-systems have been shown to improve supply-chain efficiency [5], reduce costs and waste regarding time [6], control, error detection and correction [7]. Regarding the second point, recent

© IFIP International Federation for Information Processing 2021
Published by Springer Nature Switzerland AG 2021
D. J. Powell et al. (Eds.): ELEC 2021, IFIP AICT 610, pp. 165–171, 2021.
https://doi.org/10.1007/978-3-030-92934-3_17

publications [8, 9] state that lean fosters benefits of digitalization and emergent technologies, even that without lean thinking, firms risk to "digitalize waste" [4], an echo of Ohno's advice.

This paper addresses the relationship between lean and digitalization through a case study of a performance measurement system, which was designed and used to support continuous improvement at the shop floor. Our findings support the synergy view, with one important reservation: Unless the development of the IT-systems can keep up with the developments at the shop floor, they likely become barriers to further performance improvement. The way IT development is typically organized makes this issue an important practical problem. We encourage organizations to explore new ways of organizing the relationship between distributed continuous improvement and centralized IT development.

2   Methods

Our case study analyses the use of a digital tool for performance measurement, called Team Performance (TP), in a department of a Norwegian electrochemical plant, here called Department. Department has around 240 employees, divided into five shifts with seven teams on each shift producing liquid light-metal in large batches during continuous 24-h production. This department was chosen due to being a pioneer in using TP to support continuous improvement.

Qualitative data was collected during a two-day visit to the plant in October 2020 where we collected field notes through 7 observations on the factory floor, 7 observations of meetings and 15 recorded interviews. Subsequently, we conducted 10 additional interviews by video conference. In total, the informants represent all organizational levels: operators, employee representatives, middle managers (shift managers and process managers), division manager, and divisional support staff including head of organization development and the project manager with responsibility for design and training of the overall performance measurement system. In addition, we have secondary data from collaborating with the case organizations over several years, including visits to other plants, participation in workshops and other activities.

The themes in this paper emerged inductively from the data collection and analysis process [10]. Field notes and transcribed interviews were coded thematically [11] by two of the researchers and discussed to develop common interpretations. The overall analysis was presented to and discussed with the company for informant validation. The data were categorized in several steps where the structure in the findings section represent the top-level categories.

3   Case Description and Analysis

In 2007, the case company developed its own «company-specific production system» [12], which was implemented at the different plants and departments in the following years. The production system combines conventional lean principles for process control and continuous improvement with ideas of self-managing teams and less-hierarchical organizing. While repetitive task should be standardized, operators should also be skilled and empowered to respond to deviations in the chemical processes, see [13] for a similar case.

A small corporate staff is responsible for documenting and updating the content of the production system and producing teaching material. They also conduct training and support improvement initiatives at the plants, along with local lean-coordinators, typically one or two at each plant. In practice the different plants, and even different departments, are free to emphasize different tools and principles, depending on the nature of production and short-term priorities. A structure of parallel teams is set up to aid knowledge transfer ("best practices") between plants regarding the main technical processes.

In 2016 Department was chosen as a pilot unit for implementation of TP. In short, TP is a performance measurement system that collects and displays data from several machines and other systems, presented as selected performance indicators (PIs). The teams can monitor their own performance and analyze their development over time. They can also compare their performance to other teams. Furthermore, TP is used in Department's various meetings, for quality control, feedback, problem analysis and learning. Hence, TP supports local improvement and learning, and also the company's overall lean approach as explained by the head of the corporate staff:

> TP supports execution of standard operational procedures, (...), it gives important input to our improvement meetings (...), it has an impact on the optimization of flow (...), it is important for the team and individual competence development (...), [and] it helps managers focus and pay attention to tasks in a good dialogue with the teams.

3.1 TP Supports and Motivates Performance and Improvement

TP has been well received and utilized in Department. Although our informants report variation in enthusiasm, TP was generally described as a highly effective tool for quality control and improvement. Our observations confirm that TP was evident in the working routines and regular meetings. The performance measures were also discussed informally during breaks.

TP seemed to motivate workers to perform and take part in improvement activities. Both managers and operators confirmed that TP enhanced the workers' motivation to perform and learn as they could see their own results and analyze the causes and effects directly. One union representative said:

> What we have achieved is a system that shows us that changed effort leads to improvement, more efficiency and improved profit. And at least, I should mention, TP is an opportunity to participate, which means a lot to people. Having a chance to improve your workplace.

The high level of utilization can be explained by TP being tailor-made for certain teams and processes in Department. As one of the process managers told us: *I would claim that the standard dashboards are tailored to the needs we had back then.* On the other hand, work-groups in Department which had been less involved in the design of TP found it relatively harder to relate and employ TP.

3.2   TP Needs to Keep up with Development to Support the Processes

Department had a 13-year history of systematic continuous improvement at the time of our visit. We found the learning-and-development mindset integrated in the informants' reflections, also towards a performance measurement system like TP, as illustrated by one of the managers:

> *Ideas have to grow from the organization. Understanding what to measure is a learning process, this is not initially apparent. (...) There will appear suggestions from the organization on what to measure to improve along the process.*

TP had been designed and implemented at one point in time, and the risk of the system lagging behind the developed needs for digital support of process improvement was apparent in the interviews. We will highlight three aspects here:

Firstly, concerning **adjustment of standard threshold values**: TP has a traffic-light function showing red, yellow and green depending on deviation from defined target values of the PIs. After improving their accuracy for years, the teams' actual targets had become more nuanced, leading to the demand of a more detailed palette. One of the process managers had programmed an excel spreadsheet supporting different shades of green and yellow, but she did not have access to change the standard threshold value settings in the TP system. Another manager explained the situation to us:

> *The challenge now is that some teams perform so well on the existing measures that they may get demotivated. They need something new to strive for, more than to make the green numbers even greener.*

Secondly, **development in what to measure**. Standard PIs were seen useful for the managerial teams at the corporate level as they were able to benchmark results and spread best practice within the company. Nonetheless, managers and operators at Department had a different view here. While they had moved forward in their learning process, Department called for different PIs as their problem areas had shifted. Furthermore, various teams had different needs for improvement and control. This was a concern of the union representative:

> *We have talked about finding new things to measure. Being stuck in the mud is less interesting for operators.*

An update in PIs in TP would nevertheless be unsatisfactory as the expressed need at Department was constantly improving and changing PIs to support the development areas for production. One of the mangers stated:

> *Managers have reported that we need dynamic PIs, a dynamic portal with relevant PIs. Presently TP is static, at risk of being stranded. The energy drains out and we need to get the* [central] *tech team onboard.*

Thirdly, our informants did not only call for moving threshold values or focus areas, but to **expand the whole TP system** to provide more detailed information and further support for statistical analysis. One of the process managers acknowledged that TP is not yet complete:

We have significant operations to which TP does not give feedback, operations that are important for the overall process stability. We acknowledge that we have lost some things on the way; issues become degraded as they are invisible on the dashboard.

3.3 Access to Fix and Improve Digital Solutions

Successful digitalization is about more than software. This was clearly exemplified at the time of our visit at the plant. The head of Department told us:

We used to have screens like iPads, but when you have 50 operators and 2 devices for each team you hardly have time to log on during a 15 min brake. Thus, we bought the large touch screens on the walls in the team rooms.

At the time we visited the plant, the screens were out of order to the operators' frustration:

Some teams struggle with the motivation. Normally we have touch screens in all the team rooms where they can check and discuss their performance during breaks. These screens have been broken for a while now, and then it's hard to keep the good spirit.

However, at the time we presented our findings to the case organization the hardware problem was solved. Software changes in the TP system seemed more arduous having a long decision-making chain from a local operator to his manager, coordinated at Department and then to the software team if suggestion reached up on the corporate list of priorities. One shift manager explained the gap in time and space:

Well, then we have to address problems during daytime and they [central unit] need to agree upon our requests. We do not even always agree on changed measures across shifts. It is a cumbersome process, as for everything we want to change. It is time consuming.

Even the coordinating manager at Department found this path frustrating:

I'm in charge of the digital part. If they want changes in the dashboard, I handle it. Not technically, but I report it further. (...) I struggle to get through and get help from those on digital (central unit) (...) They do not have the capacity to prioritize our needs. (...) When they are not engaged it is hard to get the operators on board.

4   Discussion and Conclusion

Our findings on the relationship between lean and digitalization is paradoxical. On one hand, TP visualizes the production process and results in a way that supports performance management and problem solving. Hence our findings show that digital tools can support lean production [4, 14]. On the other hand, our findings demonstrate that the digital systems can appear as barriers to performance and process improvement if they cannot keep up with the pace of improvement at the shop floor.

Previous research has addressed a concern about digitalization as "technology push" where processes are modified to fit new technology rather than the opposite that is central in lean [1, 15]. Further, one has argued that digitalization often enhances complexity whereas lean stands for reducing complexity [4]. Our findings show that lean and process improvement can push the need for new and more complex digital solutions whereas existing digital solutions can lag behind.

Ashrafian et al. [16] considered organizational aspects to explain both synergies and challenges between the logics and practices of lean and digitalization. The lean work organization, characterized by shop-floor empowerment, might be counteracted by digitalization. As we have seen in this case, digitalization is often driven by IT-experts organized in central units, creating or configuring solutions based on corporate or marked standards. Increased digitalization can lead to a redistribution of power in the organization. As continuous improvement becomes more data-driven and the complexity of problem analysis increases, responsibility and influence move from the operators to experts on statistics and technology at central staff divisions. Central IT-experts are challenged to deliver updated solutions to several different production units and are simultaneously striving for standardization for cost efficiency and data integration. When diverse units improve their practices and experience different IT needs, organizations might experience an increased time-space-gap [17] between central developers of digital solutions and various local users of digital solutions.

Solutions to keep the speed of process and technology development in balance could be found in designing a digitalized lean organization. Local lean activities should be coordinated by managers on all levels in the organization to increase the effects [18], whereas digitalization teams and experts should be organized in ways that close the distance in time and space to the local users on the shop-floor level. We call for more research on how to align strategies and organizational structures for lean and digital solutions in production development. Digitalization and lean should be handled less as two opposing concepts in organizational practice.

References

1. Ohno, T.: Taiichi Ohnos Workplace Management. McGraw-Hill Education, London (2013)
2. Liker, J.K., Franz, J.K.: The Toyota Way to Continuous Improvement. McGraw-Hill Education, New York (2011)
3. Pinho, C., Mendes, L.: IT in lean-based manufacturing industries: systematic literature review and research issues. Int. J. Prod. Res. 55(24), 7524–7540 (2017)
4. Lorenz, R., Buess, P., Macuvele, J., Friedli, T., Netland, T.H.: Lean and digitalization— contradictions or complements? In: Ameri, F., Stecke, K.E., von Cieminski, G., Kiritsis, D. (eds.) APMS 2019. IAICT, vol. 566, pp. 77–84. Springer, Cham (2019). https://doi.org/10.1007/978-3-030-30000-5_10
5. Ward, P., Zhou, H.: Impact of information technology integration and lean/just-in-time practices on lead-time performance. Decis. Sci. 37(2), 177–203 (2006)
6. Hoellthaler, G., Braunreuther, S., Reinhart, G.: Digital lean production an approach to identify potentials for the migration to a digitalized production system in SMEs from a lean perspective. Procedia CIRP 67, 522–527 (2018)
7. Deuse, J., Dombrowski, U., Nöhring, F., Mazarov, J., Dix, Y.: Systematic combination of lean management with digitalization to improve production systems on the example of Jidoka 4.0. Int. J. Eng. Bus. Manag. 12 (2020). https://doi.org/10.1177/1847979020951351

8. Buer, S.-V., Semini, M., Strandhagen, J.O., Sgarbossa, F.: The complementary effect of lean manufacturing and digitalisation on operational performance. Int. J. Prod. Res. **59**, 1976–1992 (2020)

9. Ciano, M.P., Dallasega, P., Orzes, G., Rossi, T.: One-to-one relationships between Industry 4.0 technologies and Lean Production techniques: a multiple case study. Int. J. Prod. Res. **59**, 1386–1410 (2021)

10. Corbin, J.M., Strauss, A.: Grounded theory research: procedures, canons, and evaluative criteria. Qual. Sociol. **13**(1), 3–21 (1990)

11. Braun, V., Clarke, V.: Using thematic analysis in psychology. Qual. Res. Psychol. **3**(2), 77–101 (2006)

12. Netland, T.H.: Exploring the phenomenon of company-specific production systems: one-best-way or own-best-way? Int. J. Prod. Res. **51**(4), 1084–1097 (2013)

13. Hekneby, T., Benders, J., Ingvaldsen, J.A.: Not so different altogether: putting lean and sociotechnical design into practice in a process industry. J. Ind. Eng. Manag. **14**(2), 219–230 (2021)

14. Meissner, A., Müller, M., Hermann, A., Metternich, J.: Digitalization as a catalyst for lean production: a learning factory approach for digital shop floor management. Procedia Manuf. **23**, 81–86 (2018)

15. Liker, J.K.: The Toyota Way: 14 Management Principles from the World's Greatest Manufacturer. McGraw-Hill, New York (2004)

16. Ashrafian, A., et al.: Sketching the landscape for lean digital transformation. In: Ameri, F., Stecke, K.E., von Cieminski, G., Kiritsis, D. (eds.) APMS 2019. IAICT, vol. 566, pp. 29–36. Springer, Cham (2019). https://doi.org/10.1007/978-3-030-30000-5_4

17. Orlikowski, W.J.: The duality of technology: rethinking the concept of technology in organizations. Organ. Sci. **3**(3), 398–427 (1992)

18. Holmemo, M.D.Q., Ingvaldsen, J.A.: Bypassing the dinosaurs?–How middle managers become the missing link in lean implementation. Total Qual. Manag. Bus. Excell. **27**(11–12), 1332–1345 (2016)

Discovering Artificial Intelligence Implementation and Insights for Lean Production

Bassel Kassem[(✉)], Federica Costa[(✉)],
and Alberto Portioli Staudacher[(✉)]

Politecnico di Milano, via Lambruschini 4/b, Milan, Italy
{bassel.kassem, federica.costa,
alberto.portioli}@polimi.it

Abstract. The research aims to understand how the implementation of Artificial Intelligence AI in Manufacturing Operations takes place. This paper will feed wider research on the interaction between Lean Production and AI, after understanding the implementation process of AI. A Systematic Literature Review (SLR) has been performed. A set of more than 2300 documents has been extracted and screened to produce a list of 90 highly selected and classified articles and conference papers dealing with the research question. After a first meta-level analysis, a structured discussion has been presented over the documents. Three macro use-cases for implementing AI into manufacturing systems have been identified. The first two use cases have been deeply analyzed by the SLR, while the third one has been left for further researches. For the first two use cases, the main applications have been presented through a comprehensive categorization (for stand-alone solution) and a clear explanation of the different paradigms (for I4.0 related implementation). Furthermore, for each case, the available frameworks have been presented. The main challenges and issues that managers should consider while implementing this kind of technology were presented. Possible consequences that AI innovations might have were also indicated. The article ends with insights for Lean production and future research.

Keywords: Artificial Intelligence · Lean · Implementation

1 Introduction

The concept of "Artificial Intelligence" (AI), later defined as "The science and engineering of making intelligent machines" was firstly introduced in 1956 at the famous Dartmouth College conference. Unfortunately, for many years AI has been considered academic, unreliable, or just a part of science fiction, causing the development of this subject experiencing different "raises and falls" in history. Only recently, this misconception has been completely reverted; the development of low-cost AI chips, the production of massive data, and the development of algorithms based upon deep learning have led to the current evolution that started in 2010 and continues to evolve.

The current interest in the topic of AI is undoubtfully linked to another major revolution that is happening nowadays, namely the fourth industrial revolution or Industry 4.0 (I4.0). Introduced by the German government in 2011, I4.0 symbolizes a

D. J. Powell et al. (Eds.): ELEC 2021, IFIP AICT 610, pp. 172–181, 2021.
https://doi.org/10.1007/978-3-030-92934-3_18

new era of industrial processes characterized by the implementation of new disruptive technologies in the manufacturing systems such as Service Automation, Robotics, the Internet of Things, and Additive Manufacturing and AI. This emerging phenomenon is reshaping how companies work from the roots, and AI is one of the main enabling technologies of this revolution.

There is no doubt that organizations gain many benefits by implementing AI [1], and most of them are using AI-powered systems to handle customers request or to cluster and target prospects for marketing reasons. According to a study published by McKinsey in November 2020 (Global survey: The state of AI in 2020, 2020), fifty percent of the respondents stated that their organizations had adopted AI. The business functions in which organizations adopted this kind of technologies were mainly service operations, product or service development, and marketing and sales, with manufacturing left behind (for the majority of the industries [2]. In addition, even if AI was born years ago, implementing these systems is still a complex and hard job for managers and professionals, and, unfortunately, implementation steps have not been largely standardized. The current research is trying to answer a simple but not trivial question related to the role of AI within manufacturing:

"How to approach the task of implementing AI in Manufacturing Operations?"

This research is a preliminary step in wider research aiming to explain the interaction between AI and Lean Production and how they could benefit from each other when applied in manufacturing. It was essential to first understand the implementation process of AI and deduce some possible outlook and space for Lean to fit in.

2 Research Methodology

The research addresses the main use cases and application of AI in manufacturing operations (a), explaining different frameworks to implement these systems and to understand the role of AI in I4.0 and Smart Manufacturing (b) and highlighting which are the most common issues, challenges, and pitfalls a manager usually encounters in the implementation process (c).

A Systematic Literature Review (SLR) of the articles and conference papers has been chosen since it is characterized by a replicable, scientific, and transparent process for the selection and analysis of the different papers according to the table below. This methodology allows to identification, evaluation, and integration of the findings of all

Table 1. SLR steps

Step	Substeps
Papers selection and classification	1. Creation of the article DB
	2. Article selection process
	3. Article classification process
Meta-level analysis of the papers	1. General descriptive analysis
	2. Further analysis of the "directly correlated" papers
Qualitative review	1. Assessment of each paper and appraisal
	2. Key concepts mapping
	3. Comparison and evaluation

relevant, high-quality individual studies addressing the research questions and the steps described are in line and inspired by previous studies [3, 4] (Table 1).

The keywords are: "AI" and "Artificial Intelligence", "Manufacturing" and "Production" and "Implementing" and "Investing" have been used in the following query.

TITLE-ABS-KEY ("AI" OR "artificial Intelligence") AND TITLE-ABS-KEY ("Implement*" OR "Invest*") AND TITLE-ABS-KEY ("Manufacturing" OR "Production").

In this way, 2344 documents have been selected. In the first step of the analysis, only the abstract has been considered. In the second one, the full article with a final list of 90 selected articles and conference papers. Classifying the items of the obtained list into three categories: 40 "Directly Correlated" articles, used to build the "backbone" of the research and to try to answer the research question, 25 "Partially Correlated" and 25 "With limited Correlation" papers used to enrich the discussion. Refer to Sankey diagram in Fig. 1.

"Machine Learning" is not included to ensure finding articles related to the general topic of Artificial Intelligence and not only to this sub-group of its technologies.

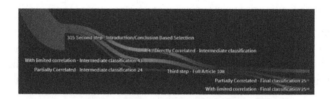

Fig. 1. Figure 2 Sankey diagram

The main statistical insights that have been obtained from the meta-level analysis have regarded the time distribution, the geographical pertinence, and the document type. The vast majority of the "Directly Correlated" articles are published after 2010.

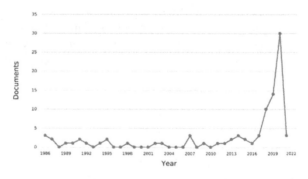

Fig. 2. Publications' timeline

To answer the research question and its subproblems, after the meta-level analysis an in-depth content review has been performed.

3 Findings and Discussion

The literature review helped to identify 4 macro areas corresponding to 4 questions that would allow answering the research question:

a. Which are the main use cases of AI in Manufacturing Operations?
b. How should the implementation process be designed in different cases?
c. Which are the most common issues, challenges, and pitfalls a manager usually encounters in the implementation process?
d. Which are the consequences of implementing AI and I4.0 in general that managers should consider when implementing these technologies?

3.1 The Main Use Cases of AI in Manufacturing Operations

From the analytical review and the first evaluation of papers, it was clear that in the literature, it was possible to identify three main use cases for the implementation of AI within the context of Manufacturing Operations. These three macro-categories were:

1. AI or Machine Learning ML applications conceived as stand-alone solutions to be implemented within the manufacturing context.
2. AI or ML applications are seen as an enabling factor for I4.0 or enabling technology for Digital Twins and Cyber-Physical Systems
3. AI or ML applications not directly used in Manufacturing operations but to design manufacturing operations within a company

The first class of articles was focused on simple AI and ML stand-alone tools applied within the manufacturing environment. In these cases, since AI can significantly improve operational effectiveness by bridging business and manufacturing models [5], companies started to use this kind of application for manufacturing Process control, Quality control, Monitoring and diagnosis, Safety evaluation, Process planning, and scheduling, Optimization of manufacturing yield, Material requirement planning, Preventive machine maintenance [6], Assistance and learning systems [7]. These applications can be considered as the simplest if compared with the ones included in the other groups, but managers willing to onboard AI in this way should consider the high variability associated with them. Indeed, within the literature, different possible classifications have been found: Classification by application type [8], by manufacturing process, by industry [9], by impact or application stage [10], and other classifications [11]. These classifications can drive managers and give them a comprehensive overview of the different AI applications available today.

The second use case presented by the literature is related to the concept of I4.0. Managers should evaluate that considering a smart factory as IT system, it is possible to implement AI at different levels [12]. Indeed, AI is one of the technologies included in the paradigm of I4.0, and it is an enabling factor for many innovations, such as for

Digital Twins and Cyber-Physical Systems [5]. According to [13], AI as an I4.0 habilitating technology will significantly impact how businesses are planned and conducted. In this case, AI and ML applications are not implemented as stand-alone solutions but interact with other innovative technologies such as Big Data technologies and the Cloud, which provide the source of the information (data) and a platform necessary to Industrial AI. The literature review highlights how AI applications (such as deep learning Image and Video processing, text and speech processing) are used to enable the intelligent functioning of Smart Manufacturing, Cyber-Physical Systems, and Digital twins [14].

The third use case, supported by a limited number of papers, was related to the use of AI in a "meta-level" way. Here, AI is used to design the system [15] or to perform technology selection [16]. Since this third use case was not sufficiently supported by the literature an in-depth analysis was not performed.

3.2   The Design of the Implementation Process

The second question was addressed in a similar but different way in the case of AI stand-alone application and the case of AI as an enabling factor for I4.0. However, in the first case, the literature presented different frameworks useful to start an effective and efficient implementation process (Stand-alone applications have limited interaction with other innovative technologies). Both for stand-alone application and I4.0 related frameworks, an incremental approach was adopted, starting from the simplest to the most complex but comprehensive one.

Referring to stand-alone applications, managers willing to onboard AI can consider different frameworks to guide their process. In particular, CRISP-DM [9], Four analytic capabilities stage models [17], and Analytic Canvas [18]. The first two frameworks are very simple and can be used as a starting point for managers, while the last one is very comprehensive and, unfortunately, complex. The Analytic Canvas is a very effective visual tool that can help managers in performing their job. Each model has a different degree of specificity and a different complexity to be managed. Indeed, professionals should decide which one (or the combination) can be the most suitable in their specific case considering the experience, the complexity, and the resources available.

Following the same approach, four different frameworks were presented to help managers to understand the role of AI within Smart Manufacturing and Cyber-Physical Systems. Three frameworks for AI in Digital Twin/Cyber-Physical Systems and one for more general implementation of AI in Smart Manufacturing. General implementation guidelines are explained by [19] or by [20]. To understand why AI is so important for Cyber-Physical Systems, [21] explained in their framework that AI has the goal of allowing managers to make more informed decisions in a shorter amount of time by processing a large amount of data coming from the production system. The same concept is presented and expanded by the data and knowledge-driven framework for digital twin manufacturing cell (DTMC) proposed by [22] and in the "framework for DT-driven industrial AI" [23]. In conclusion, managers using frameworks related to Smart manufacturing and I4.0 should be aware of the role of AI as the key enabler for Intelligent Capabilities of these systems.

3.3   Challenges and Pitfalls in the Implementation Process

After understanding which are the use cases and applications for AI within the manufacturing context and the frameworks that can be used to implement those applications, the third step each professional should go through is related to the possible issues and challenges that might arise during this kind of project.

Data Issues, since Data availability and quality represent a crucial prerequisite for even being able to use ML and AI systems, and Skills issues [6] should be carefully evaluated at the beginning of the project while the challenges related to the implementation might come out during the process.

AI implementations in I4.0 are more complex than the stand-alone applications, interacting with other technologies and with the physical environment, will most probably involve more challenges to face. Indeed, Data Issues [24], Skills Issues, Model Development Issues [22] are also frequent in the Smart Manufacturing implementation of AI. Some challenges relate to Investments Issues, Cybersecurity Issues [25] Network, and Machine-machine interaction issues [22].

In conclusion, implementing AI applications is complex, and different issues and challenges may arise. Managers and professionals willing to start these projects must carefully evaluate all these possible sources of unforeseen events. Furthermore, it is important to consider that developing AI solutions within the context of Smart Manufacturing can be even more difficult due to the double nature of this kind of innovation.

3.4   Further Consequences of Implementing AI

The final point related to AI or ML applications that any manager should always consider is linked to the consequences that this kind of innovation could bring. AI as a stand-alone application, like any other disruptive innovations, will drive companies, processes, and workflow to evolve, but, unfortunately, it is very complex to identify the consequences of this technology outside of the context of I4.0, as highlighted by the limited number of papers in the Systematic Literature Review.

For sure, AI Applications will allow companies to improve efficiency and productivity through consistent changes that will overcome some current limitations companies have by reducing cost, clarity in the production, or shortening the change-over-time [26].

The most interesting and impactful consequences that will derive from the adoption of AI and ML applications are related to Industry4.0 and Smart Manufacturing. In the literature, it was possible to identify two main groups of consequences that the implementation of AI will cause: Technical Impacts, Social Impact.

AI, together with the other I4.0 technologies, can indeed completely reshape how some processes are performed [25], creating new ways of extracting value from data [27]. Furthermore, the most drastic changes driven by AI will regard the Social sphere. The implementation of SM systems will increase demand for IT highly qualified specialists [25], causing a decreasing need for unskilled workers [28], causing a major social disruption.

The connection between people, AI, and the physical sphere [29] through the CPSs, DTs, and SMSs paradigms, but also stand-alone applications, can cause major changes, both positives and negative, that should be evaluated.

4   AI and Insights for Lean Production

Finally, within the selected articles it has been possible to find a subset of papers dealing with the outcomes coming from the combination of AI and ML techniques [30], Lean practices, knowing that the interaction of Lean with I4.0 has been the focus of the research world [31–33], In this context, AI might play a supportive role to human experts by enhancing the background to streamline the entire process and making information more accessible. According to [34], this technology can improve production control and process, help to monitor continuous flow and pull production, support early equipment failure detection while minimizing inefficiencies through the overall supply chain.

Lean has in its core the customer and its satisfaction. Though it is based on Pull production, forecasting is often used to have an idea of the trend of the customer orders. Putting customer behavior into formulas is much needed n the lean culture, and AI could help with this aspect. AI and machine learning construct patterns among the various parameters governing the relationships between the customer and the supplier. Managers could use this to build a stronger link with their customers and learn from it. Vice versa the customer who used to spend time going through the various options in the product catalog trying to personalize it or customize it according to its needs, could be assisted by AI that would learn from the customer behavior and speed up the selection process. Existing and potential integration between AI and these management practices has been a topic of discussion since the 1980s [35] and is deeply discussed by [36].

5   Conclusion and Future Research

In conclusion, implementing AI in manufacturing operations will have a long list of (mainly unpredictable) consequences on the company. It would be naïve to think that those consequences and outcomes will only be positive [37]. As described by [38], AI can also harm the value of a company and, for this reason, it should be carefully managed. Nevertheless, managers should consider all these possible consequences before starting a project of this kind.

The research has the main limitation concerning the methodology as one database is used (Scopus) and others could be included. We call for investigating "The meta-level application of AI" as future research.

As mentioned in the introduction, this paper constitutes the initial step for investigating the interaction between Lean Production and AI. This was intended to understand the implementation process of AI, and our future research will expand on the results to have systematic literature analyzing the interaction followed by real case application in manufacturing companies applying AI in their processes and adopt the lean culture as well. Future research will try to contextualize the various areas of research in this article in the lean environment. Which are the main use cases of AI in Lean Operations?

Is the implementation process any different in a lean culture? Which are the consequences of implementing AI that managers should consider when implementing these technologies in a lean environment?

References

1. Ilieva, R.Y., Nikolov, M.A.: The impact of AI & ML in agile production. In: Proceedings of 2019 10th National Conference with International Participation (ELECTRONICA), pp. 1–3 (2019). https://doi.org/10.1109/ELECTRONICA.2019.8825615
2. Stanford: Artificial Intelligence Index Report 2021 (2021)
3. Dada, O.L.: A model of entrepreneurial autonomy in franchised outlets: a systematic review of the empirical evidence. Int. J. Manag. Rev. **20**, 206–226 (2018). https://doi.org/10.1111/ijmr.12123
4. Liao, Y., Deschamps, F., Loures, E.F.R., Ramos, L.F.P.: Past, present and future of Industry 4.0 - a systematic literature review and research agenda proposal. Int. J. Prod. Res. **55**, 3609–3629 (2017). https://doi.org/10.1080/00207543.2017.1308576
5. Ghahramani, M., Qiao, Y., Zhou, M.C., Hagan, A.O., Sweeney, J.: AI-based modeling and data-driven evaluation for smart manufacturing processes. IEEE/CAA J. Autom. Sin. **7**, 1026–1037 (2020). https://doi.org/10.1109/JAS.2020.1003114
6. Wang, K.: Applying data mining to manufacturing: the nature and implications. J. Intell. Manuf. **18**, 487–495 (2007). https://doi.org/10.1007/s10845-007-0053-5
7. Fahle, S., Prinz, C., Kuhlenkötter, B.: Systematic review on machine learning (ML) methods for manufacturing processes - identifying artificial intelligence (AI) methods for field application. Procedia CIRP **93**, 413–418 (2020). https://doi.org/10.1016/j.procir.2020.04.109
8. Sharp, M., Ak, R., Hedberg, T.: A survey of the advancing use and development of machine learning in smart manufacturing. J. Manuf. Syst. **48**, 170–179 (2018). https://doi.org/10.1016/j.jmsy.2018.02.004
9. Mayr, A., et al.: Machine learning in production - potentials, challenges and exemplary applications. Procedia CIRP **86**, 49–54 (2020). https://doi.org/10.1016/j.procir.2020.01.035
10. Lwakatare, L.E., Raj, A., Bosch, J., Olsson, H.H., Crnkovic, I.: A taxonomy of software engineering challenges for machine learning systems: an empirical investigation. In: Kruchten, P., Fraser, S., Coallier, F. (eds.) XP 2019. LNBIP, vol. 355, pp. 227–243. Springer, Cham (2019). https://doi.org/10.1007/978-3-030-19034-7_14
11. Li, R., Wei, S., Li, J.: Study on the application framework and standardization demands of AI in intelligent manufacturing. In: Proceedings of 2019 International Conference on Artificial Intelligence and Advanced Manufacturing, AIAM 2019, pp. 604–607 (2019). https://doi.org/10.1109/AIAM48774.2019.00125
12. Wan, J., Li, X., Dai, H.N., Kusiak, A., Martinez-Garcia, M., Li, D.: Artificial-intelligence-driven customized manufacturing factory: key technologies, applications, and challenges. Proc IEEE **109**, 377–398 (2021). https://doi.org/10.1109/JPROC.2020.3034808
13. Ahuett-Garza, H., Kurfess, T.: A brief discussion on the trends of habilitating technologies for Industry 4.0 and Smart manufacturing. Manuf. Lett. **15**, 60–63 (2018). https://doi.org/10.1016/j.mfglet.2018.02.011

14. Engelsberger, M., Greiner, T.: Software architecture for cyber-physical control systems with flexible application of the software-as-a-service and on-premises model. In: Proceedings of the IEEE International Conference on Industrial Technology, pp. 1544–1549 (2015)

15. Renzi, C., Leali, F., Cavazzuti, M., Andrisano, A.O.: A review on artificial intelligence applications to the optimal design of dedicated and reconfigurable manufacturing systems. Int. J. Adv. Manuf. Technol. **72**(1–4), 403–418 (2014). https://doi.org/10.1007/s00170-014-5674-1

16. Evans, L., Lohse, N., Summers, M.: A fuzzy-decision-tree approach for manufacturing technology selection exploiting experience-based information. Expert Syst. Appl. **40**, 6412–6426 (2013). https://doi.org/10.1016/j.eswa.2013.05.047

17. Eriksen, L.: Industrial analytics revolutionizes big data in the digital business (2021)

18. Kühn, A., Joppen, R., Reinhart, F., Röltgen, D., Von Enzberg, S., Dumitrescu, R.: Analytics canvas - a framework for the design and specification of data analytics projects. Procedia CIRP **70**, 162–167 (2018). https://doi.org/10.1016/j.procir.2018.02.031

19. Dossou, P.E.: Development of a new framework for implementing industry 4.0 in companies. Procedia Manuf. **38**, 573–580 (2019). https://doi.org/10.1016/j.promfg.2020.01.072

20. Sjödin, D.R., Parida, V., Leksell, M., Petrovic, A.: Smart factory implementation and process innovation: a preliminary maturity model for leveraging digitalization in manufacturing moving to smart factories presents specific challenges that can be addressed through a structured approach focused on people, processes, and technologies. Res. Technol. Manag. **61**, 22–31 (2018). https://doi.org/10.1080/08956308.2018.1471277

21. Kharchenko, V., Illiashenko, O., Morozova, O., Sokolov, S.: Combination of digital twin and artificial intelligence in manufacturing using industrial IoT. In: Proceedings of 2020 IEEE 11th International Conference on Dependable Systems, Services and Technologies, DESSERT 2020, pp. 196–201 (2020). https://doi.org/10.1109/DESSERT50317.2020.9125038

22. Zhang, C., Zhou, G., He, J., Li, Z., Cheng, W.: A data- and knowledge-driven framework for digital twin manufacturing cell. Procedia CIRP **83**, 345–350 (2019). https://doi.org/10.1016/j.procir.2019.04.084

23. Alexopoulos, K., Nikolakis, N., Chryssolouris, G.: Digital twin-driven supervised machine learning for the development of artificial intelligence applications in manufacturing. Int. J. Comput. Integr. Manuf. **33**, 429–439 (2020). https://doi.org/10.1080/0951192X.2020.1747642

24. Lee, J., Davari, H., Singh, J., Pandhare, V.: Industrial artificial intelligence for industry 4.0-based manufacturing systems. Manuf. Lett. **18**, 20–23 (2018). https://doi.org/10.1016/j.mfglet.2018.09.002

25. Waibel, M.W., Steenkamp, L.P., Moloko, N., Oosthuizen, G.A.: Investigating the effects of smart production systems on sustainability elements. Procedia Manuf. **8**, 731–737 (2017). https://doi.org/10.1016/j.promfg.2017.02.094

26. Hansen, E.B., Bøgh, S.: Artificial intelligence and internet of things in small and medium-sized enterprises: a survey. J. Manuf. Syst. **58**, 362–372 (2021). https://doi.org/10.1016/j.jmsy.2020.08.009

27. Kusiak, A.: Smart manufacturing. Int. J. Prod. Res. **56**, 508–517 (2018). https://doi.org/10.1080/00207543.2017.1351644

28. Rymarczyk, J.: Technologies, opportunities and challenges of the industrial revolution 4.0: theoretical considerations. Entrep. Bus. Econ. Rev. **8**, 185–198 (2020). https://doi.org/10.15678/EBER.2020.080110

29. Pathak, P., Pal, P.R., Shrivastava, M., Ora, P.: Fifth revolution: applied AI & human intelligence with cyber physical systems. Int. J. Eng. Adv. Technol. **8**, 23–27 (2019)

30. Yadav, N., Shankar, R., Singh, S.P.: Impact of Industry 4.0/ICTs, Lean Six Sigma and quality management systems on organisational performance. TQM J. **32**, 815–835 (2020). https://doi.org/10.1108/TQM-10-2019-0251

31. Rossini, M., Cifone, F.D., Kassem, B., Costa, F., Portioli-Staudacher, A.: Being lean: how to shape digital transformation in the manufacturing sector. J. Manuf. Technol. Manag. **32**, 239–259 (2021). https://doi.org/10.1108/jmtm-12-2020-0467

32. Kassem, B., Portioli, A.: The interaction between lean production and industry 4.0: mapping the current state of literature and highlighting gaps. In: Proceedings of the Summer School Francesco Turco, pp. 123–128 (2019)

33. Rossini, M., Costa, F., Tortorella, G.L., Portioli-Staudacher, A.: The interrelation between Industry 4.0 and lean production: an empirical study on European manufacturers. Int. J. Adv. Manuf. Technol. **102**(9–12), 3963–3976 (2019). https://doi.org/10.1007/s00170-019-03441-7

34. Perico, P., Mattioli, J.: Empowering process and control in Lean 4.0 with artificial intelligence. In: Proceedings of 2020 3rd International Conference on Artificial Intelligence for Industries (AI4I 2020), pp. 6–9 (2020). https://doi.org/10.1109/AI4I49448.2020.00008

35. Manivannan, S.: Intelligence (1989)

36. Shahin, M., Chen, F.F., Bouzary, H., Krishnaiyer, K.: Integration of Lean practices and Industry 4.0 technologies: smart manufacturing for next-generation enterprises. Int. J. Adv. Manuf. Technol. **107**(5–6), 2927–2936 (2020). https://doi.org/10.1007/s00170-020-05124-0

37. Di Vaio, A., Palladino, R., Hassan, R., Escobar, O.: Artificial intelligence and business models in the sustainable development goals perspective: a systematic literature review. J. Bus. Res. **121**, 283–314 (2020). https://doi.org/10.1016/j.jbusres.2020.08.019

38. Castillo, D., Canhoto, A.I., Said, E.: The dark side of AI-powered service interactions: exploring the process of co-destruction from the customer perspective. Serv. Ind. J. **41**, 1–26 (2020). https://doi.org/10.1080/02642069.2020.1787993

Breaking Out of the Digitalization Paradox

Sourav Sengupta[(⊠)], Heidi Dreyer, and Daryl John Powell

Department of Industrial Economics and Technology Management,
Norwegian University of Science and Technology,
Gløshaugen, Trondheim, Norway
{sourav.sengupta, heidi.c.dreyer,
daryl.j.powell}@ntnu.no

Abstract. Despite the acclaimed potential of industry 4.0 for efficiency and growth, statistics show that the majority of firms' digital transformation programs fail to meet their objectives. We provide a plausible explanation of this understudied phenomenon through theoretical discussions on the four overlooked paradoxical characteristics found between digitalization activities. Further, using the lens of lean and dynamic capabilities theory, we propose strategies for firms to transcend the paradoxes and in turn, realize their expectations of the transformation initiatives.

Keywords: Digitalization · Paradox · Dynamic capability · Lean

1 Introduction

Being driven by the fear of missing out, many firms today tend to jump into this new, and trending bandwagon of digitalization without adequate understanding of the meaning, relevance, and implications. This of course leads to unwarranted failures despite the potential of revenue growth [1]. The findings of a survey conducted by Accenture strategy involving Scandinavian business leaders revealed that they need to "rethink their interpretation, approach and outlook on digital to maintain their competitive edge and guard against disruption"[1]. Further, a more recent article finds that 70% of the digital transformation initiatives fail to reach their goals [2]. Literature refers to this phenomenon as "digitalization paradox", as the investments on digitalization does not always produce the obvious or expected positive outcomes and contradicts the premise of revenue growth [3, 4]. While much has been written about Digitalization use cases and potentials, relatively little has been written about the paradox or the underlying causes. Understanding and addressing the paradox is becoming important as more and more organizations are investing on digitalization. Further, little is known about the intersection of digitalization and lean or about "Lean Digital" or "Digital Lean" paths to revenue growth. In this research, we aim to address these gaps. Through theoretical discussions on the intersection of digitalization and lean approaches and using the lens of dynamic capability theory we elaborate on: what causes the phenomenon of digitalization paradox and how firms should address it?

---

[1] https://www.accenture.com/no-en/insight-digital-nordic-wake-up-call-norwegian-businesses.

© IFIP International Federation for Information Processing 2021
Published by Springer Nature Switzerland AG 2021
D. J. Powell et al. (Eds.): ELEC 2021, IFIP AICT 610, pp. 182–190, 2021.
https://doi.org/10.1007/978-3-030-92934-3_19

The study contributes to this growing body of knowledge by exposing the inconsistencies in the conventional reasoning or assumptions at the intersection of lean and digitalization. The discussions on the overlooked paradoxical characteristics and the dynamic capabilities approach in context of the digitalization paradox provides new insights to this debate. Further, building on the contemporary lean literature, including the identified patterns of digital transformation in the context of lean production, namely "sustaining pattern" (of firms that are highly committed to lean production) and "disruptive pattern" (of firms that show low commitment to lean production) [5], and using our two-by-two framework, we elaborate on why two distinct strategic approaches are needed depending on the expectations of revenue growth. The first is problem driven with a focus on the existing customers and improving value, while the second is radical, with a focus on creating opportunities of new value and customers.

2 Digitalization and the Paradox

Digitalization to date emerge as a vague concept which at times is used interchangeably with 'digitization' or 'digital' and at times viewed under the label of industry 4.0 [6]. Nonetheless, digitalization is "more than digitizing operational processes" and is broader in meaning than industry 4.0 [7]. It can be viewed as a capability that is developed to exploit digital opportunities for operational efficiency and revenue growth [8–10]. Capabilities represents the capacity of a firm to deploy a combination of resources or a set of skills and proficiencies required to achieve the desired outcome [11, 12]. Firms differ in their "competence" and "capabilities" to engage in new technologies in their specific environments or context [6].

Digitalization can be a challenging activity and may require heavy investments and integration of various decision-making platforms and tools [13]. Nevertheless, its relationship with performance is considered complex, or even non-linear. Despite the potential for revenue generation, firms struggle to attain the expectations [4, 13]. Digitalization paradox refers to this phenomenon in which firms "invest in digitalization but struggle to earn the expected revenue growth" [3]. Studying such paradoxes become important because they expose inconsistencies in our reasoning or assumptions and present problems in fundamentally different ways [14]. In the following subsections, we theoretically elaborate on the four overlooked paradoxical characteristics found between digitalization activities that contribute to the digitalization paradox.

2.1 What is Most Obvious is Most Hidden

In the rush to leverage the opportunities, or being overly enthusiastic about the potential of a new technology, firms ignore systematically assessing the true value, or even implementing the necessary processes for delivering that value. Firms are often so engrossed in the technological marvels, that they lose sight of what is it really for and how it truly makes a difference in a specific business context for customer value addition. The benefit becomes too obvious to even conceive. This is similar to when watching a movie at home: we tend to lose sight of the screen and the surrounding environment as we get engrossed in the movie. This is a natural process, as otherwise, the consciousness about the background or the medium would be utterly distracting for us to focus. But that does takes us away from the obvious reality of where we are for the moment [15]. The implications of this phenomenon in context of digitalization could be associated with what Linde et al., [4] identified as the common digital traps leading to failures, namely, "pushing out a digital business model without understanding customer value" or value proposition, "promising additional gains without understanding the value delivery process", or "getting sold on the digital opportunity without understanding the profit formula".

2.2 Simplification Complicates

Organizational problems can be "messy" and "ill-structured" and to address them, the top management often tend to focus too much on big data and digitalization initiatives [16, 17]. Nevertheless, purely quantitative analysis could be ineffective [18]. Following the previous discussions on the paradox of obvious, overreliance on such quantitative approaches at the strategic level attenuates managers' attention to the basic questions concerning which problems to address, kind of information needed and how it could be interpreted. Nell et al., [19] refers to such issues also as digital traps, and observes "easy access to seemingly hard, concrete, and particularly insightful data, especially if it seems comparable across divisions or over time, can tempt managers to oversimplify complex problems and discount experience and other sources of knowledge." For effective implementation, firms require to build the analytics culture or capabilities [20] and be conscious of the data driven decision making biases [21, 22]. The paradox of complexity suggests simplification complicates [15]. Moreover, technology transfer often starts with decontextualizing or abstracting the useful aspects and recontextualizing it to the new settings. The recontextualizing process may not be completely successful and can lead to unexpected problems. Context becomes highly important for technology transfer. Blindness towards it could lead to complications of different forms, from social, to environmental, to economic.

2.3 Outcomes of Path Dependence are Compared with Options of the Past

Like a tapering process, path dependence leads to decisions taken at one stage of development in a firm narrowing the scope of actions in the next stage and so on. The firm eventually get trapped in a predetermined path towards a specific trajectory of growth and loses its ability to change [23, 24]. The outcomes of the path dependent sequences contradict the predictions of prevailing theoretical framework for implementation only with respect to options that were available in the past and not those presently available. This leads to what is referred to as the paradox of path dependence. For example, when a firm realizes the inefficiencies of the path or framework of technology adoption, technology reversal being an extremely costly affair, may not be an alternative option at all.

2.4   Focusing Relentlessly on Efficiency Hinders Growth

Firms today often approach digitalization from the capabilities lens and benefit majorly from operational or cost efficiency through elimination of waste or achieving excellence in individual functions [1, 25]. The efficiency path affects relative cost positions and is an important source of profitability and competitive advantage. Further, it is easier to measure returns on the investments for cost efficiency and operational excellence [1]. This path typically adopts a holistic or systems view that requires all the individual elements of the system to be aligned and stresses on internal stability over time to achieve efficiency. This helps in maintaining homogeneity across similar systems. Nonetheless, "relentless pursuit of efficiency in operational tasks can drive out the capacity to change" [26].

3   Breaking Out of the Paradox

3.1   Dynamic Capability Approach to Digitalization

Largely, a holistic systems view is adopted to develop digital (zero order) capabilities that focusses on operational efficiency (or simplification) [1]. The systems approach with its biological orientation of *homeostatic equilibrium* of living organisms and its reactive stance seeking to remain aligned with the "survival requirements of the mega-system in which it is embedded" [26], matches only with the path dependent vision of firms. This approach to digitalization does not recognize the significance of bottom-up innovation. Dynamic capabilities aim for the systemic change to start from within [26] and realizes the relevance of bottom-up innovation, especially for non-evolutionary knowledge creation or breakthrough (or proactive entrepreneurial actions towards new designs). Contrary to systems approach, it does not advocate equifinality, the idea that different systems can follow different combinations of paths and conditions to reach an identical outcome [26]. The notion of "Equifinality" is similar to the ideas of "one best way" or "standardized work" that are traditionally considered central to Lean thinking [27]; this perspective however, limits the idea of lean only to *efficiency* enhancement while ignoring its potential for *growth*. Dynamic capabilities view, by contrast, emphasize on heterogeneity in outcomes to support competitive differentiation [26]. Through this approach, digitalization typically focusses on outmaneuvering competitors through a combination of technologies and strategy. In this current era of Schumpeterian competition, uncertainties, and disruptions, manufacturers must comprehensively take control of their digitalization efforts to embrace new growth opportunities for competitiveness and value creation [1] as well as for maintaining evolutionary fitness (or the ability to respond to opportunities) over time [26, 28].

Given this background and upon synthesizing the various aligned definitions (refer to [8, 13, 29, 30], and Gartner report[2]) we reconceptualize digitalization as: a dynamic

---

[2] https://www.gartner.com/en/information-technology/glossary/digitalization.

capability (or a meta-process[3]) that evolves over time and may vary upon purposes, developed to exploit digital opportunities for revenue growth by continuously improving on and finding new sources for operational efficiencies or by changing, reinventing or creating new business models to generate new revenue sources and value producing opportunities. Digitalization when approached from dynamic capabilities perspective, combines both the evolutionary and the entrepreneurial strands necessary for revenue growth.

3.2   Combining Lean and Digitalization

The discussions concerning the paradoxical characteristics in the preceding section provides a set of plausible explanations for the digitalization paradox. For example, missing the obvious, or in other words lack of due diligence on the value propositions affects the decisions on how digitalization is to be approached. We posit that following lean principles in digitalization initiatives can significantly address these identified concerns or paradoxes. Combined adoption of Digitalization and Lean is found to have positive impact on performance [31]. In fact, Lorenz et al. [32] found that "lean is needed as a foundation for successful digitalization" and the two concepts are "complementary, not contradictory". Digitalization and Lean ideas could be combined through two approaches: the "*Digital Lean*" or the "*Lean Digital*".

Digital Lean. Digital Lean approach in the recent years have started gaining attention of the academics and practitioners [7, 31–33]. By Digital Lean we refer to digitalization for problem solving and continuous improvement, with a mindset of customer value enhancement, and in turn, revenue growth. Here, the goal is to do more with less, that is, to explore growth opportunities following an evolutionary learning path of reducing wastes and inefficiencies. Contrary to the traditional idea of lean as a path-dependent way of looking only inwards for waste elimination and efficiency, the digital lean growth path, following the dynamic capabilities view of digitalization, promotes an outward approach of continuous improvement through discovery and learning. The organizations today should not only be able to adapt to changing business environment, but also try to shape it. Moreover, for complex systems it is not easy to infer the characteristics of the whole; digitalization or the systemic change should instead start from within considering the interactions of the individual elements with a focus on value (and that becomes the basis of heterogeneity across firms) [26, 34]. To remain competitive, manufacturing firms today must consider breaking out from the path dependent "efficiency trap" to embrace the growth or transformation opportunities offered by digitalization [1].

Lean Digital. The Lean Digital path is distinct from the digital lean path that follows a lean-first approach to digitalization, in which digitalization facilitates lean implementation for continuous improvement. The lean digital path instead follows the

---

[3] A meta-process is "that orchestrates a number of processes, best practices or competencies to manage comprehensively and systemically, something that is strategically imperative, including the strategy development and execution process itself" [25].

digitalization-first approach but in a lean way, that is, avoiding digitalized wastes. By digitalized wastes [32], we refer to components of a digital initiative that does not directly or indirectly create value for customers or other stakeholders of the business. Here the goal is to do new for more, that is, creating new revenue sources through new value propositions, business models, services, or products, but with a mindset of avoiding digitalized wastes (or in other words, to pursue purposeful digitalization). Lean digital approach follows a non-evolutionary path that supports breakthrough innovation or proactive entrepreneurial actions that embrace new growth opportunities.

Synthesis. We develop a two-by-two framework (see Fig. 1) to categorize firms based on the Lean and Digitalization initiatives and the paths they may follow and show how there exists two growth paths. In this paper, we primarily focus on Quadrant 3 (Q3a and Q3b). Q3a represents the Lean Digital zone, while Q3b represents the Digital Lean zone.

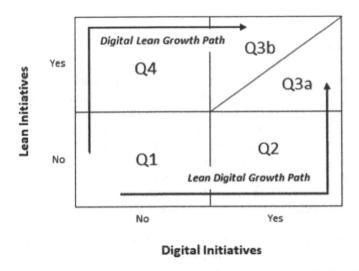

Fig. 1. Two-by-two framework for understanding lean and digital interactions

Given the above discussions on the digital lean and lean digital approaches, we observe several contradictions between them. The recent literature as we have described earlier, have found lean and digitalization to be "complementary, not contradictory". While we largely agree to this conclusion, we argue that the path to complementarity is distinct for the lean digital and the digital lean approaches and summarize them in Table 1.

Table 1. Digital lean versus lean digital

Digital-lean	Lean-digital
• Lean-first approach to digitalization (Digitalization for Lean)	• Digitalization-first approach but in a lean way (Lean for Digitalization)
• A problem solving/improvement initiative	• A breakthrough innovation initiative
• Goal is to explore opportunities to reduce waste (or Economizing) with a focus on revenue growth	• Goal is to explore opportunities for creating new revenue sources by creating new business models, services, or products, using technology and data
• For creating needed value for the existing customers	• For creating new/additional value or for targeting a new customer segment altogether
• Doing more with less	• Doing new for more
• Pursue an evolutionary learning and discovery path	• Pursue non-evolutionary path supporting entrepreneurial actions for breakthrough

4   Conclusions

This research concludes that for revenue growth, or to address the phenomenon of digitalization paradox, a dynamic capabilities approach to digitalization is more suited than top-down holistic approach. Further, we identify and present four paradoxical characteristics that also affects revenue growth through digitalization. A strategic approach to purposeful digitalization or reducing digitalized wastes showed to be vital to address these concerns. Combining lean approaches with digitalization thus, emerged as the key solution. The two strategies of combining lean and digitalization discussed in the paper, namely, digital lean and lean digital, are suited for different outcomes and expectations and hence firms should be conscious about which strategy to choose.

Further research is needed to validate and refine the framework proposed in this study, as well as to uncover any inconsistency or exception. Such research could adopt the case study method comparing and contrasting between (a) practicing lean firms that underwent digitalization of their lean processes or activities, and (b) firms that were not practicing lean, and underwent digitalization. Another plausible extension to this study is to empirically identify the more contextually embedded (and possibly paradoxical) tensions emerging of the specific digital lean initiatives.

Acknowledgements. The authors would like to acknowledge the support of the Research Council of Norway and the Norwegian research project Lean Digital.

References

1. Björkdahl, J.: Strategies for digitalization in manufacturing firms. Calif. Manage. Rev. **62**(4), 17–36 (2020)

2. Tabrizi, B., Lam, E., Girard, K., Irvin, V.: Digital transformation is not about technology. Harv. Bus. Rev. **13**(March), 1–6 (2019)
3. Gebauer, H., Fleisch, E., Lamprecht, C., Wortmann, F.: Growth paths for overcoming the digitalization paradox. Bus. Horiz. **63**(3), 313–323 (2020)
4. Linde, L., Sjödin, D., Parida, V., Gebauer, H.: Evaluation of digital business model opportunities: a framework for avoiding digitalization traps. Res. Technol. Manag. **64**(1), 43–53 (2020)
5. Rossini, M., Cifone, F.D., Kassem, B., Costa, F., Portioli-Staudacher, A.: Being lean: how to shape digital transformation in the manufacturing sector. J. Manuf. Technol. Manag. **32**(9), 239–259 (2021)
6. Zangiacomi, A., Pessot, E., Fornasiero, R., Bertetti, M., Sacco, M.: Moving towards digitalization: a multiple case study in manufacturing. Prod. Plann. Control **31**(2–3), 143–157 (2020)
7. Buer, S.-V., Semini, M., Strandhagen, J.O., Sgarbossa, F.: The complementary effect of lean manufacturing and digitalisation on operational performance. Int. J. Prod. Res. **59**(7), 1976–1992 (2021)
8. Rachinger, M., Rauter, R., Müller, C., Vorraber, W., Schirgi, E.: Digitalization and its influence on business model innovation. J. Manuf. Technol. Manag. **30**(8), 1143–1160 (2019)
9. Bogner, E., Voelklein, T., Schroedel, O., Franke, J.: Study based analysis on the current digitalization degree in the manufacturing industry in Germany. Proc. Cirp **57**, 14–19 (2016)
10. Bienhaus, F., Haddud, A.: Procurement 4.0: factors influencing the digitisation of procurement and supply chains. Bus. Process Manag. J. **24**(4), 965–984 (2018)
11. Amit, R., Schoemaker, P.J.: Strategic assets and organizational rent. Strateg. Manag. J. **14**(1), 33–46 (1993)
12. Barreto, I.: Dynamic capabilities: a review of past research and an agenda for the future. J. Manag. **36**(1), 256–280 (2010)
13. Kohtamäki, M., Parida, V., Patel, P.C., Gebauer, H.: The relationship between digitalization and servitization: the role of servitization in capturing the financial potential of digitalization. Technol. Forecast. Soc. Change **151**, 119804 (2020)
14. Schad, J., Lewis, M.W., Raisch, S., Smith, W.K.: Paradox research in management science: looking back to move forward. Acad. Manag. Ann. **10**(1), 5–64 (2016)
15. Feenberg, A.: Ten paradoxes of technology. Tech.: Res. Philos. Technol. **14**(1), 3–15 (2010)
16. Wilson, H., Daugherty, P.: Small data can play a big role in AI. Harv. Bus. Rev. (17 Feb 2020). https://hbr.org/2020/02/small-data-can-play-a-big-role-in-ai
17. Redman, T., Hoerl, R.: Most analytics projects don't require much data. Harv. Bus. Rev. (3 Oct 2019). https://hbr.org/2019/10/most-analytics-projects-dont-require-much-data
18. Baer, M., Dirks, K.T., Nickerson, J.A.: Microfoundations of strategic problem formulation. Strateg. Manag. J. **34**(2), 197–214 (2013)
19. Nell, P.C., Foss, N.J., Klein, P.G., Schmitt, J.: Avoiding digitalization traps: tools for top managers. Bus. Horiz. **64**(2), 163–169 (2021)
20. Schlegel, A., Birkel, H.S., Hartmann, E.: Enabling integrated business planning through big data analytics: a case study on sales and operations planning. Int. J. Phys. Distrib. Logist. Manag. **51**(6), 607–633 (2021)
21. Daniel, K.: Thinking, fast and slow (2017)
22. Tversky, A., Kahneman, D.: Judgment under uncertainty: heuristics and biases. Science **185**(4157), 1124–1131 (1974)
23. Schreyögg, G., Sydow, J.: Organizational path dependence: a process view. Organ. Stud. **32**(3), 321–335 (2011)

24. Sydow, J., Schreyögg, G., Koch, J.: Organizational path dependence: opening the black box. Acad. Manag. Rev. **34**(4), 689–709 (2009)
25. Feiler, P., Teece, D.: Case study, dynamic capabilities and upstream strategy: supermajor EXP. Energ. Strat. Rev. **3**, 14–20 (2014)
26. Teece, D.J.: Dynamic capabilities as (workable) management systems theory. J. Manag. Organ. **24**(3), 359–368 (2018)
27. Hopp, W.J., Spearman, M.S.: The lenses of lean: visioning the science and practice of efficiency. J. Oper. Manag. **67**(5), 610–626 (2021)
28. Teece, D.J., Pisano, G., Shuen, A.: Dynamic capabilities and strategic management. Strateg. Manag. J. **18**(7), 509–533 (1997)
29. Gölzer, P., Fritzsche, A.: Data-driven operations management: organisational implications of the digital transformation in industrial practice. Prod. Plann. Control **28**(16), 1332–1343 (2017)
30. Caiado, R.G.G., Scavarda, L.F., Gavião, L.O., Ivson, P., de Mattos Nascimento, D.L., Garza-Reyes, J.A.: A fuzzy rule-based industry 4.0 maturity model for operations and supply chain management. Int. J. Prod. Econ. **231**, 107883 (2021)
31. Cifone, F.D., Hoberg, K., Holweg, M., Staudacher, A.P.: 'Lean 4.0': how digital technologies support lean practices? In: EurOMA Conference 2019, pp. 1–10 (2019)
32. Lorenz, R., Buess, P., Macuvele, J., Friedli, T., Netland, T.H.: Lean and digitalization—contradictions or complements? In: Ameri, F., Stecke, K.E., von Cieminski, G., Kiritsis, D. (eds.) APMS 2019. IAICT, vol. 566, pp. 77–84. Springer, Cham (2019). https://doi.org/10.1007/978-3-030-30000-5_10
33. Solheim, A.B., Powell, D.J.: A learning roadmap for digital lean manufacturing. In: Lalic, B., Majstorovic, V., Marjanovic, U., von Cieminski, G., Romero, D. (eds.) APMS 2020. IAICT, vol. 592, pp. 417–424. Springer, Cham (2020). https://doi.org/10.1007/978-3-030-57997-5_48
34. Nonaka, I.: A dynamic theory of organizational knowledge creation. Organ. Sci. **5**(1), 14–37 (1994)

Lean 4.0

Mapping the Terrain for Lean Six Sigma 4.0

Jiju Antony[1], Olivia McDermott[2(✉)], Daryl John Powell[3,4],
and Michael Sony[5]

[1] Khalifa University, Abu Dhabi, UAE
[2] National University of Ireland, Galway, Ireland
`olivia.mcdermott@nuigalway.ie`
[3] SINTEF Manufacturing AS, Horten, Norway
[4] Norwegian University of Science and Technology, Trondheim, Norway
[5] Namibia University of Science and Technology, Windhoek, Namibia

Abstract. Lean Six Sigma is a powerful methodology that integrates the best of two distinct approaches to business excellence: Lean and Six Sigma. While Six Sigma focuses on the systematic reduction of variation within processes, the Lean approach aims for business growth through waste elimination and the continuous development of people. Given the advent of Industry 4.0, digitalization now presents the next frontier of industrial improvement. Lean Six Sigma can be integrated with Industry 4.0 to optimize process efficiency. This study is a systematic literature review on the integration of Industry 4.0 and Lean Six Sigma. The findings are that while this topic is an emerging area of study, there are benefits, motivations, critical success factors, and challenges to integrating Lean Six Sigma and Industry 4.0.

Keywords: Lean Six Sigma · Industry 4.0 · Continuous improvement

1 Introduction

The advent of the first Industrial Revolution transitioned production from craft production into mass production and from mass production into Lean production and Lean supply chains. Lean production is a philosophy that focusses on understanding the deeper underlying causes of waste to create more customer value. It has also been recognized as a methodology for operational excellence globally. Lean has been integrated with Six Sigma in recent years firstly by George in 2002 and subsequently adopted in its first evolution as Lean Six Sigma (LSS) by industry and academics (Lean Six Sigma 1.0). The integration has benefits with the combination resulting in both waste reduction and variation reduction [1].

In recent years Lean has been coined with "Green", and this has progressed Lean thinking into the 2nd iteration of its evolution (Lean Six Sigma 2.0). Historically, profitability and efficiency objectives have been the prevailing interest for organisations in deploying Lean Six Sigma. However, the move towards green operations has forced companies to seek alternatives to combine these with green objectives and initiatives [2, 3]. While some literature in relation to Lean and Green can be traced back to 1994 in which Davids wrote about Lean and Green [4], the Lean Six Sigma 2.0 "Green"

© IFIP International Federation for Information Processing 2021
Published by Springer Nature Switzerland AG 2021
D. J. Powell et al. (Eds.): ELEC 2021, IFIP AICT 610, pp. 193–204, 2021.
https://doi.org/10.1007/978-3-030-92934-3_20

evolution did not really take off in practice until after 2001 [2, 5–7]. LSS can contribute toward greater environmental sustainability [7].

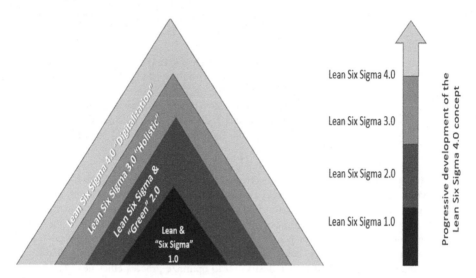

Fig. 1. Progressive development of the Lean Six Sigma 4.0 concept "sand cone" - authors own derivation.

The *"holistic"* approach of Lean Six Sigma was first touted by Snee and Hoerl as early as 2005–2007 [8–11] in various journal and conferences. However their new paradigm for Lean Six Sigma as a holistic approach was first put forward in their book *"Leading Holistic Improvement with Lean Six Sigma 2.0"* in 2018 [12]. Snee and Hoerl called the holistic approach the 2nd evolution of LSS in their book. The authors of this research paper have put forward that as the holistic approach has come after the LSS Green iteration and evolution of Lean Six Signa; the holistic approach is therefore LSS 3.0. The definition of the LSS 3.0 approach put forward by this research is based on Snee and Hoerls definition is *"An Improvement system that can successfully create and sustain significant improvements of any type, in any culture and in any business"* [12]. Further the holistic approach allows for 1. the integration of a wide set of methods so that the most appropriate methods can be adapted to tackle a problem and 2. a deployment system like LSS to create and sustain improvements [8].

However, as traditional manufacturing evolves into digitalization with the implementation and advent of Industry 4.0 (I4.0) technologies, Lean has become digitally enabled [13]. Thus, there is a new evolution of Lean 4.0 - digitalization. The advance of new Industry 4.0 technologies such as the Internet of Things (IoT), big data and data analytics, additive manufacturing, 3D printing, advanced robotics, augmented and virtual reality, cloud computing, simulation, machine learning and artificial intelligence are increasing production and service complexity [14, 15]. Both paradigms of LSS and I4.0 are promising to solve future challenges in manufacturing. The integration of I4.0 technologies into the guidelines for designing a lean value stream raises a distinct approach that benefits from the simplicity and efficiency of Lean Production with ease and agility of the technologies typical of the Fourth Industrial Revolution [4].

The Sand Cone model (Fig. 1) which describes the cumulative enhancement of capabilities as described by Ferdows and De Meyer in 1990 demonstrates the four Lean stages or evolutions [16]. Each hierarchical evolution enhances and leads to a cumulate evolution of Lean Six Sigma capabilities from Lean Six Sigma 1.0 right through to Lean Six Sigma 4.0. However, the integrated effect of Industry 4.0 technology and Lean manufacturing practice on SOP has not been empirically investigated [17, 18].

Hence, the question arises if and how these developments can possibly support each other. Thus, this paper aims to contribute to this research area.

Thus, a series of research questions arise:

- How does LSS integrate with I4.0?
- What are the motivations for integrating LSS and I4?
- What are the benefits of integrating LSS & I4.0?
- What are the critical success factors (CSF's) to integrating LSS and I4.0?

2 Methodology – Systematic Literature Review

A systematic literature review (SLR) was utilized. The authors searched systematically for articles published between 2011 and 2021, using the following full academic databases Web of Science and Scopus. The search strategy followed Tranfield et al.'s (2003) approach, which seeks to create a reliable knowledge stock by synthesizing the relevant body of literature [19]. Firstly, the emphasis on the systematic process of literature search, extraction and synthesis are higher in SQLRs than in other forms of review, making the work more scientific and replicable [19, 20]. The following search string was applied to search all the databases mentioned above: "Lean" AND "Industry 4.0", "Six Sigma" AND "Industry 4.0" and "Lean Six Sigma" AND "Industry 4.0". Table 1 provides a detailed listing of the inclusion/exclusion criteria. The references of the selected studies were manually checked to identify additional relevant studies that were missed in the database search. Grey literature (conference papers, magazine-related articles, workshops, books, editorials, prefaces) were excluded.

Table 1. Inclusion & Exclusion criteria for the SLR

Inclusion criteria	Exclusion criteria
Academic peer-reviewed journal articles books, magazine-related articles, etc.) related to Lean, Six Sigma and/or LSS and I4.0	Grey literature (conference proceedings, dissertations, text
Articles published in high quality journals ABS Ranked 3 and 4*	Articles published in languages other than English
Articles published from 2011 to 2021	Articles published before 2011 as the term Industry 4.0 was only coined in 2011
	Articles published in non-refereed journals

A flowchart was utilized to map the SLR steps, as outlined in Fig. 2. The initial search identified 2,357 articles. Duplicates were firstly removed, and the full text was retained if the abstracts stated that the study was related to Lean, Six Sigma and LSS and its application in an Industry 4.0 context. The three authors acted as reviewers and independently assessed the inclusion eligibility of the retrieved studies based on the search criteria [21]. Inclusion agreement was solved by discussions and consensus among reviewers.

Additionally, studies that were not published in peer-reviewed journals or were under 3 or 4 in the ABS journal ranking of 2018 [22] were also excluded. This process yielded 355 studies for final inclusion at this stage of the review.

The data collation was managed utilizing Zotero and MS Excel spreadsheets to record information concerning the articles selected; the authors independently reviewed each paper and coded them using a meta-framework. Once the final articles were extracted and recorded, coding was completed in order for errors to be minimized. The analysis proceeded under various characteristics in response to the research questions, including the year of publication, countries of origin, authors, journals, research methods, benefits of LSS & I4 integration, motivations for LSS & I4 integration, challenges of LSS & I4 integration, and finally the CSFs for LSS & I4 integration. With the help of the proposed methodology, the authors filtered out 19 relevant research papers for in-depth exploratory analysis of LSS & I4 integration. The insights of these publications were summarized through reviewing patterns of publications and emerging themes.

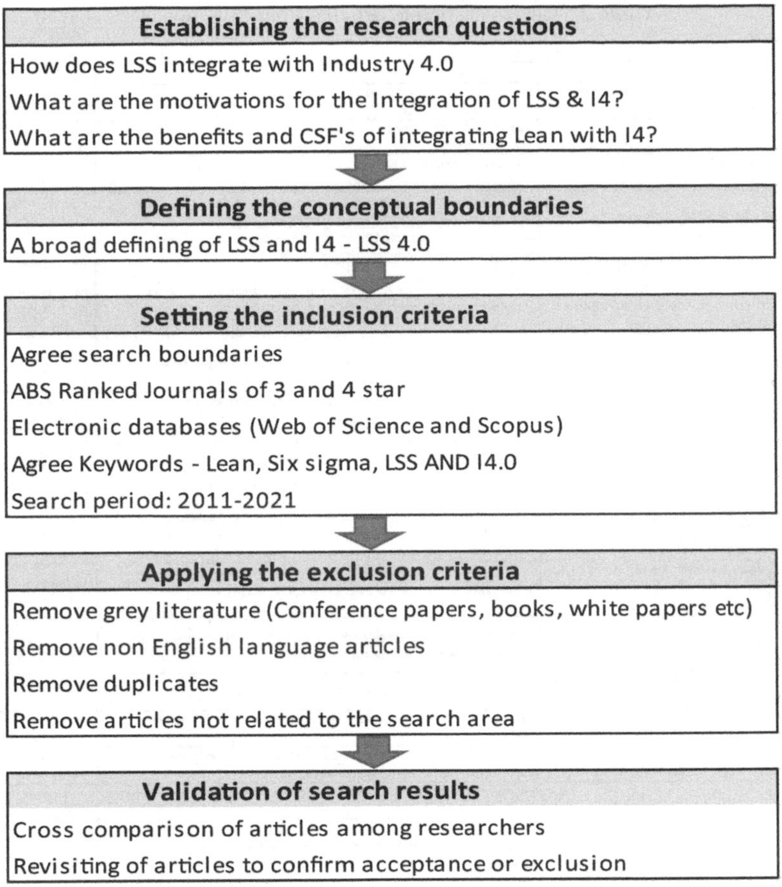

Establishing the research questions
How does LSS integrate with Industry 4.0
What are the motivations for the Integration of LSS & I4?
What are the benefits and CSF's of integrating Lean with I4?

Defining the conceptual boundaries
A broad defining of LSS and I4 - LSS 4.0

Setting the inclusion criteria
Agree search boundaries
ABS Ranked Journals of 3 and 4 star
Electronic databases (Web of Science and Scopus)
Agree Keywords - Lean, Six sigma, LSS AND I4.0
Search period: 2011-2021

Applying the exclusion criteria
Remove grey literature (Conference papers, books, white papers etc)
Remove non English language articles
Remove duplicates
Remove articles not related to the search area

Validation of search results
Cross comparison of articles among researchers
Revisiting of articles to confirm acceptance or exclusion

Fig. 2. A summary of evidence search and selection

3 Results

The final selected journals were analyzed by journal type and years of publication. Lean Six Sigma and I4.0 are still considered an emerging research area, and this fact was verified by the relatively small number of 19 final selected articles (Table 2).

Table 2. Final selection of articles from SLR review

Year	Authors	Titles	Journal
2020	Buer, S.-V., Semini, M., Strandhagen, J.O., Sgarbossa, F.⧫	The complementary effect of lean manufacturing and digitalisation on operational performance	International Journal Of Production Research
2018	Buer, S.-V., Strandhagen, J.O., Chan, F.T.S.	The link between industry 4.0 and lean manufacturing: Mapping current research and establishing a research agenda	International Journal of Production Research
2020	Ciano, M.P., Dallasega, P., Orzes, G., Rossi, T.	One-to-one relationships between Industry 4.0 technologies and Lean Production techniques: a multiple case study	International Journal of Production Research
2020	Chiarini, Belevdere, Grando	Industry 4.0 strategies and technological developments. An exploratory research from Italian manufacturing companies	Production Planning & Control
2020	Chiarini & Kumar	Lean Six Sigma and Industry 4.0 integration for Operational Excellence: evidence from Italian manufacturing companies	Production Planning & Control
2019	Felsberger, A., Qaiser, F.H., Choudhary, A., Reiner, G. Andreas Felsberger, Fahham Hasan Qaiser, Alok Choudhary & Gerald Reiner	The impact of Industry 4.0 on the reconciliation of dynamic capabilities: evidence from the European manufacturing industries	Production Planning & Control
2019	Laurie Hughes, Yogesh K. Dwivedi, Nripendra P. Rana, Michael D. Williams & Vishnupriya Raghavan	Perspectives on the future of manufacturing within the Industry 4.0 era	Production Planning & Control
2018	Hannola, L., Richter, A., Richter, S., Stocker, A.	Empowering production workers with digitally facilitated knowledge processes–a conceptual framework	International Journal Of Production Research
2020	Kamble, S., Gunasekaran, A., Dhone, N.C.	Industry 4.0 and lean manufacturing practices for sustainable organisational performance in Indian manufacturing companies	International Journal Of Production Research
2021	Giulio Marcucci, Sara Antomarioni, Filippo Emanuele Ciarapica & Maurizio Bevilacqua	The impact of Operations and IT-related Industry 4.0 key technologies on organizational resilience	Production Planning & Control
2021	Rossella Pozzi, Tommaso Rossi & Raffaele Secchi	Industry 4.0 technologies: critical success factors for implementation and improvements in manufacturing companies	Production Planning & Control
2017	Alexandre Moeuf, Robert Pellerin, Samir Lamouri, Simon Tamayo-Giraldo & Rodolphe Barbaray	The industrial management of SMEs in the era of Industry 4.0	International Journal of Production Research
2020	Núñez-Merino, M., Maqueira-Marin, J.M., Moyano-Fuentes, J., Martínez-Jurado, P.J.	Information and digital technologies of Industry 4.0 and Lean supply chain management: a systematic literature review	International Journal Of Production Research
2020	Rosin, F., Forget, P., Lamouri, S., Pellerin, R.	Impacts of Industry 4.0 technologies on Lean principles	International Journal of Production Research
2020	Sahu A.K., Padhy R.K., Dhir A.	EPEC 4.0: an Industry 4.0-supported lean production control concept for the semi-process industry	Production Planning & Control
2020	Tortorella, G.L., Pradhan, N., Macias de Anda, E., (...), Sawhney, R., Kumar, M.	Designing lean value streams in the fourth industrial revolution era: proposition of technology-integrated guidelines	International Journal Of Production Research
2018	Tortorella, G.L., Fettermann, D.	Implementation of Industry 4.0 and lean production in Brazilian manufacturing companies	International Journal Of Production Research
2019	Tortorella, Gl; Giglio, R; van Dun, DH	Industry 4.0 adoption as a moderator of the impact of lean production practices on operational performance improvement	International Journal Of Operations & Production Management
2017	Yin, Y., Stecke, K.E., Li, D.	The evolution of production systems from Industry 2.0 through Industry 4.0	International Journal Of Production Research

The majority of the final selection came from 3 main journals (Fig. 3) – the International Journal of Production Research (42%), Production Planning and Control (37%) and the International Journal of Operations & Production Management (5%). It was not surprising that LSS 4.0 is still very much an emerging area of research as several authors who have written about the topic have had similar findings [17, 23]. Also, article analysis by year of publication demonstrated that LSS 4.0 has emerged as a topic of researcher interest only from 2017 onwards.

Very few articles discussed the impact of Industry 4.0 on (1) 'soft' lean practices, (2) the facilitating effect of lean manufacturing on Industry 4.0 implementations,

(3) empirical studies on the performance implications of an Industry 4.0 and lean manufacturing integration, (4) the effect of environmental factors on the integration of Industry 4.0 and lean manufacturing, and (5) implementation framework for moving toward an Industry 4.0 and lean manufacturing integration [17, 23].

Some authors under the SLR study and had more than one publication on the theme of LSS & Industry 4.0 integration. These authors included Tortorella (Brazil), who had three articles, Buer (Norway), Strandhagan (Norway), Chiarini (Italy) and Kumar (UK), whom all had published two research articles.

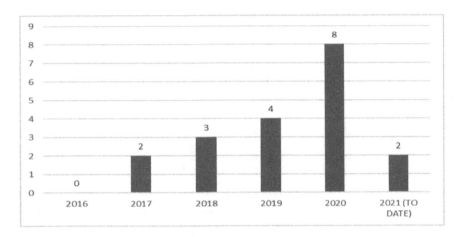

Fig. 3. Breakdown of publication by year in the LSS & I4 area

The articles selected were screened and themes noted under various categories and themes such as the benefits of LSS & I4 integration, the motivations for LSS & I4 integration, the challenges of LSS & I4 integration, and finally, the CSFs for LSS & I4 integration. These common themes were summarized in Table 3. All articles selected were unanimous in the support or the benefits of integrating Lean and Industry 4.0.

There were fewer articles that discussed the CSF's for and challenge to integrating LSS and Industry 4.0.

Table 3. Themes emerging around Lean Six Sigma 4.0 after SLR of the final selection of articles

Cited article	Benefits of intregrating Lean Six Sigma & Industry 4.0	Motivations for intregrating Lean Six Sigma & Industry 4.0	Challenges of intregrating Lean Six Sigma & Industry 4.0	Critical Success Factors for integrating Lean Six Sigma & Industry 4.0
Sven-Vegard Buer, Marco Semini, Jan Ola Strandhagen & Fabio Sgarbossa (2021)	X	X	X	X
Sven-Vegard Buer, Jan Ola Strandhagen & Felix T. S. Chan (2018)	X	x	X	X
Maria Pia Ciano, Patrick Dallasega, Guido Orzes & Tommaso Rossi (2021)	X			
Andrea Chiarini, Valeria Belvedere & Alberto Grando (2020)	X	X		
Andrea Chiarini & Maneesh Kumar (2020)	X	X		X
Andreas Felsberger, Fahham Hasan Qaiser, Alok Choudhary & Gerald Reiner (2020)	X	X		
Laurie Hughes, Yogesh K. Dwivedi, Nripendra P. Rana, Michael D. Williams & Vishnupriya Raghavan (2020)	X		X	
Lea Hannola, Alexander Richter, Shahper Richter & Alexander Stocker (2018)	X	X		
Sachin Kamble, Angappa Gunasekaran & Neelkanth C. Dhone (2020)	X	X	X	X
Giulio Marcucci, Sara Antomarioni, Filippo Emanuele Ciarapica & Maurizio Bevilacqua (2021)	X			
Rossella Pozzi, Tommaso Rossi & Raffaele Secchi (2021)	X			
Alexandre Moeuf, Robert Pellerin, Samir Lamouri, Simon Tamayo-Giraldo & Rodolphe Barbaray (2017)	X	X		
Miguel Núñez-Merino, Juan Manuel Maqueira-Marín, José Moyano-Fuentes & Pedro José Martínez-Jurado (2020)	X	X	x	X
Frédéric Rosin, Pascal Forget, Samir Lamouri & Robert Pellerin (2020)	X	X	X	X
Philipp Spenhoff, Johan C. (Hans) Wortmann & Marco Semini (2021)	X			
Guilherme Luz Tortorella, Ninad Pradhan, Enrique Macias de Anda, Samuel Trevino Martinez, Rupy Sawhney & Maneesh Kumar (2020)	X	X	X	
Guilherme Luz Tortorella & Diego Fettermann (2018)	X			
Tortorella, G.L., Giglio, R. and van Dun, D.H. (2019)	X	X	X	X
Yong Yin, Kathryn E. Stecke & Dongni Li (2018)	X	x		

4 Discussion

According to the literature reviewed as part of this SLR, there is a strong rationale for integrating LSS and I4.0 in terms of benefits for integration and motivations for integration. There are also challenges to integrating and critical success factors (CSF's)

of integrating LSS and I4.0. Lean practices are positively associated with Industry 4.0 technologies, and their concurrent implementation leads to larger performance improvements [14]. The integration of I4.0 technologies into the guidelines for designing a lean value stream raises a distinct approach that benefits from the simplicity and efficiency of Lean Production with the ease and agility of the technologies typical of the Fourth Industrial Revolution [12, 13]. I4.0 and Lean practices complement I4.0 technologies [12], and there is a synergistic effect between both to target operational excellence [12, 25].

Lean manufacturing tools can also be facilitators or even prerequisites for a move towards Industry 4.0. Industry 4.0's technologies are improving the implementation of Lean. Lean principles, depending on the technologies' capability levels. I4.0 technology can support strong integration of Just-in-time and Jidoka [26, 27]. While technologies from Industry 4.0 do not seem to cover the integrated nature of Lean principles overall, they can reinforce the efficiency of these principles. Lean practices tend to be more impactful with Industry 4.0 as it allows a better understanding of customers' demands and accelerates information sharing processes. Real-time integration with minimal waste generation is an advantage [18]. Buer et al. especially highlighted that total productive maintenance (TPM), Kanban, production smoothing, autonomation, waste elimination, and Kanban and Andon could be improved by I4.0 technology [24]. However, I4.0 demonstrates little or no support for waste reduction via people and teamwork. Softer Lean principles such as these, which are more focused on communication between employees and creativity, do not seem to be subject to improvements by Industry 4.0 technologies at this time [26]. In terms of Industry 4.0 alignment with the Lean Toyota production system, flow line, job shop, cell, flexible manufacturing system, and seru – I4.0 can be integrated and applied to the different production types to match different demand dimensions over time [14]. Big data collection and evaluation for Lean production and TPS should become easier because of IoT. KPI data generated from I4.0 will enable Lean [28].

There are challenges to integrating Lean and I4.0. While there are synergies between the two, I4.0 technologies may not contribute to improved operational excellence if implemented as a standalone application in the absence of Lean practice, according to Kamble et al. in his 2020 study [17]. The study findings indicate that while I4.0 technology has a direct and positive effect on operational performance, this effect is magnified in the presence of Lean manufacturing practices as a mediating variable. In summary, the successful implementation of Lean practices are crucial for overcoming I4.0 technology implementation barriers. Industry 4.0 and Lean have convergent and divergent characteristics [15, 24]. The synergistic and complementary nature of LSS and I4.0 might change according to the context in which the deployment is taking place - for example, a developed vs developing economy [18]. Integration and implementation of these concepts can therefore be slower. Lean is focused on developing human integration, which is crucial for overcoming the I4.0 technology implementation barriers. Thus the successful implementation of Lean enables organizations to initiate the implementation process for Industry 4.0 technology [17].

Due to Industry 4.0 and Lean's convergent and divergent characteristics. Lean entails socio-cultural changes that are stimulated daily through fast and simple workfloor experimentations, which may conflict with the high levels of capital expenditure and technological expertise demanded by Industry 4.0 [18].

Industry 4.0 combines many technologies such as sensors, automation, robots, and cyber-physical systems. These technologies may require changing the operational procedures of a production system and Lean SOP's, and standardized work. How a production system adapts to an environment with new technologies and customer demand dimensions has to be investigated [27].

The CSF's to integrating Lean and I4.0 are many. There is a clear need to pursue the deployment of Lean management while improving certain Lean principles using Industry 4.0 technologies [26]. The roadmap towards Industry 4.0 is complex and multifaceted, as manufacturers seek to transition towards new and emerging technologies whilst retaining operational effectiveness and a sustainability focus [28].

Lean practices help to install organizational habits and mindsets that favor systemic process improvements. While implementing Lean, companies must intelligently weigh the trade-offs when introducing novel technologies instead of simple standard operating procedures [23]. While I4.0 will enhance knowledge management, workers on the shop floor need to be highly skilled in decision-making as the separation of dispositive and executive work diminishes with the implementation of I4.0 in a Lean environment [29]. The integration of Lean and I4.0 needs reinvented mapping tools and implies a horizontal integration and a vertical, end-to-end integration within an organization [30].

The articles selected were more supportive and discursive in terms of the motivation of Lean and I4.0 integration. It is clear that there is a dearth of literature around Lean and I4.0 CSF's and challenges as the area is still an emerging and unknown one.

5 Conclusion

In light of the upcoming fourth industrial revolution, the researchers find that Lean is not obsolete but rather is more important than ever to reap the benefits from emerging Industry 4.0 technologies and integrate with them for improved operational performance. A limitation of the study is the lack of research in this area as it is an emerging area. Also, within the SLR conference papers were excluded and may perhaps have been a source of further research around Lean and Industry 4.0 integration. Future research opportunities include more longitudinal studies on organizations implementing Lean, LSS and Industry 4.0 type technologies to ascertain and benchmark the benefits, CSF's and other learnings. Qualitative and quantitative studies with professionals working on Lean and organizational digitalization programs would be an opportunity to leverage further learnings around this new evolution of Lean 4.0.

References

1. George, M.L.: Lean Six Sigma: Combining Six Sigma Quality with Lean Production Speed. McGraw-Hill, New York (2002)
2. Garza-Reyes, J.A.: Lean and green – a systematic review of the state of the art literature. J. Clean. Prod. **102**, 18–29 (2015). https://doi.org/10.1016/j.jclepro.2015.04.064

3. Cherrafi, A., Elfezazi, S., Govindan, K., Garza-Reyes, J.A., Benhida, K., Mokhlis, A.: A framework for the integration of green and Lean Six Sigma for superior sustainability performance. Int. J. Prod. Res. **55**, 4481–4515 (2017). https://doi.org/10.1080/00207543. 2016.1266406
4. Davids, M.: Environmental strategies: lean and green. J. Bus. Strateg. **15**, 18–20 (1994). https://doi.org/10.1108/eb039621
5. Garza-Reyes, J.A., Winck Jacques, G., Lim, M.K., Kumar, V., Rocha-Lona, L.: Lean and green – synergies, differences, limitations, and the need for six sigma. In: Grabot, B., Vallespir, B., Gomes, S., Bouras, A., Kiritsis, D. (eds.) APMS 2014. IAICT, vol. 439, pp. 71–81. Springer, Heidelberg (2014). https://doi.org/10.1007/978-3-662-44736-9_9
6. Yang, M.G., Hong, P., Modi, S.B.: Impact of lean manufacturing and environmental management on business performance: an empirical study of manufacturing firms. Int. J. Prod. Econ. **129**, 251–261 (2011). https://doi.org/10.1016/j.ijpe.2010.10.017
7. Powell, D., Lundeby, S., Chabada, L., Dreyer, H.: Lean Six Sigma and environmental sustainability: the case of a Norwegian dairy producer. Int. J. Lean Six Sigma **8**, 53–64 (2017). https://doi.org/10.1108/IJLSS-06-2015-0024
8. Snee, R.D.: Lean Six Sigma – getting better all the time. Int. J. Lean Six Sigma **1**, 9–29 (2010). https://doi.org/10.1108/20401461011033130
9. Snee, R., Hoerl, R.: Leading Six Sigma: A Step-by-Step Guide Based on Experience with GE and Other Six Sigma Companies. FT Press, Upper Saddle River (2002)
10. Snee, R.D: Making another world: W. Edward Deming and a holistic approach to performance improvement. In: Joint Statistical Meetings, Seattle (2007)
11. Hoerl, R.W., Snee, R.: Statistical thinking and methods in quality improvement: a look to the future. Qual. Eng. **22**, 119–129 (2010). https://doi.org/10.1080/08982112.2010.481485
12. Snee, R., Hoerl, R.: Leading holistic improvement with Lean Six Sigma 2.0: Ron Snee: 9780134288888 (2018)
13. Calabrese, A., Dora, M., Ghiron, N.L., Tiburzi, L.: Industry's 4.0 transformation process: how to start, where to aim, what to be aware of. Prod. Plann. Control 1–21 (2020). https://doi.org/10.1080/09537287.2020.1830315
14. Chiarini, A.: Industry 4.0, quality management and TQM world. A systematic literature review and a proposed agenda for further research. The TQM J. **32**, 603–616 (2020). https://doi.org/10.1108/TQM-04-2020-0082
15. Chiarini, A., Belvedere, V., Grando, A.: Industry 4.0 strategies and technological developments. An exploratory research from Italian manufacturing companies. Prod. Plann. Control **31**, 1385–1398 (2020). https://doi.org/10.1080/09537287.2019.1710304
16. Ferdows, K., De Meyer, A.: Lasting improvements in manufacturing performance: in search of a new theory. J. Oper. Manag. **9**, 168–184 (1990). https://doi.org/10.1016/0272-6963(90) 90094-T
17. Kamble, S., Gunasekaran, A., Dhone, N.C.: Industry 4.0 and lean manufacturing practices for sustainable organisational performance in Indian manufacturing companies. Int. J. Prod. Res. **58**, 1319–1337 (2020). https://doi.org/10.1080/00207543.2019.1630772
18. Tortorella, G.L., Giglio, R., van Dun, D.H.: Industry 4.0 adoption as a moderator of the impact of lean production practices on operational performance improvement. Int. J. Oper. Prod. Manag. **39**, 860–886 (2019). https://doi.org/10.1108/IJOPM-01-2019-0005
19. Tranfield, D., Denyer, D., Smart, P.: Towards a methodology for developing evidence-informed management knowledge by means of systematic review. Br. J. Manag. **14**, 207–222 (2003). https://doi.org/10.1111/1467-8551.00375
20. Yang, E.C.L., Khoo-Lattimore, C., Arcodia, C.: A systematic literature review of risk and gender research in tourism. Tour. Manage. **58**, 89–100 (2017)

21. Parameswaran, U.D., Ozawa-Kirk, J.L., Latendresse, G.: To live (code) or to not: a new method for coding in qualitative research. Qual. Soc. Work. **19**, 630–644 (2020). https://doi.org/10.1177/1473325019840394

22. Academic Journal Guide 2018. https://charteredabs.org/academic-journal-guide-2018/. Accessed 11 June 2021

23. Tortorella, G.L., Pradhan, N., de Anda, E.M., Martinez, S.T., Sawhney, R., Kumar, M.: Designing lean value streams in the fourth industrial revolution era: proposition of technology-integrated guidelines. Int. J. Prod. Res. **58**, 5020–5033 (2020)

24. Buer, S.-V., Semini, M., Strandhagen, J.O., Sgarbossa, F.: The complementary effect of lean manufacturing and digitalisation on operational performance. Int. J. Prod. Res. **59**, 1976–1992 (2021). https://doi.org/10.1080/00207543.2020.1790684

25. Calabrese, A., Ghiron, N.L., Tiburzi, L.: 'Evolutions' and 'revolutions' in manufacturers' implementation of industry 4.0: a literature review, a multiple case study, and a conceptual framework. Prod. Plann. Control **32**, 213–227 (2021). https://doi.org/10.1080/09537287.2020.1719715

26. Rosin, F., Forget, P., Lamouri, S., Pellerin, R.: Impacts of industry 4.0 technologies on Lean principles. Int. J. Prod. Res. **58**, 1644–1661 (2020). https://doi.org/10.1080/00207543.2019.1672902

27. Yin, Y., Stecke, K.E., Li, D.: The evolution of production systems from industry 2.0 through industry 4.0. Int. J. Prod. Res. **56**, 848–861 (2018). https://doi.org/10.1080/00207543.2017.1403664

28. Hughes, L., Dwivedi, Y.K., Rana, N.P., Williams, M.D., Raghavan, V.: Perspectives on the future of manufacturing within the industry 4.0 era. Null 1–21 (2020). https://doi.org/10.1080/09537287.2020.1810762

29. Hannola, L., Richter, A., Richter, S., Stocker, A.: Empowering production workers with digitally facilitated knowledge processes – a conceptual framework. Int. J. Prod. Res. **56**, 4729–4743 (2018). https://doi.org/10.1080/00207543.2018.1445877

30. Chiarini, A.: Corporate Social Responsibility strategies using the TQM - hoshin kanri as an alternative system to the balanced scorecard. TQM J. **28** (2016). https://doi.org/10.1108/TQM-03-2014-0035

Learning Through Action: On the Use of Logistics4.0 Lab as Learning Developer

Mirco Peron[✉], Erlend Alfnes, and Fabio Sgarbossa

Department of Mechanical and Industrial Engineering,
Norwegian University of Science and Technology, Richard Birkelands vei 2B,
7034 Trondheim, Norway
mirco.peron@ntnu.no

Abstract. The concept of learning factory is taking more and more hold as teaching method, especially after the advent of Industry 4.0 (I4.0). Learning factories have in fact proved to be effective in developing knowledge and skills necessary for students to master the potentialities of adopting new I4.0 technologies in several aspects of production and logistics systems. Driven by these potentialities and aiming to create and spread new knowledge on the use of I4.0 technologies in production and logistics systems, the Production Management group at NTNU, with the support of the Department of Mechanical and Industrial Engineering, has established in 2018 the Logistics 4.0 (Log4.0) Lab. Since then, the Log4.0 Lab has been used to develop state-of-the-art research investigating the impact of I4.0 technologies on production and logistics systems and to transfer the developed knowledge to the students to render them ready for and attractive to the job market. In this paper, we provide some examples of the use of the Log4.0 Lab for teaching purposes, and specifically we focus on its use with respect to the Lean 4.0 concept, i.e., the integration of I4.0 technologies with Lean practices and concepts.

Keywords: Logistics 4.0 Lab · Action learning · Learning factory · Teaching

1 Introduction

Traditional teaching methods show limited effects in the development of skill sets for industrial applications. Applied sciences such as manufacturing and logistics cannot, in fact, be learned effectively only inside a classroom, and new teaching approaches enabling training in realistic manufacturing environments are hence needed [1]. Modernizing the learning processes and bringing it closer to industrial practice allow to better educate students and to provide them the skills required by the job market [2]. In this perspective, learning factories have emerged as a very effective solution.

Learning factories are facilities with an authentic factory environment that is used for research, training, and/or learning purposes, and often have a multiple of machines and equipment that can be used to enable a changeable setting for problem- and action-oriented learning [3]: they hence offer the possibility of realistic representation of a factory (sub) system with the necessary products, processes, and resources in an experience-orientated, participative, digital as well as realistic learning environment [4].

D. J. Powell et al. (Eds.): ELEC 2021, IFIP AICT 610, pp. 205–212, 2021.
https://doi.org/10.1007/978-3-030-92934-3_21

The learning factory concept was developed in 1990s, when Penn State University established a large facility equipped with machines and equipment to do industry-sponsored projects. This center is still ongoing and is doing projects that involve students and employees. Recently, the use of learning factories has increased, focusing on many different topics, among which also the Lean concept. However, it is after the advent of Industry 4.0 (I4.0) that the use of learning factories has reached the peak, taking many forms, from the traditional factories to smaller lab environments [3]. There are in fact now more than 120 learning factories worldwide, with more than 50 in Germany, the homeland of I4.0 [4].

Learning factories have in fact proved to contribute to strengthen the transfer of new skills and ways of working needed for adoption of new I4.0 technologies through the use of student projects and research projects with industry [5, 6]. Driven by the benefits associated with learning factories, the Production Management group at NTNU, with the support of the Department of Mechanical and Industrial Engineering, decided to establish the Logistics 4.0 (Log4.0) Lab with a twofold goal. First, to identify new benefits associated with the use of emerging technologies in production and logistics systems, and second, to transfer the developed knowledge to students.

In this paper, we will provide some main examples of the use of the Log4.0 Lab as learning factory for bachelor and master students as NTNU. Specifically, we will describe its use as learning factory only within the concept of Lean 4.0, where students are taught how I4.0 technologies can be integrated with Lean practices and concepts.

2 Logistics4.0 Lab

In 2018, the Production Management group at NTNU, with the support of the Department of Mechanical and Industrial Engineering, established the Logistics 4.0 Lab with the main purpose to create new knowledge on the use of emerging technologies in production and logistics systems. Specifically, through the test and study of the impact of emerging technologies on logistics systems, done in collaboration with companies, either as suppliers of such technologies or as case applications, new methods and models for designing and management of future logistics systems are being developed. Moreover, the Logistics 4.0 Lab is an important platform and arena where the knowledge developed through learning games, hackathons, project-works and master projects is transferred to bachelor and master students.

To do this, a replication of real-life operations and material handling activities in production systems has been made, including assembly workstations, a storage area and material handling systems, and a material management support system. These elements of a conventional production and logistics system have been integrated with a wide range of emerging technologies, such as indoor positioning system, motion capture system, augmented, virtual and immersive reality, visual interactive board, real time control and advanced simulation tools, 3D mapping, mobile robots, smart material handling systems, assistive devices and tools for smart operators & managers.

Among them, it is worth mentioning those which will be further discussed in the following sections. The first one is the motion capture (mocap) system. The researchers in the Logistics 4.0 Lab have developed, in collaboration with the supplier, the integration of such system with a virtual reality (VR) platform in order to replicate the operations in a virtual 3D environment. The motion capture system is a suit composed by 29 inertial sensors which are used to create a digital twin of the operator who is wearing the suit. In the VR platform, the virtual environment can be recreated using 3D models and, through the integration with the suit, the operator can see her/himself in such virtual environment using a VR set. The combination of the mocap system with the VR has been studied in the Log4.0 Lab to virtual design the assembly workplace, showing the students how this solution can highly reduce the time and resources required by traditional workplace design procedures.

The second example of technology studied in the Log4.0 Lab is the integration of photogrammetry with several camera devices and supporting systems. In this way, the pictures taken through different cameras (action camera, 360 camera, smartphone camera) can be merged thanks to some photogrammetry software in order to create a virtual environments (3D scanning). These solutions have been developed for multiple purposes. They have been integrated with CAD systems to support re-layout design and they can be also interconnected to digital twin in order to replicate the real environment into the virtual environment. Moreover, the researcher at Log 4.0 Lab have developed the use of such solution for virtual factory tours, Gemba walks and 5S analysis.

Then, another example of technology studied in the Log4.0 Lab is the Augmented Reality (AR). In this case, the researchers have implemented Microsoft Hololens for assisting workers in various operations, from assembly to order picking. These AR glasses have a specific software development kit which can be used to develop functionalities based on the requirements of the users. In this way, the information and instructions can be given to the user through the glasses, and they are dynamically adapted to the user's point of view. In the Log4.0 Lab, the students have been taught how AR can act as Poka-Yoke in order picking operations in a warehouse and how it can support operators' cognitive tasks.

Another technology implemented in the lab to convey and give information to operator is conventional projector integrated with motion sensing input device (in specific, Microsoft Kinect). The projector is used to send the information to the desk where the operator is working, and the motion sensing input device performs gesture control as input for deciding which information to be sent to the operator. As before, this application has been shown to be an effective Poka-Yoke solution for assembly operations, and students have had the possibility to study it in first person.

3 Action Learning at the Log4.0 Lab: Some Examples

Since the establishment of the Log4.0 Lab in 2018, the Production Management group has pushed its vocation to action learning even further by using the Log4.0 Lab for teaching and research purposes. Many activities have in fact been carried out, involving more than 20 students between summer jobs and master theses, resulting in the development of new teaching materials and/or in the publication of the results in both conference papers and international journals. In the following we will focus on the results of these activities in terms of new teaching materials. The topics covered by the new teaching materials have been many, but in this work we will focus only on the teaching materials relevant for the Lean4.0 concept (i.e., the integration of Lean practices and concepts and I4.0 technologies), and specifically on those showing how the design and the operational level can be affected by the integration of Lean and I4.0 technologies. More in details, the new teaching materials show that the design level is positively affected by the possibility provided by I4.0 technologies to reduce wastes (waste of resources, time, ...) and by the increased efficiency, and that, similarly, the operational level is positively affected by I4.0 technologies since, when serving as support for the operators, they limit the probability of the operators to do mistakes and they facilitate cognitive tasks. In the following we will describe separately the main content of each of these teaching materials, starting from those related to the design level and finishing with those related to the operational level. It is worth mentioning that the teaching materials herein considered are the results of four master theses.

3.1   Teaching Materials on Lean4.0 in the Design Level

Immersive Virtual Mock-Up Approach for Workstation Design
The following teaching material shows how the use of mocap system and VR can assist and improve the design of assembly workstations. The combined use of mocap system and VR allows to design assembly workstations in a virtual environment, without the need to build a physical mock-up [7, 8]. Thanks to the developed teaching materials, students at both bachelor and master level have the possibility to learn how to design an assembly workstation in such a way. Moreover, they learn the theoretical foundations necessary to evaluate the design considering either productivity, operators' wellbeing, or both. Students are shown how to create different workstation layouts for assembling a jet pump, as well as to test the different layouts virtually for then choosing the best solution considering both the productivity and the operators' wellbeing aspect. An example of the virtual assembly operation is reported in Fig. 1.

a)
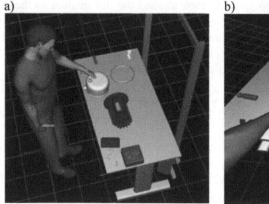
b)
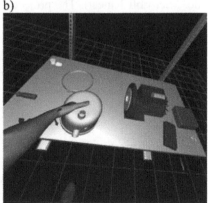

Fig. 1. Birdseye (a) and operator's (b) view of the assembly operation

Use of 3D Scanning for Manufacturing Layout Redesigns

The following teaching material shows how Industry 4.0 technologies can improve the redesign of the facility layout, and this can greatly contribute to the success of a company by acting on the reduction of the Lean waste of unnecessary transportation and handling activities (a well-designed manufacturing facility layout is reported to potentially reduce operating expenses by 50% [9]). Thanks to this teaching material, students at both bachelor and master level learn how the most suitable approach for solving facility layout designs, i.e. systematic layout planning (SLP) [10], can be supported by an innovative technology like the 3D scanning [11]. Specifically, students learn to operate the two most common 3D scanning types (i.e., photogrammetry and structured light) to determine the detailed and accurate information of the layout required by SLP. Students are shown how to redesign the shopfloor of a workshops (Fig. 2) using the two different 3D scanning, highlighting the pros and cons of both technologies.

a)

b)

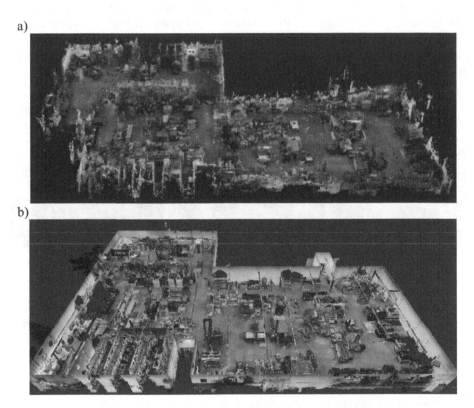

Fig. 2. Areal view of the shopfloor with 3D scanning photogrammetry (a) and structure light (b) type

Moreover, it is worth mentioning that 3D scanning has been used in a master level course to allow the students to have a digital Gemba of a warehouse during pandemic times and to provide digital safety tours for new lab users, but this won't be discussed in this work.

3.2   Teaching Materials on Lean4.0 in the Operational Level

Use of Smart Glasses in Order Picking Operations

Thanks to this teaching material, students at bachelor level have the possibility to learn how smart glasses can serve as Poka-Yoke for error proof order picking operations, which represent the most time-consuming, labor-intensive and expensive activities for most warehouses [12]. Specifically, they have the possibility to reflect on the advantages of a pick-by-vision solution compared to the traditional paper-based order picking system. Moreover, thanks to this learning material, students learn also how to carry out cost-benefit analyses. Students are in fact shown through a case study where different customer demands are simulated how to compare the pick-by-vision solution to other innovative order picking systems (i.e., are barcode handheld, RFID tags handheld, pick-by-voice, pick-by-light and RFID pick-by-light systems) from both productivity and economic perspectives in order to determine when it is convenient to adopt the pick-by-vision solution. An example of a student using the pick-by-vision solution for order picking operations is reported in Fig. 3.

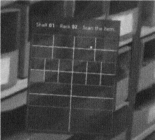

Fig. 3. Details of the pick-by-vision

Kinect-Projector-Based Assistance Technology for Manual Assembly

Thanks to this learning material, students at bachelor and master level are taught how digital assistance technologies can support an error proof manual assembly. Students, in fact, are able to experience firsthand that digital technologies can provide clear and easy-to-read assembly instructions, as reported in literature [13, 14], and they are shown quantitatively the benefits of adopting such technologies. Specifically, students can experience firsthand how the digital technology developed in the Log4.0 Lab (consisting of a projector coupled with a Microsoft Kinect motion sensing device, Fig. 4) decreases the number of assembly errors and the cycle time for the assembling of a LEGO model compared to paper-based instructions.

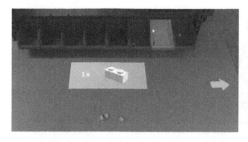

Fig. 4. Examples of assembling operations projected onto the workplace

4   Conclusions

In this paper we reported some examples of the use of the Logistics 4.0 Lab as learning factory by the Production Management group at the MTP Department at NTNU. Specifically, due to space limitations, we focused only on the developed teaching materials relevant for the Lean4.0 concept (i.e., the integration of Lean practices and concepts and I4.0 technologies). In particular, we show how the design and the operational levels can be affected by the integration of Lean and I4.0 technologies, allowing students to understand thanks to tangible examples how I4.0 technologies can assist Lean practices and concepts.

Students revealed to be enthusiastic for the new teaching materials. Some of their comments collected by the responsible of the courses during the periodic students-teacher meeting are here reported:

"It was an amazing experience to cherish and I learnt a lot of useful information on how Industry 4.0 technology could be used to make our lives easier and automate the manual tasks."

"Working with these technologies was really interesting at a personal and academical level. I was able to apply the 3D scanning on a real case and it was fascinating to see its practical implementation and benefits."

"I did like it! I feel like I gained a lot of useful knowledge and it helped me expand view within the field."

However, although these benefits achievable by using the Log4.0 Lab as learning factory, there were also some drawbacks associated with it. The main drawback is the necessity for the researchers in the Log4.0 Lab to master the new technologies. This requires highly skilled and highly motivated researchers, that are willing to keep pace with the continuous technological advancements. However, these are characteristics that are intrinsic in persons who have decided to follow the academic career, where producing state-of-the-art research and teaching is the order of the day, and the Production Management group at the MTP Department has proved over the year to be extremely successful in this.

References

1. Rentzos, L., Doukas, M., Mavrikios, D., Mourtzis, D., Chryssolouris, G.: Integrating manufacturing education with industrial practice using teaching factory paradigm: a construction equipment application. Procedia CiRP **17**, 189–194 (2014)
2. Gento, A.M., Pimentel, C., Pascual, J.A.: Lean school: an example of industry-university collaboration. Prod. Plan. Control **32**(6), 473–488 (2021)
3. Abele, E., et al.: Learning factories for research, education, and training. Procedia CiRp **32**, 1–6 (2015)
4. Pittich, D., Tenberg, R., Lensing, K.: Learning factories for complex competence acquisition. Eur. J. Eng. Educ. **45**(2), 196–213 (2020)
5. Baena, F., Guarin, A., Mora, J., Sauza, J., Retat, S.: Learning factory: the path to Industry 4.0. Procedia Manuf. **9**, 73–80 (2017)
6. Leal, L.F., Fleury, A., Zancul, E.: Starting up a learning factory focused on Industry 4.0. Procedia Manuf. **45**, 436–441 (2020)
7. Battini, D., Calzavara, M., Persona, A., Sgarbossa, F., Visentin, V., Zennaro, I.: Integrating mocap system and immersive reality for efficient human-centred workstation design. IFAC-PapersOnLine **51**(11), 188–193 (2018)
8. Simonetto, M., Peron, M., Sgarbossa, F.: A methodological framework to integrate motion capture system and Virtual Reality for assembly system 4.0 workplace design. Saf Sci. (under review)
9. Tompkins, J.A., White, J.A., Bozer, Y.A., Tanchoco, J.M.A.: Facilities Planning. Wiley, Hoboken (2010)
10. Heragu, S.S.: Facilities Design. CRC Press, Boca Raton (2018)
11. Peron, M., Fragapane, G., Sgarbossa, F., Kay, M.: Digital facility layout planning. Sustainability **12**(8), 3349 (2020)
12. Battini, D., Calzavara, M., Persona, A., Sgarbossa, F.: A comparative analysis of different paperless picking systems. Ind. Manag. Data Syst. **115**(3), 483–503 (2015)
13. Peron, M., Sgarbossa, F., Strandhagen, J.O.: Decision support model for implementing assistive technologies in assembly activities: a case study. Int. J. Prod. Res. **0**(0), 1–27 (2020)
14. Simonetto, M., Peron, M., Fragapane, G., Sgarbossa, F.: Digital assembly assistance system in Industry 4.0 era: a case study with projected augmented reality. In: Wang, Y., Martinsen, K., Yu, T., Wang, K. (eds.) IWAMA 2020. LNEE, vol. 737, pp. 644–651. Springer, Singapore (2021). https://doi.org/10.1007/978-981-33-6318-2_80

Blended Network Action Learning – A Digital Lean Approach to Solving Complex Organizational Problems Across Space and Time

Daryl John Powell[1,2(✉)], Melanie de Vries[3], Mitchell van Roij[3], and Jannes Slomp[3]

[1] SINTEF Manufacturing, Horten, Norway
daryl.powell@sintef.no
[2] Norwegian University of Science and Technology, Trondheim, Norway
[3] HAN University of Applied Science, Arnhem, The Netherlands

Abstract. Network Action Learning has emerged as an innovative development of Action Learning and has been described as a lean approach to collaborative strategic improvement with problem-solving at its core, be it either within- or across organizational boundaries. Virtual Action Learning is also presented as an emergent variety of Action Learning, bringing together geographically dispersed individuals within and across organizations in an online, virtual environment. Given the onset of new, innovative digital technologies – particularly in response of the Covid-19 pandemic – Blended Learning has also emerged as an educational platform that represents some combination of face-to-face and online learning using mobile technologies and cloud-based resources. Though Virtual Action Learning has been discussed as neither better than nor second best to face-to-face Action Learning, in this paper, we suggest that a blended approach may be the most effective method. Thus, the purpose of this paper is to construct a Blended approach to Network Action Learning, where intra- and inter-firm Network Action Learning can take place using a hybrid, physical-virtual approach to promoting collaborative strategic improvement and gemba-based problem-solving.

Keywords: Organizational learning · Blended learning · Network action learning · Lean strategy · Digital lean

1 Introduction

Network Action Learning (NAL) has emerged as an innovative development of [1]'s theories of Action Learning (AL) and has been presented as a useful and useable means of enabling intra- and inter-firm learning to achieve sustainable, collaborative strategic and operational improvement, specifically in the context of buyer-led lean transformations [2]. For example, [3] present NAL as an approach to collaborative strategic improvement, be it either intra-firm or inter-firm improvement. As such, in this paper, we explore a blended approach to NAL for addressing complex organizational

D. J. Powell et al. (Eds.): ELEC 2021, IFIP AICT 610, pp. 213–224, 2021.
https://doi.org/10.1007/978-3-030-92934-3_22

problems across space and time – assisting organizations in deploying a bottom-up, gemba-led lean strategy.

[4] present Virtual Action Learning (VAL) as an emergent variety of AL, bringing together geographically dispersed individuals within and across organizations in an online, virtual environment. Given the onset of new, innovative digital technologies – particularly in response of the Covid-19 pandemic – Blended Learning (BL) has also emerged as an educational platform that represents some combination of face-to-face (F2F) and online learning using mobile technologies and cloud-based resources.

Though Dickenson et al. argue that VAL is neither better than nor second best to F2F AL, we suggest that given the current situation with social distancing measures and limited international travel due to Covid-19, a blended approach may actually be a superior approach. We also suggest that the solving of complex problems demands broad knowledge and an extensive skills base, thus virtual mobility of international experts and problem owners allows for the development of learning networks, in which peers can challenge each other to reflect over concurrent challenges and the underlying assumptions that may be causing them. This will help advance the state-of-the-art, particularly given the onset of Industry 4.0 and the emergence of key enabling digital technologies that allow for greater virtual mobility. Therefore, the purpose of this paper is to construct and present a Blended Network Action Learning (BNAL) methodology, where intra- and inter-firm NAL can take place both physically and virtually to promote collaborative strategic improvement to solve complex organizational problems across space and time.

2   Theoretical Background

2.1   Action Learning

Action Learning (AL) has emerged as a radical process for increasing organizational knowledge and capacity for better adapting to change [3]. It can be considered as a lever for developing, improving, and assimilating learning in organizations. [1] outlines the following assumptions that underpin AL:

- Learning is cradled in the task and formal instruction is not sufficient,
- Solving problems requires insightful questions,
- Learning involves doing, is voluntary, spurred by urgent problems or enticing opportunities and is measured by the results of action.

[1] formulated his action learning concept around the formula $L = P + Q$, where L stands for learning, P for programmed knowledge and Q for questioning insight. In his theory of action, [5] presented his science of praxeology of cyclical systems – alpha, beta and gamma. System alpha focuses on investigating a problem. System beta focuses on solving the problem, and the negotiation cycles required to implement the solution. System gamma focuses on the learning as experienced by participants, and involves self-awareness, reflection and questioning. It is important to emphasize that the three systems (alpha, beta and gamma) are neither linear nor sequential, nor entirely

discrete. The three are best understood as a holistic system of interlocking yet over-lapping parts which deserve differing emphases at different times [3].

At the heart of AL is a distinction between different kinds of issues. As such, [1] distinguishes between puzzles and problems. Puzzles are those difficulties for which a solution exists, and which are amenable to expert advice. For example, experts may swiftly and simply select a solution to "puzzles", thus solving puzzles is not amenable to AL. Problems, on the other hand, are difficulties where no single solution can possibly exist. Most complex organizational change projects fall into the category of a problem – where there is no single solution and where there are many opinions as to what the course of action might be. Problems are amenable to AL as, in response, different people can advocate different courses of action in accordance with their own value systems, past experiences and intended outcomes.

2.2   Network Action Learning

[3] suggests that collaborative strategic improvement requires developing a capacity to learn within and across a network, for example, in both the intra- and inter-organizational context. Not just as individuals in organizations, but especially within and between organizations. With roots in action learning, the authors propose Network Action Learning (NAL) as a useful and usable approach to collaborative strategic improvement:

"Continuous and collaborative improvement are, in essence, processes of action and learning: problems are identified; solutions are created, analyzed, selected and implemented; resulting not only in improved operational performance but also in improved capability (through learning)".

They extend the action learning formula and define NAL as $L = P + Q + O + IO$. This formulation captures the action learning process in the context of both intra- and inter-organizational learning. Here, P refers to the established knowledge of collaborative improvement, Q relates to the questioning process, and O and IO relate to emerging insights in the organizational and inter-organizational contexts. As such, *"the action learning by the network is built on exposing programmed knowledge to questioning, combined with organizational- and inter-organizational insights created in action"* (p. 69). To increase competitive advantage, however, the network must be capable of exploiting this learning – by demonstrating an *absorptive capacity*. As such, participants within and across organizations in the network must engage in appropriate learning interventions in a structured way, consistent with [6] who argue that organizational design is critical to building learning mechanisms that develop and sustain learning capabilities.

3   Towards a Philosophy for Blended Network Action Learning

[7] suggests that researchers are typically encouraged to ground their research in a research philosophy consisting of an ontology (reflecting the researcher's under-standing of self, own experience, the nature of the relational world and the nature of

knowledge and theory), an epistemology (expressing how the researcher seeks to know), a methodology (articulating the set of ideas justifying the approach which the researcher adopts for the process of inquiry), and finally a method (for planning enacting, evaluating and understanding research).

In terms of a philosophy for BNAL, ontology is reflected in [1] in that *"there can be no learning without action, and no action (sober and deliberate) without learning"* (p.83). The classic formulation (equating learning and knowing) $L = P + Q$ provides an epistemological basis. Most significant for this paper is that of methodology, which we base on [5]'s theory of action and science of praxeology of cyclical systems - alpha, beta and gamma:

- *System Alpha*: In BNAL, system alpha frames the complex organizational problem to be solved. It focuses on identifying and analysing a real organizational problem including analysing the external environment, current organizational performance, and management values (what the managers want to achieve).
- *System Beta*: Revans' scientific method presents us with a structured means for investigating, understanding, and solving problems, in action. In BNAL, system beta concerns the deployment of the scientific method and involves exploring the problem-solving process, through multiple cycles of action and reflection. Action learners uses appropriate theoretical perspectives to frame the results of the action and reflection cycles, with a view to identifying emergent actionable knowledge.
- *System Gamma:* The (individual and collective) learning is the focus of system gamma. In BNAL, the active participation of action learners in developing and executing systems alpha and beta has implications for the scope of system gamma. The action learners' involvement in system gamma exposes the process of how their engagement with the problem has challenged their own thought processes, to further inquiry. The interpretation and evaluation of each action learner's own involvement underpins the emergent actionable knowledge, ensuring the quality of BNAL.

As such, the remainder of this paper is structured as follows: we first present a framework for problem evaluation based on the Find, Face, Frame, Form (4F) framework as presented in *The Lean Strategy* [8]. Secondly, we present a framework for BNAL as well as guidelines for its application.

4   The Lean Strategy Approach to Problem Resolution: Find, Face, Frame, Form

[8] distinguishes between two types of strategy – the traditional, "Porter" style approach strategy formulation and execution, and the "superior" lean strategy. They suggest that lean-thinking executives abandon all preconceptions of traditional management reasoning. For example, *defining* "problems" in the board room, *deciding* what must be done to resolve them, *driving* execution through action plans, and then *dealing* with unexpected consequences (4D) is not an effective means to *grow* a business. Lean leaders must *find* problems by going to the "Gemba" to see the problems faced by workers and customers with their own eyes. This lets them develop a clear understanding of what factors are preventing them from hitting current targets.

Armed with first-hand, specific knowledge, lean leaders then *face* the main challenges (the "elephants" in the room/the obvious problem(s) no one wants to discuss) by creating key operational indicators such as improving quality, speeding up delivery, or reducing incidents. Next, they *frame* the challenges and goals in such a way that everyone in the company can understand them and know how they can contribute - lean leaders will propose lean solution types to problem types, such as pulling (instead of pushing) the workflow to create value faster for clients or by applying value analysis/value engineering (VA/VE) to conceive and deliver products that clients love, over and over again. Finally, lean leaders support and develop people to enable them to *form* their own solutions, so that the sum of all local solutions and ideas forms an effective, collective response to the main challenges.

This forms the basis for BNAL (and a critical success criteria) – where the organization's leaders must adopt Gemba-leadership to encourage and guide people in their improvement activities, and must begin by accepting the workplace-based, ground-up strategic thinking of finding and facing problems at the Gemba, framing those problems with pre-defined conditions (e.g., just in time, zero defects etc.), and facing them together with the teams themselves (4F).

5 BNAL Methodology: Framework and Guidelines

Similar to [9], we identify six main components that make up the BNAL framework:

- The problem
- The network
- The questioning and reflective process
- The commitment to taking action
- The commitment to learning
- The learning facilitator

All six core components must be formalized at the beginning of a BNAL initiative, but as BNAL is dynamic and a process of discovery, the elements may evolve as the process develops (for example, the problem may be reframed, and new members may be added to the network, etc.).

5.1 The Problem

The starting point for BNAL is the problem encountered in an organization (also referred to as the task, the project, the challenge, or the opportunity). Without a problem, there can be no BNAL. The problem should be important and should provide an opportunity for learning (the best BNAL projects provide rich learning opportunities). Such examples could be:

- Improving quality
- Reducing waste in operations
- Increasing productivity
- Better leveraging technology to create value for customers

5.2 The Network

The core entity on BNAL is the network (also known as the "set"). The individuals that make up the BNAL team are those who are responsible for framing the problem, constructing action, planning action, taking action and reflecting over action – in a blended manner (e.g., a mix of physical and virtual mobility).

5.3 The Questioning and Reflective Process

The questioning and reflective process is a structured means of guiding strategic improvement within the network. For example, network participants can act and learn together in a blended (hybrid physical-virtual) way. The organizations in the network may have similar problems and therefore similar interventions/actions can be taken. They can share the outcome of these results in the network and learn from one another, by asking each other challenging questions about the interventions. This provides the basis for problem solving while satisfying the learning imperative that is central to BNAL. This process demands that a safe and open environment be created so that organizations become comfortable to share their problems, assumptions, and ideas. Together the individuals try to progress further by challenging one another.

BNAL recognizes that problem solving must begin by first diverging through the use of inquiry, before converging on a solution. Some example questions include:

- What is the real issue?
- What evidence do we have?
- What do we need to learn?
- Why are we doing this?
- What are the risks?
- Which stakeholders should we engage?
- How do we ensure that the outcomes are achieved?
- Would the problem be solved after we implement this?
- How do we know we have achieved the results?
- What have we learned?

5.4 The Commitment to Action

The members of the BNAL set are committed to implementing change in one or more of the participating organizations. This means that members must be committed to working together to improve the collaborative relationship (be it temporary or long-term) and must commit to solving the problem through participation in action.

5.5 The Commitment to Learning

The learning that occurs in BNAL is of greater strategic value to the participating organizations than the immediate tactical value of solving the problem at hand. Learning to learn is a key component of BNAL and emerges through combining action with questioning and reflection. It is important to create a safe environment where reflection and learning can occur. Failures in the group must be seen as opportunities to learn rather than events to be hidden or ignored.

5.6 The Learning Facilitator

The learning facilitator, coach, or indeed *sensei* [10], is the catalyst in BNAL. The learning facilitator must enhance the network's ability to learn and take meaningful action. The characteristics of the facilitator should be more in the areas of group facilitation and learning than in the technical expertise required to solve the problem.

6 A Theory of Action and Science of Praxeology

Given that the six core components are in place, BNAL is guided further by [5]'s theory of action and science of praxeology of cyclical systems - alpha, beta and gamma:

6.1 System Alpha – Finding, Facing, and Framing (or Re-framing) the Problem

System alpha concerns the process of constructing action in BNAL. This subtask aims to provide a set of guidelines for constructing a BNAL project to address a problem, including recruitment and initial contact of network participants, selecting the type of participation/mobility (physical or virtual), and arriving at a (broad) definition of learning and improvement needs.

Gemba Visit. The BNAL approach begins with a process of reflection and questioning insight at the gemba ("the real place") to locate the problem in practice. The gemba visit should be carried out at least by the company representatives (project owner/-sponsor/-manager) and the learning facilitator (sensei). Other representatives from the network can be involved where applicable.

Find and Face the Problem. Participants in the gemba visit have the potential to discover many problems. Some can be solved with existing solutions and programmed knowledge (these problems are referred to as puzzles and, though amenable to experts, such problems are not amenable to action learning), while others require a great deal of reflection and insightful questions (solving such complex, organizational problems are the primary goal of BNAL). Finding and facing problems effectively often requires the local management team to be challenged by the facilitator (learning coach) to think differently about the observed situation. Facing the main issues of the business by starting with the management team's own misconceptions and taking a helicopter view to find the challenges which impede customer and/or employee satisfaction and limit organizational growth is a critical part of this phase.

Frame the Problem. Framing the problem can often mean aligning the entire organization (or indeed network) around compelling learning goals. As such, the facilitator will help company/network representatives to frame the problem and identify the necessary learning and improvement needs (the next step).

Define Learning and Improvement Needs. Though the participants in the BNAL initiative may not have prior experience of either blended- or network action learning, they may be familiar with the Deming cycle [11]: plan-do-check-act (PDCA). This well-established cycle of action and reflection is often referred to as the *learning* cycle.

For companies engaging in BNAL, all improvement actions must be rooted in shared concerns – and a shared understanding of the problem(s) where:

1. Improvement and learning go together, with the share objective of overcoming a problem for which there is no single solution.
2. Simply treating the problem as a puzzle and attempting to solve it with (existing) commercial solutions is not a solution. Rather, if seen as a means and rational for engaging with the problem, the puzzle provides a vehicle for engagement with the real problem.
3. Knowledge gaps present the set with learning needs, where the group must engage in action learning. Simply assigning a reading task or a lecture would be to introduce P only. The plan is to take action, thus questioning insight (Q) from the action must be combined with P in order to solve the problem. This process emphasizes the important role of the learning facilitator – who will help the problem-owner to identify whether the organization has the necessary skills and knowledge to solve the problem alone, or indeed whether external parties should be engaged in the action learning process. This then leads to the identification and construction of the network (see the following section).

Identify and Construct Network. The degree of complexity of the problem and the available resources in the organization determine whether the problem can be solved within the own organization or if other actors should be involved. In the latter case, the learning facilitator should assist the organization in sourcing the relevant expertise externally – acting as a knowledge broker to create ties with external stakeholders. Such ties can be formulated both through physical and virtual (blended) communication. Assuming the problem is significantly complex that it cannot be solved by the organization in isolation, the first step for the learning facilitator is to assess the knowledge, competency and capacity of the existing network of the organization. This is because existing ties require little effort to build the mutual trust which is beneficial for knowledge transfer and learning interventions in BNAL. Also, as BNAL is focused on problems with a high degree of complexity that often cannot be solved in the organization due to lack of available resources, the organization should reach out to actors beyond the network to start an alliance. By bundling the knowledge and resources of the actors in the network the complex problem can be more easily solved.

Thereafter, the BNAL set is tasked both with action on the initiative as well as with extracting learning from the experience of action towards a solution for the wider problem. As such, the network needs to include an appropriate mix of levels, affiliations, disciplines, functions, responsibilities and experiences. The network also needs to interact on a regular basis throughout the BNAL initiative, where some of this interaction is through participation in scheduled meetings, each with practical, commercial and learning outcomes. A plan for such interaction is the topic of the next section – forming and implementing the solution(s) to the problem.

6.2   System Beta – Forming and Implementing the Solution(s) to the Problem

System beta concerns the process description for planning action. This subtask aims to develop a set of guidelines for selecting programmed knowledge from existing theory and the knowledge and experiences of participants to help form possible solutions to the problem defined in the previous step, and considers how blended learning approaches can be used to provide network participants with the extra fundamental knowledge required in order to address the problem at hand. Important issues to raise here are definition of network roles and responsibilities, assessment of current state, identification and discussion of existing theory, and planning for milestones and performance deadlines.

System beta also concerns the process description for taking action. This subtask will develop a set of guidelines for how the individuals in a network can effectively take action to solve the problem, also with a view to creating new knowledge and learning. Important considerations include identification of emerging issues as well as review of training and facilitation needs.

Define Network Roles and Responsibilities. A core part of BNAL is the network (also known as the "set"). The individuals that make up the network are those who are responsible for solving the problems (this might include intra- and inter-organizational representatives, e.g., managers from companies experiencing the same or similar problems) – through constructing action, planning action, taking action and reflecting over the action. After the problem is identified, the foundation for the network should be laid using the following six steps, for which we rely on the work of [12] to further conceptualize the intra- and inter-organizational networks, namely *allocation, regulation* and *evaluation*, as well as the important role of the *network administrator (see below)*.

Allocation: Once the partners for the network are selected, the resources, tasks and responsibilities should be allocated and aligned across the network partners. The partners are tied together in the network and strong cooperation is needed to solve the problem. It is important that this is all formalized.

Regulation: In this step, rules for the collaboration are formalized and implemented. All network partners should live by the rules of the game (though these rules can be both formal and informal). When a new partner enters the network, she should comply with the existing rules in the network. However, the rules of engagement may change over time as the network evolves.

Evaluation: The last step in creating an effective network to solve problems with BNAL is evaluation. The network should be evaluated regularly to see whether it is going in the right direction. The contributions of the individual partners, the performance of the whole network and the relations between the network partners are evaluated. It should be evaluated if actions should be taken to stay on track. On top of that, it is important that every partners' opinion is considered in the evaluation. Organizations weigh up the disadvantages and advantages of being part of the network and this in turn influences the effort they will make. Effort to maintain quality relationships with other partners and effort to take action and share knowledge. Thus, the effectiveness of the network depends on how the partners rate the quality of the network.

Network Administrator: A network administrator should also be appointed to facilitate the network – this is a distinctly different role to that of the learning facilitator. The network administrators' job is to administer knowledge sharing among partners, while the learning facilitator strives to enhance the network's ability to learn and take meaningful action. Regarding the evaluation, the network administrator evaluates the network from his perspective. Is the way the network facilitator sees the network equal to how the individual partners experience it? If not, it is the job of the network facilitator to find the imbalance and take action. For an effective network in which partners are willing to share their knowledge, resources and learnings, high levels of trust and reciprocity are important. In the evaluation it should be considered if the levels of trust and reciprocity are desirable or that actions should be taken.

Planning and Taking Action. Having established the roles and responsibilities within the network, the set can begin to plan and take action in order to address the problem at hand. This involves using the scientific method as follows:

1. Assess Current State
2. Agree on Target State
3. Plan for Action (Incl. Selection of Programmed Knowledge)
4. Take Action (using loops of PDCA)

Having also found and framed the problem in the previous step, A3 management is a well-known and well-documented scientific problem-solving process that presents leaders with a step-by-step approach to plan and take action, closely modelled on PDCA [13]. The term A3 in fact refers to an international standard paper size (297 × 420 mm). Toyota adopted the name A3 drawing on insight that every issue an organization faces can and should be captured on a single sheet of A3 paper. While the basic thinking for an A3 follows a common logic, the precise format and wording are flexible, and most organizations tweak the design to fit their unique requirements [14].

A3 management also serves as an important means of communication – such that countermeasures developed during the problem-solving process can be standardized and shared with others [15]. [13] present this form of "standardized storytelling" as a powerful tool to engage and empower leaders as well as front line personnel. They conclude that it is the thinking behind paper, not the A3 paper itself, that is most important.

6.3   System Gamma – Reflecting over Learning and Emergent Actionable Knowledge

System gamma concerns the process description for reflecting over action and learning, which occurs in parallel to the activities defined in system beta. This subtask aims to establish a set of guidelines for reflecting over the BNAL cycle(s), including how the experiences and new knowledge can be shared within and outside of the network using blended learning. This should also include assessments of the scale of the collaborative improvement, and a plan/review as to how the process of change has been communicated within and outside the network.

Regarding the A3 process, the effect confirmation and follow-up phases are critical for system gamma. Here, the participants in the network (set) must study the effects of the action (preferably at the Gemba) and use insightful questioning to identify important lessons learned. Here questions must be prioritized over statements.

Any emergent learning should be documented (on the A3 or otherwise) and communicated within and across the participating organizations, to share and re-apply any emergent actionable knowledge.

7   Conclusion

This paper presents BNAL as a means of guiding collaborative problem-solving and learning across space and time. We document both a means for problem evaluation (Find, Face, Frame, Form) and a framework for planning and taking action (including guidelines for reflection and learning). Further work should aim to empirically refine and validate the BNAL methodology through case-based action research. Critical success factors for facilitating BNAL should also be evaluated.

Acknowledgements. The authors would like to acknowledge support from the European Commission for financing the Erasmus+ Knowledge Alliance *LEAN4.0* (Grant number 601227) and the Research Council of Norway for the project *Lean Digital*.

References

1. Revans, R.W.: What is action learning? J. Manag. Dev. **1**(3), 64–75 (1982)
2. Powell, D.J., Coughlan, P.: Rethinking lean supplier development as a learning system. Int. J. Oper. Prod. Manag. **40**(7/8), 921–943 (2020)
3. Coughlan, P., Coghlan, D.: Collaborative Strategic Improvement Through Network Action Learning: The Path to Sustainability. Edward Elgar Publishing, Cheltenham (2011)
4. Dickenson, M., Burgoyne, J., Pedler, M.: Virtual action learning: practices and challenges. Action Learn.: Res. Pract. **7**(1), 59–72 (2010)
5. Revans, R.W.: Developing Effective Managers: A New Approach to Business Education. Praeger, New York (1971)
6. Shani, A.R., Docherty, P.: Learning by Design: Building Sustainable Organizations. Blackwell, Oxford (2003)
7. Guba, E., Lincoln, Y.: Competing paradigms in qualitative research. In: Denzin, N.K., Lincoln, Y. (eds.) Handbook of Qualitative Research, pp. 93–99. Sage, Thousand Oaks (1994)
8. Ballé, M., et al.: The Lean Strategy: Using Lean to Create Competitive Advantage, Unleash Innovation, and Deliver Sustainable Growth. McGraw Hill Professional, New York (2017)

9. Marquardt, M.J.: Optimizing the Power of Action Learning: Solving Problems and Building Leaders in Real Time. Davies-Black Publishing, Palo Alto (2004)
10. Reke, E., Powell, D., Olivencia, S., Coignet, P., Chartier, N., Ballé, M.: Recapturing the spirit of lean: the role of the sensei in developing lean leaders. In: Rossi, M., Rossini, M., Terzi, S. (eds.) ELEC 2019. LNNS, vol. 122, pp. 117–125. Springer, Cham (2020). https://doi.org/10.1007/978-3-030-41429-0_12
11. Deming, W.E.: Out of the Crisis. Massachusetts Institute of Technology, Cambridge (1982)
12. Sydow, J., Schüßler, E., Müller-Seitz, G.: Managing Inter-organizational Relations: Debates and Cases. Macmillan International Higher Education (2015)
13. Richardson, T., Richardson, E.: The Toyota Engagement Equation: How to Understand and Implement Continuous Improvement Thinking in Any Organization. McGraw Hill, New York (2017)
14. Shook, J.: Managing to Learn: Using the A3 Management Process to Solve Problems, Gain Agreement, Mentor and Lead. Lean Enterprise Institute (2008)
15. Liker, J.K., Hoseus, M.: Toyota Culture: The Heart and Soul of the Toyota Way. McGraw-Hill, New York (2008)

Lean Management

'Fake Lean'; On Deviating from an Ambiguous Essence

Jos Benders[1,2(✉)] ⓘ, Marte Daae-Qvale Holmemo[1] ⓘ,
and Jonas A. Ingvaldsen[1] ⓘ

[1] Department of Industrial Economics and Technology Management NTNU,
Alfred Getz vei 3, 7491 Trondheim, Norway
jos.benders@ntnu.no
[2] CESO, KU Leuven, Leuven, Belgium

Abstract. The term 'fake lean' is useful as it points to the various ways in which lean is put into praxis. At the same time, 'fake' in this expression condemns particular uses. Against the background of literature on organization concepts, we discuss the notion of 'fake lean'. This essay centers around the tenability of the essentialist norm inherent in 'fake lean'. We encourage users of 'lean' to reflect on how they put it into practice, and whether or not they decide to 'fake' this organization concept.

Keywords: Fake lean · Lean principles · Organization concepts · Continuous improvement

1 An Appealing Term; or is it?

The term 'fake lean' works as a powerful formulation to warn against improper uses of 'lean'. It was coined by Emiliani [1], who targets in particular the use of continuous improvement without involving shopfloor employees. More generally, 'fake lean' seems an appropriate label for what are seen as incomplete or erroneous lean applications [f.i. 2–4]. The term is intuitively appealing for drawing attention to different uses of 'lean'. In addition, Emiliani's critical stance reminds us that employee input in operational affairs is quintessential for organizations to survive or even flourish. This seems ignored when lean is used to legitimize dismissals [5], or strengthen managerial or expert authority [6]. Thus, at face value 'fake lean' appears to be a useful notion. Nevertheless, or perhaps therefore, with this paper we aim to problematize the notion of 'fake lean'. The term does two things:

1. point to the variety of uses to which 'lean' can be put; and
2. qualify some of these uses as improper.

We discuss both aspects against the more general background of 'organization concepts', as we prefer to call prescriptive ideas on organizing. The purpose of our exercise is to make users of 'lean', such as consultants, industrial engineers and managers, reflect on the design potential of lean, how they use this potential, and normative implications of their uses.

© IFIP International Federation for Information Processing 2021
Published by Springer Nature Switzerland AG 2021
D. J. Powell et al. (Eds.): ELEC 2021, IFIP AICT 610, pp. 227–234, 2021.
https://doi.org/10.1007/978-3-030-92934-3_23

2 Organization Concepts

An organization concept (OC) consists of 'prescriptive notions on how to manage or organize, promising performance improvement, meant for consumption by managers, and known by a particular label' [7]. They are characterized by what we prefer to call 'interpretative viability' [8]. In other words, for ideas to be disseminated at a large scale, they must appeal to different parties, each of which can interpret the ideas in their own way(s). The interpretive space presents a design potential: users can tailor OCs to their own preferences and situation. At the same time, this interpretative space means that concepts can be criticized as being ambiguous.

There is a substantial market for organization concepts. At any moment in time, managers tend to be faced with numerous issues to improve. Its vendors present OCs as solutions to such issues, and promise performance improvement. Books and other publications promoting OCs are not balanced academic accounts of the pros and cons of a concepts; instead, they emphasize the need, the feasibility and advantages of implementing an OC. Management consultants are the main providers of OCs. They have a vested interest in selling these ideas, earning money by assisting their customers in tailoring OCs to those customers' situation. Within organizations, the use of an OC means selecting and adapting (parts of) its content. Often staff at all hierarchical levels is involved, which may all give their own twists to how, and if, an OC is actually put into practice. The ultimate changes from such transformation processes may vary from full-fledged realization of the original content to dilution of that content and purely ceremonial adoption. Yet in all cases, tailoring the OC to a user's situation necessarily involves interpreting how the OC may be put into practice in that particular situation. Analytically, we distinguish between three different modes of interpretation:

1. Make abstract ideas concrete;
2. Use ideas selectively;
3. Ceremonial use (variants (1) impression management and (2) legitimating actions).

3 Lean-Inspired Changes in Practice

In an illustrative fashion, we now show how lean has been interpreted in the three modes distinguished above.

Make Abstract Ideas Concrete. According to Womack and Jones [9], the first lean principle is to specify customer value. The principle explicitly calls for making the abstract concrete, i.e. specifying what value is for the customers to be served. This may be quite clear if customers can specify their wishes, as may be the case when ordering a car or truck. Yet even when producing trucks, Johansson and Osterman [10] show that experienced industrial engineers found it hard to agree on specifying 'waste' and 'value' in production processes. Such interpretive difficulties abound when 'lean thinking' is applied in service industries and especially in healthcare. Whilst achieving a high 'quality of care' is of paramount importance to medical disciplines, it is nearly impossible to come up with uncontested specifications of that quality of care, let alone operationalize those. This is for instance evident in discussions on how euthanasia relates to quality of care/value for the patient. Such difficulties were the reason why within a Dutch network to promote lean healthcare this issue was skipped at the favour of a focus on continuous improvement [7].

Another issue may arise when different principles are applied simultaneously. Picture 1 shows the re-organized storage racks for incontinence diapers in a Flemish nursing home. The previously existing storage was reorganized by the nursing home staff so that they could now easily pick the right diapers and replenish timely. The picture was taken during a tour in the presence of the home's director and the director of another nursing home. The latter remarked that this was not lean in his view because of the substantial variation in different types of diapers. In his home, the choice was confined to a much smaller number of types from which the residents could choose, so that logistics handling was easier. As response, the organizing home's director pointed out that, following the first lean principle, residents were given the choice of their preference. This indeed lead to considerable variation, yet in this case it was seen as wanted variation. This example illustrates how both directors prioritized different principles, with different results (Fig. 1).

Fig. 1. Re-organized storage of incontinence diapers in Flemish nursing home (2017)

Selective Use. The use of 'lean' tends to be selective. One example was already discussed above in the deliberate choice to prioritize CI in 'lean health care' and leaving implicit what 'quality of care' is. More generally, only some lean elements rather than the entire package tend to be part of implementation efforts. The fourth principle, 'let the customer pull value', even appears hardly ever included in service industries, as it seems self-evident and therefor superfluous in services based on interacting directly with customers. The second and third principles focus on creating flows, and seem often applied in isolation. Furthermore, this 'value stream mapping' (VSM) tends to focus on single flows and generally does not take account of situations where several flows interact. In that case, compromises must be made whereby creating a perfect flow for one particular process tends to hamper the flow of other processes. Thus, even within one organization applying VSM entails selecting processes to be optimized at the expense of other processes. Perhaps the most prevailing mode of selective use concerns equating lean with CI. Below the level of principles, attempts to implement lean are often classified as 'tool based', i.e. particular tools and/or practices are used without the 'lean philosophy' or principles.

Ceremonial Use. Implementing concepts can be a long lasting process, certainly if behavioral changes are intended. It is not uncommon that high failure rates are reported in organizational change efforts, and thus also for attempts to implement a concept. In such case, the original intention easily dilutes in the process of being implemented [11]. The outcomes of such change programs tend to drift from the original intention, and as such these outcomes can be perceived as ceremonial. Another cause for ceremonial use is when a concept is fashionable, and managers feel pressured to keep up the appearance that their organizations are in tune with current developments. A strong version of such 'going with the flow' is when a concept is purely used to legitimate organizational changes, irrespective of whether or not these changes are related to the concept's content. In that respect, 'lean' is particularly powerful as it denotes both a particular way of conducting performance improvements as well as its result. The term 'lean' is also used as adjective, whereby lean stands for slim and fit and as antonym to 'fat'. Indeed, instruments have been developed to measure 'leanness' focusing on how well organizations perform rather than how they achieved these results. Such use became particularly apparent in the first half of the 1990s. *The Machine the Changed the World* happened to appear at the brink of an economic crisis, which created fertile ground for the message that organizations needed to become 'lean'. This was at a large scale interpreted as motivation for a large array of cost-cutting measures with an emphasis on downsizing and delayering [5], and gave rise to the statement 'lean is mean'. The Dutch truck manufacturer DAF is a case in point: many employees were fired in an (ultimately unsuccessful) attempt to avoid bankruptcy [12].

4 Lean Back and Beyond Toyota

At least some of the changes described above may be considered examples of 'fake lean'. Detailing which ones, however, requires specifying the content of its opposite, 'real lean'. As we argue below, that is easier said than done as over the course of time there has been considerable debate over lean's essence.

'Lean' and Toyota are closely connected, yet not identical. As of the mid-1930s, the engineer Taiichi Ohno [13] started developing what was to become the 'Toyota Production System' (TPS). Already at the start of his career, Ohno developed three insights which he found essential [14]:

1. organize along product flows rather than by functions;
2. make small rather than large batches;
3. prevent rather than repair.

He incorporated these ideas, and many others, throughout the 1940s and 1950s into a coherent whole. This lengthy trial-and-error development process amounted into the TPS. His ideas drew considerable attention from Toyota's competitors, but remained largely unknown beyond Japan until in 1977 Sugimori and three other Toyota officials published a first paper in English which sketched the main TPS outlines [15]. In the 1970s Japanese mass producers started outcompeting American and European competitors, leading to many searches into the source(s) of their success.

In retrospect, a landmark event was the reception of the book *The Machine that Changed the World* in 1990 [16]. Based on extensive quantitative research its messages were that (1) 'lean production' of passenger cars lead to far superior performance than the ways 'Western' producers worked, (2) the latter had to take over this way of producing or face bankruptcy, and (3) taking this over was feasible. The book's launch was shortly before a severe economic crisis set in, creating an ideal setting for this message. In 1996, 'The Machine..' was followed up by another book entitled Lean Thinking. Its authors Womack and Jones presented the following motivation to write this: 'many readers (authors: of 'The Machine..') [...] told us that they were anxious to give lean production a try. Their question was a seemingly simple one: How do we do it? [...] The fact was, we didn't know the answers. We had been busy benchmarking industrial performance [...] but The Machine focused on aggregated processes [...] rather than broad principles' [9, pp. 9–10]. Thus, the authors presented the following five principles of 'Lean Thinking':

1. specify customer value by specific product;
2. identify the value stream for every product;
3. create an uninterrupted value stream per product;
4. let the customer pull value;
5. pursue perfection (by improving constantly).

The high level of abstraction made it possible to apply lean principles beyond the repetitive manufacturing industries which so far had been the focus of lean-inspired changes. In the final chapter, Womack and Jones 'dreamed' about how several sectors may in the future be transformed by lean principles: long distance travel, food production and distribution, construction, and medical care. Actually, lean-inspired changes were to be initiated, and still are, far beyond the sectors about which the authors had dreamed, including service industries and the public sector. Whereas manufacturing industries were the main focus in the first 'lean production' wave, lean's scope was now broadened to, in principle, all other sectors. The abstract principles opened up many new areas of application, and the 'prehistory' meant that a large variety of different methods, tools and practices was available and could be drawn upon to make concrete changes.

In the meantime, Toyota developed further. In 2001 and after almost ten years of writing, it published the 'Toyota Way 2001' to make its corporate philosophy and corresponding underlying values and guiding principles explicit. These include continuous improvement and respect for people [17].

The relationship between Toyota and lean can be typified as 'back and beyond'. By calling the production system of Japanese car manufacturers 'lean production', Womack c.s. extended the scope of the Toyota Production System to other Japanese car manufacturers and potentially non-Japanese. A further and more important step beyond Toyota was 'Lean Thinking' with its five principles intended for application in any economic sector. At the same time, in trying to apply 'lean' Toyota is constantly taken as source of inspiration. Yet the company formulated its own principles five years after the 'Lean Principles'. Thus, for some three decades now Toyota and lean are both seen as closely related and distinct which leaves considerable interpretive space as to lean's core.

Continuous Improvement. Some special attention is needed for the principle of continuous improvement (CI), both because it appears to be the core of many lean implementations, and because employee involvement in CI is used as a key criterion to distinguish between 'real' and 'fake' implementations [1]. Toyota officials Sugimori et al. initially called the 'respect-for-human system' one of the two key pillars of the Toyota Production System. This Respect for human system contains: (1) elimination of waste movements, (2) attention for worker safety, and (3) self-display of workers' capabilities by entrusting them with greater responsibility and authority. The latter includes the authority to stop the production line if a worker feels the need, being informed about production progress and 'a system whereby workers can take part in making improvements. Any employee at Toyota has a right to make an improvement on the waste he has found' [15, p. 559]. This initial formulation is very concise, continuity is only implied, and the target of improvements is constrained to reducing waste. Later publications make clear that CI is a well-developed, systematic and disciplined approach within Toyota. The starting point for improvements are standard operating procedures (SOPs). Workers are to follow the SOPs closely. In doing so, they are best placed to observe and experience whether particular SOPs work well. If not, they can signal the issue [18]. Depending on the complexity of this issue, alternatives to the current SOP are investigated, by or in close cooperation with shopfloor employees. Such improvement work requires in-depth understanding of production operations as well as improvement practices, which may take several years to learn [19]. Furthermore, employment security is an important building block: reducing waste such as superfluous activities means eliminating work and thus jobs. Obviously, employees likely refrain suggesting improvements if they threaten to 'engineer themselves out of their jobs'. Equally important, yet hardly addressed in the prescriptive literature, is the importance of creating the right mindset. Besser [20] described the efforts to which Toyota went to create a 'Team Toyota' at its greenfield Kentucky (US) plant.

As Sugimori et al. stated it, involving staff in improvement activities is but one manifestation of 'respect for human'. Again, this principle lends itself to many interpretations and indeed, translations from Japanese into English. The principle is generally called 'respect for people'. Recently, Ljungblom & Tennerfors [17] proposed to understand and translate it as 'respect for craftmanship' (RFC) in order to capture the essence of what Toyota means.

5   Faking Forward?

As argued in the introduction, the term 'fake lean' does two things:

1. point to the variety of uses to which 'lean' can be put; and
2. qualify some of these uses as improper.

Qualifying particular uses requires specifying lean's essence in the first place. The next step is to compare a particular use to that norm. When the use in question deviates from the norm, that use may be classified as 'fake'.

Our excursion into literature about organization concepts shows that:

1. concepts, such as 'lean', are characterized by a certain interpretive space;
2. using this space tends to result in a variety of applications;
3. there are various modes through which these come about;

It also became clear that lean's interpretive space is considerable, arguably even larger than that of any other organization concept. This makes it hard to pin down its essence. Whilst 'fake lean' assumes the existence of its opposite 'real lean', 'real lean' cannot be demarcated uncontestedly. This holds at the overall level: lean production does not equal lean thinking does not equal the Toyota Production System does not equal the Toyota Way. If we use lean selectively and constrain the term to continuous improvement, a similar issue arises: in the course of time, 'respect for human/people/craftsmanship' has been the focus of interpretive struggles. Pinning down the essence of 'lean' is thus not straightforward, and unavoidably leads to contestable choices.

One choice can be to take CI within Toyota as the norm. Toyota operates a specific and highly disciplined way of involving employees in CI, of which socialization into the 'Team Toyota', extensive training and coaching, and employment security are key building elements. It remains to be seen, however, to what extent this Toyota way of involving employees is (1) desirable and (2) feasible in other organizations and sectors. Taiichi Ohno found inspiration when developing the Toyota Production System in many other organizations and sectors. In doing so, he was not concerned about staying in line with their original intention, but rather how they might fit his own vision. In that sense, he might be seen as a productive faker. In exactly the same fashion, designers who look to lean or Toyota for inspiration should be concerned how its ideas or tools may suit their own context and vision. Formulated differently, using a concept's innovative potential should prevail over a conservative concern for sticking to the original. Educators of 'lean' should pay attention to the design potential as well as the implicit normative stances to increase their audiences' reflectivity, and therewith capabilities to handle the rhetoric and contents of 'lean'.

References

1. Emiliani, B.: Real Lean: Understanding the Lean Management System. The Center for Lean Business Management LLC. Wethersfield (2007)
2. Liker, J.: Toyota Way: 14 Management Principles from the World's greatest Manufacturer. McGraw-Hill Education, New York (2004)
3. Nicholas, J.: Hoshin Kanri and critical success factors in quality management and lean production. Total Qual. Manag. Bus. Excell. 27(3–4), 250–264 (2016)
4. Thirkell, E., Ashman, I.: Lean towards learning: connecting Lean Thinking and human resource management in UK higher education. Int. J. Hum. Resour. Manag. 25(21), 2957–2977 (2014)
5. Benders, J., van Bijsterveld, M.: Leaning on lean; the reception of a management fashion in Germany. N. Technol. Work. Employ. 15(1), 50–64 (2000)
6. Alcadipani, R., Hassard, J., Islam, G.: "I Shot the Sheriff": irony, sarcasm and the changing nature of workplace resistance. J. Manag. Stud. 55(8), 1452–1487 (2018)
7. Benders, J., van Grinsven, M., Ingvaldsen, J.A.: The persistence of management ideas: how framing keeps 'Lean' moving. In: Sturdy, A., Heusinkveld, S., Reay, T., Strang, D. (eds.) The Oxford Handbook of Management Ideas, pp. 271–285. Oxford University Press, Oxford (2019)
8. Ortmann, G.: Formen der Produktion: Organisation und Rekursivität. Westdeutscher Verlag, Opladen (1995)
9. Womack, J.P., Jones, D.T.: Lean Thinking: Banish Waste and Create Wealth in Your Corporation. Simon & Schuster, New York (1996)
10. Johansson, P.E., Osterman, C.: Conceptions and operational use of value and waste in lean manufacturing–an interpretivist approach. Int. J. Prod. Res. 55(23), 6903–6915 (2017)
11. van Loenen, V.G.M.W.: Lean in (inter) actie: twee perspectieven op een ontwikkeling van Lean. Doctoral thesis. Radboud University Nijmegen (2021)
12. Benders, J., Slomp, J.: Struggling with solutions; a case study of organisation concepts at work. Int. J. Prod. Res. 47(18), 5237–5243 (2009)
13. Ohno, T.: Toyota Production System: Beyond Large-scale Production. CRC Press, Boca Raton/London/New York (1988)
14. Fujimoto, T.: The Evolution of a Manufacturing System at Toyota. Oxford University Press, New York (1999)
15. Sugimori, Y., Kusunoki, K., Cho, F., Uchikawa, S.: Toyota production system and Kanban system materialization of just-in-time and respect-for-human system. Int. J. Prod. Res. 15(6), 553–564 (1977)
16. Womack, J.P., Jones, D.T., Roos, D.: The Machine that Changed the World: The Story of Lean Production. Rawson Associates, New York (1990)
17. Ljungblom, M., Lennerfors, T.: The Lean principle respect for people as respect for craftsmanship. Int. J. Lean Six Sigma 12(6), 1209–1230 (2021)
18. Adler, P.S., Cole, R.E.: Designed for learning: a tale of two auto plants. Sloan Manag. Rev. 34(3), 85–94 (1993)
19. Spear, S.J.: Learning to lead at Toyota. Harv. Bus. Rev. 82(5), 78–91 (2004)
20. Besser, T.L.: Team Toyota: Transplanting the Toyota Culture to the Camry Plant in Kentucky. SUNY Press, Albany (1996)

Toyota Inspired Excellence Models

José Dinis-Carvalho$^{(\boxtimes)}$ ⬤ and Helena Macedo ⬤

Algoritmi Centre, University of Minho, 4800-058 Guimarães, Portugal
dinis@dps.uminho.pt

Abstract. Toyota Production System (TPS) created in the 1950s undoubtedly marked the beginning of a new era in production and economy. The paradigm shifts introduced both in terms of the management and organization of material flows and in the way the respect for people became an important pillar, brought enormous benefits to the society. From the 1970s until today, companies and organizations around the world have been implementing this new way of organizing and managing the industry and services to achieve excellence. Since the 1970s, several TPS-Inspired Models of Excellence have been created and have been competing for their academic acceptance and adoption in companies and organizations around the world. The purpose of this article is to analyze the most popular models and compare them in terms of the following criteria: Focus on Pull Flow; Focus on Process; Focus on Respect for People (or Sociotechnical scope); Existence of associated techniques; Coverage on Indirect Areas; Popularity in scientific journals; and Popularity in books. Although being "Lean" frequently referred as synonymous of TPS, according to those criteria, authors are inclined to conclude that Kaizen Model, Toyota Way, and Shingo Model are the most comprehensive excellence models considered in this study.

Keywords: Lean thinking · Shingo model · Toyota way

1 Introduction

In several types of products, the use of a brand to replace the product name is quite common in everyday life. People use terms such as "Gillette", "Jacuzzi", or "Chiclet" when referring to the corresponding generic product. In these cases, we relax the language and collectively choose to use the brand name to designate generic products. This is happening even though there is a generic product name that can be used in most cases such as "Razor blade" or "Chewing gum". There are other cases, such as "Post-it" that it may not be easy to find a product name that is shared by most people. A similar phenomenon is occurring in the Industrial Organization and Management body of knowledge where professionals and academics use brand names such as "Lean Thinking", "Kaizen", "Shingo Model", "Theory Of Constraints", "Toyota Way", "Agile Manufacturing", and others. These and other "management brand names" worked and some of them still work as management fashions as referred by Abrahamson [1]. Management fashion is defined by the author as "*a relatively transitory collective belief, disseminated by management fashion setters, that a management technique leads rational management progress*". The author argues that management fashion should not be treated as a special case of aesthetic fashion and that management

© IFIP International Federation for Information Processing 2021
Published by Springer Nature Switzerland AG 2021
D. J. Powell et al. (Eds.): ELEC 2021, IFIP AICT 610, pp. 235–246, 2021.
https://doi.org/10.1007/978-3-030-92934-3_24

fashion, far from being cosmetic and trivial, is a serious matter. A different definition for management fashion is the following: Management fashion is *"the production and consumption of temporarily intensive management discourse, and the organizational changes induced by and associated with this discourse"* [2]. It seems that managers' decisions to embrace new management concepts and ideas are more often informed by collective beliefs about rational or progressive managerial practice than from determined rationalization [3].

Although the aforementioned "management brands" can be seen as different management fashions, in reality some of them are very similar in concepts and the fashion part is only related to the specific fashion setter. Most of these management fashions have the same source but the truth is that so far there does not seem to be a consensus on the generic name that brings together all these brand alternatives. One of the challenges of this field is to find an appropriate designation for the organization and management paradigm created by Toyota to be accepted by most practitioners and academics. Maybe only time will tell but for now we will use in this article the designation "TPS Inspired Excellence Models - TIEM". The term "Lean" may even be the term that has collected more popularity and for many people, it is already accepted as the natural generic designation for these TIEM. The negative aspect of "Lean" designation is that its meaning is very connected mainly with one just one of the two parts of the socio-technical nature of organizations, the technical part. The 5 principles of Lean Thinking [4] are mainly about the focus on value, generation of pull flow and pursue perfection. The fifth lean principles "pursue perfection" is assumed in this article to be equivalent to "Continuous Improvement" since perfection is achieved by continuously removing waste and improving flow pulled by customers.

Other models such as Shingo Model [5] and Toyota Way [6] include also principles clearly oriented to the social part such as "Respect every individual", "Lead with humility", "Think systematically", "Develop exceptional people and teams who follow your company's philosophy", and "Make decisions slowly by consensus, thoroughly considering all options; implement decisions rapidly".

Toyota, contrary to most companies, always pursued the continuous improvement of its processes but at the same time, assured that the focus on its employees was being maintained. The treatment of employees with respect and consideration, and the utilization and enhancement of their plenty capacities is one of the basic concepts of Toyota Production System (TPS). Nowadays, respect for people in the organizational context became a theme of global interest, pursued by all the organizations that seek excellence. But not always was this way. During the 1980s and 1990s, most western companies and universities were more interested in the physics concerning the flow control of materials than the human, behavior, and cultural side of TPS. For that reason, TPS is one organizational excellence model that was followed by many in the past but still arouses the interest of the most competitive companies of the present.

If we look back and scrutinize the concepts and principles of organizational excellence models that appeared after TPS, such as Theory of Constraints (TOC), Kaizen Model, Lean Production, Agile Manufacturing, Lean Thinking, Toyota Way, or Shingo Model, we may say that many of them were probably inspired in TPS and follow its principles and concepts. Such is the case of continuous improvement, which besides being one of the main concepts of Toyota Way, is one of the five principles of

Lean Thinking, a set of principles (a dimension) in the Shingo Model and one of the fourteen principles of Generic Features Model of Agile Manufacturing. The same is applied for the concept "treat the workers as human being and with consideration". Followed by Toyota Way, it is generically described in the Toyota website as "Continuous Improvement and Respect for people in everything we do". In Shingo Model this concept appears inside the dimension of Culture Enablers as "Respect every individual". Moreover, in Agile Manufacturing it appears inside the Generic Features Model as "Empowerment of all the people in the enterprise".

The objective of this paper is to analyze and compare the different TIEM in terms of the following criteria: Focus on Pull Flow; Focus on Process; Focus on respect for people (or Sociotechnical scope); Existence of associated techniques; Coverage on Indirect Areas; Popularity in scientific journals; and Popularity in books.

2 Description of the Main Excellence Models Inspired in TPS

The Toyota Production System (TPS) has inspired many models of excellence not only in production but also in the organization as a whole. Since the first journal article published in 1977 about TPS [7] models have been created and evolving to the present day (see general overview in Fig. 1). *Toyota Production System: Beyond Large-Scale Production*, was the first book in English about TPS [8], published in 1988 by Taiichi Ohno, one of TPS creators, although that version is just a translation of the first Japanese version published ten years earlier in 1978.

Models of excellence are understood here as being descriptions of how to proceed to achieve a competitive advantage in the market. In other words, they are descriptions of what to do, what principles to follow, and what tools to use to be more effective and efficient than competitors.

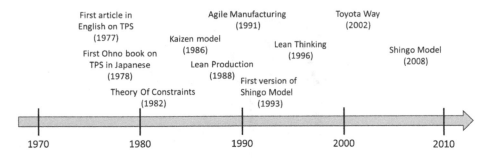

Fig. 1. Main excellence models.

Probably the first kind of excellence model inspired by TPS, published in English after the TPS itself, was presented by Eliyahu Goldratt in his famous and bestseller book "The Goal" [9]. One of the possible reasons that justify the success of this book is the fact that although it is a book with technical content it was written in a novel format. This innovative way of presenting the model made it very attractive due to the ease of its reading and understanding. The model presented and coined as Theory Of Constraints, became very popular as its Optimized Production Technology method was firstly published in 1982 [10] as well as the Drum-Buffer-Rope dispatching technique published a few years later [11].

The second excellence model inspired by TPS is most probably the one presented in a book by Masaaki Imai in 1986 [12]. In that book, the author suggests that the economic success of Japan was the result of the Japanese management practices summarized in the so-called Kaizen umbrella presented in Fig. 2. Under the umbrella, a list of concepts, principles, and tools are presented as the Kaizen model guidelines or structure. From that list, it is possible to understand that the scope of the model covers the sociotechnical nature of organizations, from the technical part to the human part as expressed in the article referred earlier from 1977 about TPS. In that article, the authors argue that TPS is based on the following two main concepts: Reducing cost from the elimination of waste and treat the workers as human being and with consideration. In the items presented under the umbrella of Fig. 2 the reader can see the technical aspects such as "robotics" and "kanban", as well as the human and behavior side as "Small-group activities" and "Cooperative labor-management relations".

Despite the existence of this very comprehensive model, during the 1980s and 1990s in the West, the terms that became popular were mainly "Just-In-Time" and "Kanban" as being the central part of TPS. Just-In-Time was referred by Sugimori, Kusonoki and Uchikawa [13] and later referred by Taiichi Ohno [8] as one of the two pillars of TPS. During these decades, most western companies and universities were more interested in the physics concerning the flow control of materials than the human, behavior and cultural side of TPS. JIT or "Just-In-Time" was accepted as a kind of operational excellence model pursued by most industrial engineering professionals and scholars.

Fig. 2. The Kaizen umbrella [12].

Both Just-In-Time and Theory of Constraints models were very much focused on just one side of the socio-technical nature of organizations, the technical side, more precisely in the material flow control. "Just-In-Time" or just "JIT" has long been connoted in the West, in a relaxed way, as if it were the materialization of TPS or simply equivalent to TPS. After the successful publication in 1990 of the book "The machine that changed the world" [14] and later in 1996 with the publication of "Lean Thinking" book [4] the term JIT was gradually replaced by the term "Lean Production", "Lean Manufacturing", or simply "Lean". Although changing the term used, the Lean Thinking model was still very much focused on only the same technical side of the TPS as JIT.

Despite the focus of Lean Thinking was on the technical part of organizations, such as value, value stream identification, and pull flow, the principle of pursuing perfection leaves some room for the social sciences' part. While the importance of teamwork, empowerment, motivation, and Bottom-Up initiatives are also briefly referred in that original book, the focus of Lean Thinking is towards value, flow and its continuous improvement. Lean Thinking was materialized as following 5 principles: (1) identification of value, (2) identification of the value stream, (3) promoting flow, (4) promoting flow pulled by demand, (5) pursue perfection (also known as continuous improvement).

Agile Manufacturing (AM) is another famous model of excellence proposed by a group of researchers at Iacocca Institute in 1991 [15]. This model comes to life shortly after the first scientific article presenting "Lean Production" [16] and the famous book "The machine that changed the world" from which Lean production became famous and just two years before the book "Lean Thinking" being published. Maybe inspired in TPS, the AM model clearly distances itself from the TPS questioning some of its concepts and never mentioning some of the classic TPS tools such as *5S*, *SMED*, *Heijunka*, *Kanban*, and *Poka-Yoke*. In this model, there is an important component of the inclusion of new technologies and in the integration of the following 3 pillars [17]: Organization, People, and Technology. The Organization pillar refers to the innovative management structures and organizations; The People pillar refers to the skill base of knowledge and empowered people, and the Technology pillar refers to the flexible and intelligent technology. The AM conceptual framework includes Competitive foundations, Core concepts and Generic features model, as described in Fig. 3.

Generic features model	Core concepts
• Integrated enterprises	• Strategy to achieve agility
• Human networking organization	• Strategy to exploit agility
• Enterprise based on natural groups	• Integration of organization, people and technology
• Increased competencies of all people	• Interdisciplinary design methodology
• Focus on core competences	
• Virtual corporations	
• An environment supportive of experimentation, learning and innovation	
• Multi-skilled and flexible people	**Competence foundations**
• Team working	
• Empowerment of all the people in the enterprise	• Continuous change
	• Rapid response
• Knowledge management	• Quality improvement
• Skill and knowledge enhancing technologies	• Social responsibility
• Continuous improvement	• Total customer focus
• Change and risk management	

Fig. 3. Conceptual framework of Agile Manufacturing (adapted from [17]).

At the beginning of the 21st century, the social sciences' side of organizations started to gain more and more recognition in many organizations around the world. One of the companies that clearly include that invisible side in the form of principles was again Toyota by creating the Toyota Way excellence model. The Toyota Way is one of the models of excellence whose principles very effectively cover the entire spectrum of the socio-technical nature of organizations [6]. The principles with grey background in Table 1 are principles more linked to the continuous improvement side of the Toyota Way while the other ones are more linked to the Respect for People side.

Table 1. The 14 principles of the Toyota Way.

Section	Principles
Long Term Philosophy	#1. Base your management decisions on a long-term philosophy, even at the expense of short-term financial goals.
The Right Process Will Produce the Right Results	#2. Create continuous process flow to bring problems to the surface.
	#3. Use "pull" systems to avoid overproduction.
	#4. Level out the workload (Heijunka). (Work like the tortoise, not the hare.)
	#5. Build a culture of stopping to fix problems, to get quality right the first time.
	#6. Standardized tasks are the foundation for continuous improvement and employee empowerment.
	#7. Use visual control so no problems are hidden.
	#8. Use only reliable, thoroughly tested technology that serves your people and processes.
Add Value to the Organization by Developing Your People and Partners	#9. Grow leaders who thoroughly understand the work, live the philosophy, and teach it to others.
	#10. Develop exceptional people and teams who follow your company's philosophy.
	#11. Respect your extended network of partners and suppliers by challenging them and helping them improve.
Continuously Solving Root Problems Drives Organizational Learning	#12. Go and see for yourself to thoroughly understand the situation (Genchi Genbutsu).
	#13. Make decisions slowly by consensus, thoroughly considering all options; implement decisions rapidly (Nemawashi).
	#14. Become a learning organization through relentless reflection (*Hansei*) and continuous improvement (*Kaizen*).

Finally, the Shingo Model started to be developed in 1988 to support the Shingo Prize, awarding the first company in 1989 [18]. The first version of the Shingo model, also referred as "1st Assessment Model" was established in 1993 [19]. Very little emphasis was given in that version to the human side of organizations and no reference was given to continuous improvement concept. A new Shingo Model was released in 2008 [19] with emphasis on principles and culture where clear relevance was given to continuous process improvement, assigning a set of principles to that dimension.

The actual version of the Shingo Model [20] is very much an enhancement of that new Shingo Model. In the point of view of scientific publications the first article found in Scopus database referring the Shingo Model was published in 2014 [21]. In that article, the authors refer the Shingo Institute website in 2012 as the source of those principles. The ten guiding principles are categorized into three dimensions - Cultural Enablers, Continuous Improvement, and Enterprise Alignment, as shown in Fig. 4. The first dimension of the guiding principles lies on the Culture Enablers principles of respect for people and lead with humility, and they are at the bottom of the pyramid because they concentrate on the foundation of an organization: the people. This class refers to the type of behaviors required in order to effectively accommodate all the other principles. The second dimension of the guiding principles pyramid – Continuous Improvement – refers to the principles related to the production processes focus and its improvement. The "Enterprise Alignment" class refers to the formal vision and purpose of the entire organization.

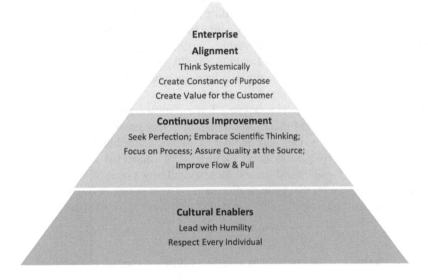

Fig. 4. Shingo model guiding principles.

The principles of this model can be assigned to each one of the sides, technical and social, in a relatively easy way. The principles in the class "Continuous Improvement" can be assigned to the technical side while the principles in the other classes, "Cultural Enablers" and "Enterprise Alignment" can be assigned to the social sciences side.

3 Publications Data Analysis

The models considered in this study are the following: TPS (Toyota Production System), TOC (Theory Of Constraints), KM (Kaizen Model), AM (Agile Manufacturing), LT (Lean Thinking), TW (Toyota Way), and SM (Shingo Model). Regarding the number of publications of the different excellence models considered in this study, the results are shown in Table 2. In the same table are also presented the first published book and first journal article for each model. The data was collected from Scopus and Mendeley databases mainly because the first one is an indexed scientific database commonly recognized as including only good quality publications. The second one (Mendeley) was chosen because it includes also other articles with less impact factor and many books that are not listed in the first one. The keywords used in the search were the following: "Toyota Production System" for TPS; "Theory of Constraints" for TOC; "Kaizen" for KM; "Agile Manufacturing" for AM; "Lean Manufacturing" OR "Lean Production" OR "Lean Thinking" for LT; "Toyota Way" for TW; "Shingo Model" OR "Shingo Prize" for SM.

In Scopus database the search was performed within Article Title, Abstract, and Keywords. In Mendeley database the search is not customizable and it can only be carried out within Article Title and Abstract. The results shown in Table 2 are not 100% accurate for many reasons. Regarding "Kaizen Model" some articles may mention "Kaizen" not about the Kaizen Model, but just using the word "Kaizen" to refer to Continuous Improvement. Regarding "Lean Thinking" the search was carried out using the three keywords "Lean Manufacturing", "Lean Production", and "Lean Thinking because in most cases the authors are referring to the same general philosophy that is assumed here as Lean Thinking. Finally, in the "Shingo Model", the search included also the "Shingo Prize" keyword since the model, although not being formally published it was already existing to support the prize.

The results from the search show that LT is the most popular model both in journal articles as in books. The AM model although being very popular in the academic word through journal articles very small number of books are published. That fact maybe shows little demand by practitioners. Although not connected with Agile Manufacturing, curiously, "Agile" word gained large popularity not in manufacturing but in software development by the Agile manifesto [22]. This popularity may result from the fact that some methodologies, such as Scrum [23], were associated with it.

Table 2. Publications Data analysis of different excellence models.

		TPS	TOC	KM	AM	LT	TW	SM
Scopus	Documents	716	5227	1596	6657	10134	68	44
	Book	12	142	21	16	82	1	5
	Journal Article	391	3275	830	2048	4710	39	18
Mendeley	Documents	1337	2867	3975	1870	15424	183	87
	Book	42	45	92	3	303	15	10
	Journal	831	1978	2577	1078	9732	99	37
Year of the 1st Book		1988	1984	1986	1991	1996	2004	2011
		[8]	[9]	[12]	[15]	[4]	[6]	[24]
Year of the 1st Journal Article		1977	1985	1986	1994	1997	2003	2013
		[4]	[25]	[26]	[27]	[28]	[29]	[30]

The TOC model has been very popular among scholars with a large number of journal publications but also with an interesting number of published books (45). The large number of books may indicate some curiosity and popularity among practitioners. That attraction may be explained by the existence of a Production Planning and Control system called Optimized Production Technology (OPT) which uses a flow control technique called Drum-Buffer-Rope (DBR). TW and SM are the most recent model and that is probably the reason why the number of publications is still quite low, especially SM.

4   Comparing the Models

The method to compare the models is based on 7 criteria listed in the first column of Table 3. For each criteria a "High", "Average" or "Low" is assigned to each excellence model according to authors' judgment from the available and relevant published material. The criteria were selected according to authors' point of view regarding their impact in organizational and management excellence. This way to analyze and compare these models does not pretend to be the best way but it covers the criteria that are most relevant according to the point of view of the authors. The reasons behind the selection of each one of them is presented in the following paragraphs.

Focus on Pull Flow was selected because it plays a key role in the overall performance of production. This concept or principle is one of the most important paradigm shifts proposed by TPS and copied in western companies. On the technical side of the TPS, this concept is responsible for breaking many beliefs and myths developed in the mass production era. This principle is clearly stated in all excellence models considered in this article with the exception of the AM model as can be seen in Fig. 3.

Table 3. Comparison between excellence models.

	TPS	TOC	KM	AM	LT	TW	SM
Focus on Pull Flow	High	High	High	Low	High	High	High
Focus on Process	High	Low	High	Average	High	High	High
Focus on Respect for People (or Sociotechnical scope)	High	Low	High	High	Low	High	High
Existence of Associated Techniques	High	Average	High	Low	High	High	High
Coverage on Indirect Areas	Low	Low	High	Low	Low	High	High
Popularity in Journals	High	High	High	High	High	Low	Low
Popularity in Books	Average	Average	High	Low	High	Low	Low

The second criterion, Focus on Process, is understood here as the process being the only responsible for its outcomes. People cannot be blamed for poor processes. Poor processes cannot produce excellent results so every process must be totally controlled and reliable. TOC does not show pieces of evidence of focus on process except for the bottleneck since its only concern is the throughput protection. The AM model shows little evidences in this respect and, apart from that, since it promotes the use of new technologies its reliability can be difficult to guaranty. The Toyota Way, for instance, states in one of its principles "Use only reliable, thoroughly tested technology that serves your people and processes".

The Respect for People criterion was included because excellence cannot be achieved without the human side of organizations. This principle is clearly stated in most excellence models considered in this article except for TOC and LT. Since the primary focus of TOC is the flow of materials little focus is naturally given to human aspects. The reality of LT is slightly different. Most Lean Thinking followers and practitioners may claim that the model is also concerned with that "respect for people" side of TPS. That seems true because "Lean" is understood by many as just a different word to refer to TPS or now the Toyota Way. The reality is that even the word "Lean" suggests ideas such as "without fat", "without waste", or very little quantities of WIP in the productions, nice production flow, and so on. The word "Lean" suggests much more the physical aspects of production than the aspects linked to the social sciences. Only one principle of LT can include some aspects of the "respect for people" side, the Pursue of perfection principle.

The criterion "Existence of Associated Techniques" was selected because professionals normally feel more comfortable when techniques are available to implement in order to achieve results. Techniques help the materialization of a principle or a concept and for that reason this criterion was considered here in this study. TPS, KM, LT, TW, and SM are highly linked to several tools and techniques while TOC holds only one specific technique for material flow management and AM has no specific connection to specific techniques or tools.

Coverage on Indirect Areas is an important criterion since more and more people work in indirect areas in companies. The competitiveness of any company is also achieved by the performance of its indirect areas. Based on that it is relevant the level at which the model can be applied in indirect areas such as office, intellectual, and research and development work. KM, TW, and SM are the only models covering those areas.

The last two criteria related to popularity are measure by the number of publications in scientific journals and books. TW and SM are not very popular maybe because they are very recent.

5   Conclusions

The objective of this paper is to analyze and compare some of the most popular TIEM (TPS Inspired Excellence Models) in terms of some specific criteria. The study used the formal information supplied by the creators of each model as well as data from published scientific articles and books. Although Lean Thinking, Lean Production, and simply "Lean" is widely accepted as a different name for TPS, the reality shows that formally the principles presented by their founders [4] do not cover some important aspects that were present in the original TPS model. Based on the principles and/or concepts formally defined for each excellence model considered in this study the authors conclude that Kaizen Model, Toyota Way, and Shingo Model are the most comprehensive excellence models considered. The interesting aspect of Kaizen Model is that it was proposed in 1986 while the Toyota Way was proposed in 2003 and Shingo Model was only formally presented in 2011. Contrary to the Theory Of Constraints, Agile Manufacturing although including the social sciences side of the technical nature of organizations, does not recognize the value of pull flow, which is an important practical principle to achieve excellence. Moreover, AM does not provide nor promote the use of practical tools and techniques to achieve excellence.

Acknowledgements. This work has been supported by FCT – Fundação para a Ciência e Tecnologia within the R&D Units Project Scope: UIDB/00319/2020.

References

1. Abrahamson, E.: Management fashion. Acad. Manag. Rev. **21**(1) (1996). https://doi.org/10.5465/AMR.1996.9602161572
2. Benders, J., Van Veen, K.: What's in a fashion? Interpretative viability and management fashions. Organization **8**(1) (2001). https://doi.org/10.1177/135050840181003
3. Perkmann, M., Spicer, A.: How are management fashions institutionalized? The role of institutional work. Hum. Relat. **61**(6) (2008). https://doi.org/10.1177/0018726708092406
4. Womack, J., Jones, D.: Lean Thinking: Banish Waste and Create Wealth in Your Corporation. Fee Press, New York (1996)
5. The_Shingo_Institute. MyEducator - The Shingo Institute (2020). https://app.myeducator.com/reader/web/1705a/foreword/re0xx/. Accessed 18 Aug 2020
6. Liker, J.: Toyota Way: 14 Management Principles from the World's Greatest Manufacturer. McGraw-Hill Education, New York (2004)
7. Sugimori, Y., Kusunoki, K., Cho, F., Uchikawa, S.: Toyota production system and Kanban system: materialization of just-in-time and respect for human systems. Int. J. Prod. Res. **15**, 553–564 (1977). https://doi.org/10.1080/00207547708943149
8. Ohno, T.: Toyota Production System: Beyond Large-Scale Production, 3ª Edição. Productivity, Inc., New York (1988)
9. Goldratt, E., Cox, J.: The Goal: A Process of Ongoing Improvement. North River Press, Great Barrington (1984)
10. Fox, R.E.: MRP, Kanban, and OPT - What's Best? (1982)
11. Goldratt, E.: Computerized shop floor scheduling. Int. J. Prod. Res. **26**(3) (1988). https://doi.org/10.1080/00207548808947875
12. Imai, M.: KAIZEN – the key to Japan's Competitive Success (1986)
13. Sugimori, Y., Kusunoki, K., Cho, F., Uchikawa, S.: Toyota production system and Kanban system Materialization of just-in-time and respect-for-human system. Int. J. Prod. Res. **15**(6), 553–564 (1977). https://doi.org/10.1080/00207547708943149
14. Womack, J., Jones, D., Roos, D.: The Machine that Changed the World. Free Press, New York (1990)

15. Nagel, R.N., Dove, R.: 21st Century Manufacturing Enterprise Strategy, An Industry-Led View. Diane Publishing Co., Bethlehem (1991)
16. Krafcik, J.F.: Triumph of the lean production system. Sloan Manag. Rev. **30**(1), 41–52 (1988). https://doi.org/10.1108/01443570911005992
17. Kidd, P.: Agile Manufacturing: Forging New Frontiers. Addison-Wesley, Boston (1994)
18. Shingo Institute. The Shingo Model (2020). https://shingo.org/shingo-model/. Accessed 08 June 2020
19. Shingo Institute. Shingo Institute: Our history (2106). https://www.cob.calpoly.edu/centralcoastlean/wp-content/uploads/sites/6/2017/07/Shingo-California-presentation-Feb-2016-edited.pdf. Accessed 06 Sept 2021
20. Miller, R.D.: Hearing the Voice of the Shingo Principles: Creating Sustainable Cultures of Enterprise Excellence. Routledge, Milton Park (2018)
21. Bhullar, A.S., Gan, C.W., Ang, A.J.L., Ma, B., Lim, R.Y.G., Toh, M.H.: Operational excellence frameworks-Case studies and applicability to SMEs in Singapore. In: IEEE International Conference on Industrial Engineering and Engineering Management, vol. 2015-January (2014). https://doi.org/10.1109/IEEM.2014.7058722
22. Beck, K., et al.: Agile Manifesto, Software Development (2001). http://agilemanifesto.org/
23. Sutherland, J., Schwaber, K.: The Scrum Guide, the Definitive Guide to scrum: The Rules of the Game (2017). https://www.scrumguides.org/scrum-guide.html
24. Rizzo, K.: Operational and Performance Excellence: The Shingo Model. Printing Industries Pr (2011)
25. Politou, A., Georgiadis, P.: Production planning and control in flow shop operations using drum buffer rope methodology: a system dynamics approach, pp. 1–18 (1985)
26. Schneiderman, A.M.: Optimum quality costs and zero defects: are they contradictory concepts? Qual. Prog. **19**(11), 28–31 (1986)
27. Ross, E.M.: Twenty-first century enterprise, agile manufacturing and something called CALS. World Cl. Des. Manuf. **1**(3), 5–10 (1994). https://doi.org/10.1108/09642369210056593
28. Piercy, N.F., Morgan, N.A.: The impact of lean thinking and the lean enterprise on marketing: threat or synergy? J. Mark. Manag. **13**(7), 679–693 (1997). https://doi.org/10.1080/0267257X.1997.9964504
29. Fane, G.R., Vaghefi, M.R., Van Deusen, C., Woods, L.A.: Competitive advantage the Toyota way. Bus. Strateg. Rev. **14**(4), 51–60 (2003). https://doi.org/10.1111/j..2003.00286.x
30. Naftanaila, I., Radu, C., Cioana, G.: Operational excellence - a key to world-class business performance. Stud. Bus. Econ. **8**(3), 133–141 (2013)

Quantum Lean: The Next Step in Lean Systems

Sean Fields[(⊠)] and Michael Sanders[(⊠)]

BeehiveFund (a Non-profit 501c3 Corporation),
The Woodlands, TX 77380, USA
{s.fields,michael}@beehivefund.org

Abstract. Based on the Toyota Production System (TPS) business model, Lean Systems were popularized in the 1990s and have helped many businesses achieve significant gains in profitability and competitiveness. Despite this, a majority of organizations that attempt to adopt this system fail to do so. One reason this can occur is an inability to adapt lean principles to specific environments. To address this problem, alternative methods like Scrum, Agile, and Quick Response Manufacturing (QRM) were developed. As a further advance to lean approaches, an alternative system called "Quantum Lean" (QL) was developed to build on the work of lean innovators and apply continuous improvement to lean systems. Compared to other lean approaches, QL offers a combination of greater simplicity, efficiency, speed, and comprehensiveness. The key to this system is a framework that is:

Time-Focused – Instead of addressing waste like traditional lean does, QL attacks time. This results in much greater simplicity by leveraging the fact that time is the fuel that feeds all waste. Instead of attacking waste in its 8 forms, addressing time is simpler.

Product-Centric – Instead of tracking time from the view of the customer/resource/owner, time is tracked and measured from the standpoint of the product and only the product. Relentlessly maintaining this framework assures consistency in conclusions and reduces the conflicting objectives (e.g., owner vs. employee vs. customer) that come with other continuous improvement approaches.

Simple and Comprehensive – QL employs a comprehensive and structured analytic methods including Product Path Diagramming (PPD). Compared to other methods like Value Stream Mapping (Traditional Lean) or Manufacturing Critical-Path Time Mapping (QRM), PPD offers greater simplicity, refined prioritization, and/or an improved ability to identify and minimize every waste and contributor to time-in-fulfillment.

Although there are additional benefits, QL's simplicity is arguably its greatest virtue and makes additional advantages possible.

Keywords: Lean · Lean Systems · Lean manufacturing · Quick Response Manufacturing (QRM) · Quantum Lean · Continuous improvement · Value Stream Mapping (VSM) · Manufacturing Critical-Path Time Mapping (MCT) · Product Path Diagramming (PPD) · Operational excellence

© IFIP International Federation for Information Processing 2021
Published by Springer Nature Switzerland AG 2021
D. J. Powell et al. (Eds.): ELEC 2021, IFIP AICT 610, pp. 247–258, 2021.
https://doi.org/10.1007/978-3-030-92934-3_25

1 Introduction to Lean Systems [1]

Due to global competition, many Western companies were investigating ways to preserve markets against low-cost countries. As a result of this, one of the movements that sustained was lean systems. Lean is an approach that's based on Toyota's business techniques (aka "Toyota Production System" (TPS)). It includes a variety of methods to achieve quality, efficiency, and quick delivery and has been deployed at many companies to great effect. It's been so successful that these techniques have been extended to services and government.

Although Henry Ford and the Japanese pioneered lean ideas, the term "lean production" was popularized by James Womack, Daniel Jones, and Daniel Roos in their 1990 book" The Machine That Changed the World". In this work, the authors articulated the methods that the Japanese used to achieve dominance in the auto industry. From there, Womack and Jones published a book called "Lean Thinking" that propounded the principles of what is known as "Lean Systems" [1].

From this point, businesses worldwide started implementing the concepts. At the same time, practitioners developed variations with the goal of addressing lean deficiencies and improving results. One example of this is Quick Response Manufacturing (QRM), which is rooted in the concept of time-based competition pioneered by the Japanese in the 1980's and first formulated by George Stalk, Jr. [2] To expand and clarify QL principles, QL will be compared and contrasted with traditional lean systems and QRM.

2 Overview of Lean Systems, Quick Response Manufacturing, and Quantum Lean

2.1 Traditional Lean (TL) [1]

While lean deployments sometimes depart from the ideas originally espoused by Womack and company, there are common principles and practices that are common to typical lean programs:

- Value-Centric – According to the book "Lean Thinking", value is defined as "the capability provided to customer at the right time at an appropriate price, as defined in each case by the customer". This idea of value is the starting point for many lean efforts.
- Precepts - According to the book "Lean Thinking", the precepts of lean are:
  - Identifying Value Streams – The actions required to bring a product from concept to completion are documented. Often, a flowcharting technique called Value Stream Mapping (VSM) is used to identify activities and wastes. 26 + symbols are used to distinguish the different types if activities, events, and elements.
  - Waste-Based Analysis – After the value stream is identified, activities are typically characterized as value-added (activities that clearly create value) and non-value added (activities that clearly create no value). Non-value added work

is also called "waste". To aid in identifying inefficiencies, waste is often classified according to the following 8 categories (aka, "8 Wastes"):

- People – Not realizing the full potential of a workforce's talent, skills, and knowledge
- Motion – Unnecessary people movements
- Transportation – Material and product movement
- Defects – Efforts as a result of scrap, rework, and non-conformity
- Overprocessing – Unnecessary or non-value added processes
- Overproduction – Manufacture of product that is not needed
- Inventory – Excess product and material not being processed
- Waiting – Delay in waiting for the next step of a process
- Establishing Flow – A state where the product is fulfilled with no stoppages, scrap or backflows
- Pull – Product fulfillment only initiates at the request of a customer
- Perfection – The pursuit of complete waste elimination.

2.2   Quick Response Manufacturing (QRM) [2]

As part of lean's refinement, other systems were developed to widen lean's applicability and to simplify the improvement process. One example of this is Quick Response Manufacturing (QRM). Key principles and practices of this approach include:

- Time-Focus – Quick Response Manufacturing (QRM) emphasizes the reduction of internal and external lead times to achieve sustainable competitive advantage.
- Time-Based Analysis – Activities are evaluated based on their effect on lead time. In addition, unique metrics are used to track performance.
- Manufacturing Critical-Path Time (MCT) Mapping – To help determine where improvement is needed, MCT Maps are used to define when a product is being processed and when it is waiting.
- Points of Emphasis – Frequently, QRM's points of emphasis include:
  - QRM Mindset – Instilling an understanding of the link between waste and lead time along with the creation and training of a steering committee to select and guide implementation projects.
  - Organization – QRM emphasizes the establishment of QRM Cells for "Focused Target Market Segments" where products with shorter lead times can yield significant market benefits. QRM Cells use cellular production principles to expedite the production of these items.
  - System Dynamics – An emphasis on system dynamics (e.g. queueing theory) to guide lead reduction efforts.

2.3   Quantum Lean (QL) [3]

To simplify the improvement process and widen lean's applicability, Quantum Lean was developed to make lean more accessible to beginners and to address incompatibilities between conventional techniques and environments that aren't mass-production. As one example of mismatched settings, there is a high probability that a VSM will fail to identify significant wastes in environments like job-shops, offices, and services. After several implementations where such a situation occurred, QL's originators concluded that a new angle was essential and developed an alternative approach to lean systems. It is easy to learn, prioritizes opportunities, and indicates corresponding solutions. As an additional advantage, it can be used in any business environment including mass production, job shops, and services.

QL Approach - Quantum Lean documents a product's in fulfillment by capturing the actions, events, and occurrences that a product undergoes. To facilitate analysis and prioritization, the time that constitutes these activities and events are classified as conversion, non-conversion, and delay and are defined as:

- Conversion – When the product is being transformed by a resource into a configuration that is closer to finished form.
- Non-Conversion - When the product is being processed by a resource but is not brought closer to the finished form. For simplicity, QL limits this category to moving, handling, rework, and inspection.
- Delay – When no man, machine, or any other resource is expending effort on the product.

The QL symbols look like the following (See Fig. 1):

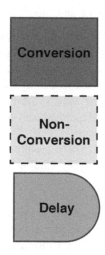

Fig. 1. Quantum lean diagramming symbols

An additional and critical provision of QL is all time is strictly tracked from the standpoint of the product, and only the product. For example, if a worker spends 10 min finding tools, QL characterizes this as a delay for the product instead of as a non-conversion time due to a worker's excess motion.

Once the activities that comprise a product's time in fulfillment are documented and defined, QL prioritizes the activities for improvement based on the category (delay, conversion, non-conversion) that occupies the plurality of a product's time. On an assembly line where delay and non-conversion are insignificant, the first concern will be streamlining conversion activities. In a typical job shop, the priority will be minimizing delays.

From this point, each event/occurrence that falls within the prioritized category is analyzed using QL heuristics to pinpoint causes of excess time and identify solutions to eliminate or minimize it. By addressing these factors in this manner, inefficiencies are addressed without the need to identify particular wastes (i.e. "the 8 wastes").

Over time, the QL system targets and streamlines every aspect and stage of an organization's fulfillment. In addition, QL can be applied to any type of organization including mass production, job shops, offices, and services.

Overall, QL offers a simplified, comprehensive, and thorough system for systematically eliminating inefficiency in ways that other systems don't. This point will be elaborated on in the following sections comparing and contrasting QL against other approaches.

3 Comparison of Quantum Lean (QL) to Other Lean Systems

While all lean systems share a goal of minimizing time in fulfillment, each school of thought deploys different frameworks, strategies, and methods. This section compares and contrasts QL's approach with QRM and TL and will illustrate key differences with an example of each approach's analytic methods (VSM, MCT, and PPD).

3.1 Comparison of Quantum Lean (QL) to Traditional Lean (TL)

Major differences between Quantum Lean (QL) and Traditional Lean (TL) include:

- Waste-based vs. Time-Based – As a way to minimize time in fulfillment, a major precept of TL is that flow and pull need to be established. However, TL lacks an accessible and systematic framework to achieve this. For this reason, many TL implementations deemphasize flow/pull and focus on waste-reduction. Although this can be effective in mass-production environments where flow/pull is already established, a major opportunity for increasing efficiency through increased flow can be lost in other environments like job-shops, offices, and service entities.
  QL's time-based technique avoids this pitfall by providing a systematic and accessible way to maximize flow for any organization that has not yet achieved flow. In addition, QL provides a mechanism for systematically eliminating waste once flow has been established.
- Multiple Frames of Reference vs. Product-Centricity – In the field, TL implementations often use the 8 waste framework to classify activities and this approach lacks a consistent frame of reference. This results in activities being analyzed from a variety of viewpoints that include:
  - Resource – Examples include reducing labor costs, increasing resource utilization, and decreasing equipment usage (e.g. forklift travel).
  - Customer – Examples include increasing on-time delivery and customer satisfaction.

- Money – Examples include increasing profitability, inventory turns, decreasing capital costs, and minimizing supply usage.
- Organization – Example would include increasing market share
- When a variety of viewpoints are used, contradictory results occur. QL achieves consistent results by strictly maintaining a product-centric framework when analyzing time-in-fulfillment, prioritizing targets, and determining improvements.

Practically, these differences between TL and QL manifest themselves in the following ways:

TL is complicated compared to QL – Although the principles are simple, traditional lean methods are complicated and workforces are frequently slow to understand them. For example, Value Stream Mapping (VSM) utilizes 26+ symbols. In addition, the concept of 8 wastes further complicates implementation.

As many find it difficult to internalize waste-centric and value-centric frameworks, shortcuts are often taken and can manifest themselves into insignificant pursuits like minimizing travel or take the form of a tool-obsession where a lean tool like Kanban is adopted whether or not it is really needed.

TL is prone to inconsistent conclusions – Everyone is familiar with the idea of value. At the same time, no two opinions on it are alike. No matter how carefully value is explained, preconceived notions tend to override the most precise definition. This often results in difficulties with harmonizing stakeholders. As sound analysis requires a consistent frame of reference, this key requirement can't be met due to value's ambiguity.

In addition to value-driven contradictions, waste-based techniques compound the confusion with a constantly shifting framework. Depending on the waste, the frame of reference can center on a product (overproduction, transportation, inventory), people (motion, waiting), both people and product (defects, overprocessing), or an organization (people).

When frames of reference shift, this leads to conflicting objectives and decisions. A classic example of this is a buyer awarding a contract to a low bidder and causing production costs to skyrocket due to poor part quality. While the buyer made an optimal decision in a purchasing-centric framework, his conclusions are otherwise faulty.

TL is divisive – Classifying activity as "non-value added" carries a connotation that can alienate employees. In addition, a common understanding of goals is a must for harmonizing a workforce. With the inconsistencies associated with value and waste frameworks, it's challenging to coordinate stakeholders. Some of the potential conflicts that may result include:

- Department vs. Department - An example of this would be a purchasing department that sources lower-cost parts, but creates significant downstream quality problems for production.
- Customer vs. Company – From automated customer service to self-check at the grocery store, companies inconveniencing customers to save money is common.
- Employer vs. Employee – In the path to "lean", many companies penalize employees with cost-cutting efforts.

- Employee vs. Employee – Conventional lean approaches offer nothing to mitigate the innate friction that can occur among people whose perceived interests are in conflict.

Although conflict is inevitable in any organization, QL avoids the conflicts that are embedded in TL's structure.

3.2 Comparison of Quantum Lean (QL) to Quick Response Manufacturing (QRM)

While QL and QRM both emphasize lead time reduction, there are significant differences between Quantum Lean (QL) and Quick Response Manufacturing (QRM) that include:

Application – QRM is largely restricted toward custom manufacturers that have not established flow. Because its framework and techniques are comprehensive in scope, QL is applicable to any business at any stage of a lean transformation.

Scope – QRM emphasizes the elimination of delays and deemphasizes the streamlining of non-conversion and conversion activities. QL provides a framework for systematically addressing delays, conversions, and non-conversions in a manner that accomplishes maximum impact in the shortest time.

Granularity – The QRM technique of Manufacturing Critical Path (MCT) identifies time components of a product's time in fulfillment and subdivides them according to whether a product is being processed or not being processed. The QL product path diagram (PPD) categorizes the time in fulfillment and performs a finer classification of the time components according to the nature of the delays (e.g. WIP, batch), conversions, and non-conversions.

Practically, these differences between QRM and QL manifest themselves in the following ways:

QL is more versatile – By design, QRM relegates itself to custom organizations. By contrast, QL is highly applicable to any kind of operation including mass production, job shops, offices and services.

QL is more comprehensive – QRM's greatest utility is reserved to organizations that have not established pull or flow. Once flow/pull has been achieved, QRM's usefulness fades. By contrast, QL works equally well in all phases of a lean transformation.

QL is easier to apply – While an MRT and a PPD are similarly easy to develop, the additional granularity of a PPD and straightforward QL heuristics allow inexperienced users to quickly to prioritize targets, identify solutions, and develop sound action plans. By contrast, an MCT provides limited insight into the causes of excess time-in-fulfillment and little guidance is provided for identifying corresponding solutions.

4   Example – Product Path Diagram (PPD) and Value Stream Mapping (VSM) and Manufacturing Critical Path (MCT) Map

To compare and contrast the PPD, VSM, and MCT Map, a common scenario for a fabrication shop will be used:

- An operator waits 30 min to get access to a crane so he can remove sheet metal from a rack.
- Using the crane, sheet metal is removed from a shelf and moved to a computer-controlled cutting table in 10 min.
- Due to a lack of coordination between production control and the shop floor, the program needed to run the table will not be available for an hour.
- Once a program is available to run the cutting table, setup operations require 30 min.
- After setup, the sheet metal is cut into 30 pieces at a rate of two minutes per piece.
- After cutting is finished, a forklift arrives in 15 min and the parts are moved to assembly in five minutes.
- Since assembly is working on another order, the parts are placed in queue.
- After 100 min in queue, assembly is ready to process the cut pieces.

4.1   Value Stream Mapping (VSM) - Fabrication

For the fabrication shop example, a VSM would look like this (See Fig. 2):

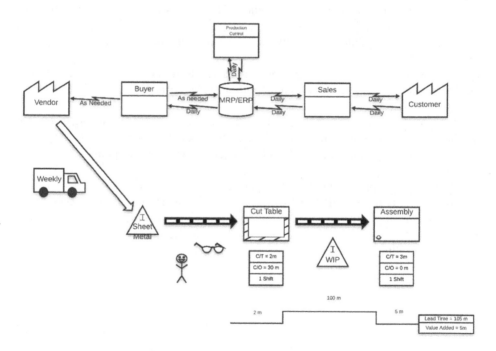

Fig. 2. Value Stream Map (Fabrication)

Despite abundant opportunities to improve, the VSM gives few indications of this:

- The VSM contains several symbols, but little insight can be gleaned about what to work on and to what priority order.
- The only potential targets that can be identified from this diagram include:
  - WIP – Assy – This can be identified from the timeline at the bottom of the VSM
  - Setup – This can be identified from the information box associated with cutting table operation.
- Missing waits and wastes include:
  - Wait for Program
  - Batch
  - Wait for Crane
  - Wait for Forklift
  - Move to Table
  - Move to Assy

4.2 Manufacturing Critical Path (MCT) Map - Fabrication

For the fabrication shop example, an MCT map would look like this (See Fig. 3):

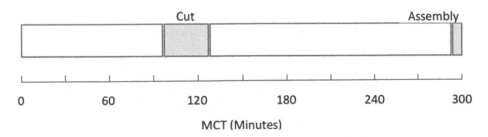

Fig. 3. – MCT Map (Fabrication)

The MCT prioritizes areas to improve but does not indicate potential causes for wait times. These causes will need to be identified in a subsequent investigation. In addition, no wastes are specifically identified. Some points:

- An MCT Map is virtually identical to a VSM timeline. An MCT Map could possibly be characterized as a VSM without any VSM icons.
- Once the time between Cut and Assembly is prioritized for reduction, the task of determining what to work on and to what priority remains.
- With QRM almost solely focused on addressing the contribution of system dynamics to lead time, significant impacts for lead time reduction (like non-conversions and their associated delays) may be overlooked.
- Since wait times between conversion operations often consist of multiple events and actions, it is possible that the priority established by an MCT Map may result in addressing targets that offer suboptimal impact.

4.3   Product Path Diagram (PPD) - Fabrication

The PPD for the fabrication shop example looks like this (See Fig. 4):

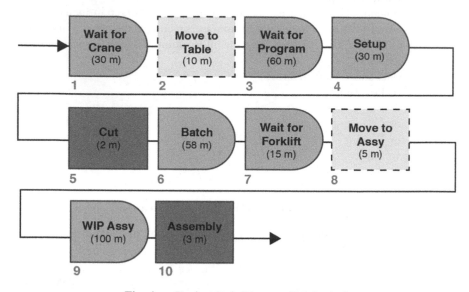

Fig. 4. – Product Path Diagram (Fabrication)

The PPD accurately reveals a target-rich environment. As each element is addressed, QL heuristics guide the user in prioritization, determining potential reasons for the time contributor, and identifying appropriate lean tools to minimize or even eliminate the time. Since the bottleneck is not in the portion of the shop being diagrammed, the PPD indicates that the following targets should be addressed in the following priority order:

- WIP – Assy
- Wait for Program
- Batch
- 4a) Wait for Crane
- 4b) Setup
- Wait for Forklift
- Move to Table
- Move to Assy
- Assembly
- Cut

5   Advantages of Quantum Lean Systems

Benefits afforded by Quantum Lean include:

- Simplicity – QL methods for analyzing and optimizing processes are much simpler than waste-based techniques. Where the QL diagramming method uses three symbols, Value Stream Mapping (VSM) employs 26+. Also, QL's focus on minimizing a product's time in fulfillment sidesteps the challenge of identifying 8 (or 12) types of wastes. Reducing the number of flowcharting symbols from 26 to 3 and the things employees have to look at from 8 to 1 correspondingly reduces complexity.
- Versatility – Without modification, QL can be applied in any setting including product development, production, and office. By contrast, other lean systems vary their approach based on environment. For example, where there are 8 wastes assigned for manufacturing, there are 12 for product development. In addition, QL diagramming techniques can be used for improving job methods and single minute exchange of dies.
- Comprehensiveness – QL seamlessly addresses every phase of a lean implementation including the implementation of pull, flow, and waste elimination. QRM only addresses the establishment of pull and flow. Frequently, TL only addresses waste elimination.
- Speed – With enhanced versatility and simplicity, QL techniques accelerates a workforce's learning curve. In addition, as QL is easier to understand and remember, the odds of QL being misapplied is much lower than when other approaches are used.
- Buy-In – With QL's product-centricity, personnel aren't put on the defensive. For example, if an employee has to walk several hundred feet to get tools, typical lean efforts will zero in on a person's excess travel. Conversely, a product-centric approach looks at the same situation as the product having to wait for tools to become available. Approaching lean the latter way deflects undeserved blame for problems away from employees. In addition, employees appreciate QL's simplicity.
- Cost – QL's versatility and simplicity reduces training costs, minimizes misfires, and accelerates deployment. Overall, this results in lower implementation costs and a quicker return on investment.

Although there are additional benefits, QL's simplicity is arguably its greatest virtue and what makes the other advantages possible.

6   Summary

Lean systems have accomplished much and offers a compelling and positive vision with the possibility of wins for owners, employees, and customers alike. As it has made such a significant difference, the world owes a tremendous debt to its founders. Like any school of thought, improvement is possible and Quantum Lean (QL) offers next-level performance for lean systems that includes greater simplicity, increased workforce buy-in, quicker deployment, and lower cost of implementation. With a similarly positive and compelling vision, QL offers the same benefits as TL, but is formulated to deliver them more efficiently and effectively. Compared to QRM, QL is applicable to all phases of a lean implementation, offers benefits more comprehensively, and allows an entire workforce (instead of a steering committee) to identify targets and solutions.

References

1. Weigel, A.: Web.mit.edu (2000). https://web.mit.edu/esd.83/www/notebook/WomackJones. PDF
2. Suri, R.: It's About Time. The Competitive Advantage of Quick Response Manufacturing. Productivity Press (2010)
3. Fields, S., Sanders, M.: Quantum Lean: Taking Lean Systems to the Next Level, 1st edn. J Ross Publishing, Ft, Lauderdale (2020)

What Comes After the Transformation?

Characteristics of Continuous Improvement Organizations

Ton van Kollenburg$^{(\boxtimes)}$ (iD) and Alinda Kokkinou (iD)

Avans University of Applied Science, Breda, The Netherlands
`ajc.vankollenburg@avans.nl`

Abstract. Continuous Improvement initiatives are diverse, encompassing, amongst others, lean management, six sigma, TPM and TQM. Striking is that the Continuous Improvement literature focuses mainly on the transformation process itself, with little attention given to its outcome. This study addresses this gap by examining the characteristics of organizations after the continuous improvement transformation. A review of scientific papers and management books yielded eight elements characterizing a continuous improvement organization. These elements were described using 25 main characteristics, thus bringing clarity to academics and practitioners about the definition and attributes of becoming a continuous improvement organization.

Keywords: Continuous improvement · Lean management · Organization

1 Introduction

According to Gupta and Sharma multiple academics and practitioners have attempted to define lean management [1]. Some focused on the customer perspective, others on the waste reduction perspective. However, there is no apparent consensus on the definition of lean. Partly, this is caused by its evolution over a long period, partly because of its mistaken equivalence with other quality-related approaches [1]. On top of that, based on their literature review, Bhuiyan & Bagchel [2] found no theoretical basis for continuous improvement (CI), so CI should be used as a general term that has acquired many of its attributes from other quality initiatives. A clear definition of continuous improvement is thus missing.

Literature study from Jurburg, Viles, Tanco and Mateo [3] shows that, from the second half of the twentieth century, companies worldwide have started to adopt CI systems, with many benefits. An analysis of 1090 papers on CI topics shows that 76% of the articles about lean management, continuous improvement, six sigma, and so forth are about implementation. The other 24% discuss philosophies, culture, concept, and innovation [4]. Despite all the focus on implementation, implementing CI remains difficult. For organizations, it seems evident that implementation is hard for them, but do they know where they are going? What kind of organization do they want to become?

© IFIP International Federation for Information Processing 2021
Published by Springer Nature Switzerland AG 2021
D. J. Powell et al. (Eds.): ELEC 2021, IFIP AICT 610, pp. 259–268, 2021.
https://doi.org/10.1007/978-3-030-92934-3_26

Many companies are trying to adopt CI due to changes in the business environment, the emergence of new management systems, and the importance of quality management [4]. At the same time, a proper definition within organizations is missing, and the design of a CI organization is unclear. Therefore, this study aims to find out what a CI organization is, to help academics and practitioners. Academics need a clear definition of CI and CI organization characteristics to avoid empirical testing of vague and imprecise concepts leading to a body of research that examines a different aspect of the same underlying constructs, masked by different terminology [5]. Practitioners need to know where they are heading to maintain focus in their organizational transformation, e.g., by creating or using a so-called maturity model. Knowing what a CI organization is, also helps process improvement teams, for they must understand the definitions of the methodology, tools, and change vehicles of quality tools to avoid mismatches which can be fatal to a fledgling improvement program [6].

The main question in this study is: "What are the characteristics of a continuously improving organization as emerging from literature? From this question, several subquestions arise:

1. What is the definition of CI?
2. What are the improvement approaches related to CI?
3. What are the CI characteristics according to these approaches, and how can they be combined into a model?

2   Methodology

2.1   Methods

Two literature reviews were conducted to find the characteristics of a CI organization. The first one focused on the definition of CI and improvement approaches that match with CI. This review was used to select the literature for the second review, focusing on finding the characteristics of a CI organization. A problem with literature study is that most research is carried out in developed economies [1], possibly leading to biased results.

2.2   Definition of CI and Related Improvement Approaches

Articles reviewing CI literature were searched, using databases like Emerald, ScienceDirect, and Springer, and looking for keywords like 'continuous improvement' and 'lean management'. In addition, the so-called 'snowball effect' was used, in which the articles found are used as a source to find others. Based on these articles, a theoretical definition of CI was created. Practitioners play an essential role in the development of CI [4]. Therefor this definition was challenged with those from CI practitioners, that were obtained from websites created by practitioners.

In combination with the definition of CI, the findings of the first literature review were used to define and scope the second literature review. A problem with this approach is that areas that are not mainstream or emerged too recently are missed. To use as many related approaches as possible, the improvement approaches found were discussed in the research group Improving Business of Avans University of Applied Science and complemented with related approaches.

2.3   Characteristics of a CI Organization

In the search for scientific papers and relevant management books, to find the characteristics of a CI organization, the following keywords were used: 'TPM', 'TQM','six sigma', 'theory of constraints', 'QRM', 'agile', 'scrum' and 'lean'. This search may result in a list of irrelevant publications, while the influential ones are missed. Therefore, another search with Google was done, using the following keywords: 'best lean books', 'most sold lean books' and 'top 50 bestselling management books of all time'. The sources identified were tested against the definition of CI.

Selected books and articles were studied to find characteristics in a sequence of steps. First, the sources were scanned. If they were found to include characteristics, they were added to the literature list. After training and briefing, the research team members examined each source in the list, inventorying characteristics of a CI organization after the implementation or transformation. Their findings were inventoried in a list. As the literature review progressed, the list was evaluated regularly to see if every team member had correctly identified the characteristics. Differences of opinion were discussed and resolved in the team. After the reading phase, all characteristics were coded. For every code, the characteristics were analyzed to remove double characteristics and come to each code's core: one or a few short descriptions summarizing the code. Finally, these descriptions were iteratively recombined into a model of a CI organization consisting of eight elements.

3   Results

3.1   Definition of CI and Related Improvement Approaches

Definition of CI

CI tends to be used as a general term. It has acquired many of its attributes from other improvement approaches such as Total Quality Management (TQM) and lean manufacturing, and it is often defined as a culture of continuous improvements of any size that includes all different levels in an organization [2]. Based on a literature review of three decades of CI, Sanchez & Blanco conclude that every author has their definition, but they highlight three characteristics [4]:

- Continuous improvement as a cycle; not as an only act;
- All people from the organization should participate;
- The aim is to improve by focusing on eliminating waste and identifying new areas of improvement.

Singh & Singh also mention the relevance of sustained improvement and improvement in all organizational systems [7]. Zollo & Winter similarly mention two different elements in their definition, namely the systematic generation and modification of operating routines and the pursuit of improved effectiveness [8]. Finally, Jurburg, Viles, Tanco & Mateo mention a systematic approach in the whole organization with everyone achieving greater business productivity, quality, safety, ergonomics, and competitiveness [3].

Based on these definitions, CI is defined as *a systematic and cyclic approach to eliminate waste by involving everyone in improving operating routines in all the systems of an organization and come to sustained improvements regarding effectiveness and efficiency.*

This academic definition was challenged by contrasting it with practitioners' definitions. A casual Google search came up with 404,000,000 results regarding 'definition continuous improvement'. By focusing on definitions from more reputable practitioners (thus excluding websites like Wikipedia and retaining only websites with named authors), further analysis revealed that after seven websites, definitions started to repeat. The keywords derived from the definitions of practitioners on these websites were [9–15]:

- never-ending, long term;
- strive for perfection in everything you do;
- continuous (happening all the time) versus continual (not going on all the time);
- adding more value by improvement of products, services, or processes;
- raising performances regarding efficiency, effectiveness, quality, speed, flexibility, cost, and sustainability;
- perfecting on-the-go instead of one-off initiatives;
- improvements can be incremental (over time) and breakthrough (all at once);
- responsibility of everyone in the company;
- not something you "do", but a way a company operates;
- condition to become an agile company.

Testing the definition of CI resulted in de following definition (the changes are underlined): *a never-ending, systematic, and cyclic approach that is happening all the time, striving for perfection by adding more value and eliminating waste, where everyone is involved in improving products, services or processes in all the systems of an organization, thus coming to sustained improvements regarding effectiveness and efficiency, e.g., quality, speed, flexibility, cost, and sustainability.*

Related Improvement Approaches

Several authors discuss possibly related approaches regarding CI, e.g., kaizen, lean, and TQM [1, 16], or the similarity of various lean production models [17]. Similar is the discussion about lean, six sigma, and lean six sigma: While they have different objectives, together or separately, they can improve business processes [18, 19].

Studying variants of the Toyota Production System (TPS), Netland [20] determines that companies do not develop these variants from scratch; they are influenced by existing best practices in their industry, resulting in similar variants when it comes to content. Therefore, even differences across different industries are hardly present.

This phenomenon also seems to apply to CI. For example, companies merge different CI initiatives, resulting in a combined CI program, of which lean six sigma is the most well-known hybrid methodology [2]. Similarly, companies using total quality management introduced six sigma to be able to prioritize quality projects. Other examples are Imai [21], combining total quality control, and kaizen and Shirose [22], emphasizing the commonality of just in time (JIT), total quality control (TQC), and total productive maintenance (TPM).

CI is part of a family of related approaches and concepts. Therefore, the related approaches will be used to find the characteristics of a CI organization. The following approaches related to CI emerge from the literature [1, 3, 16, 19, 20, 22]: balanced scorecard, business excellence, business process re-engineering, improvement methodology, JIT, kaizen, lean manufacturing, lean six sigma, lean thinking, organizational excellence, quality management systems, six sigma, theory of constraints, TPM, TQC, TQM, TPS, and world-class manufacturing.

However, this list stills lacks modern approaches that seem to be related, namely quick response manufacturing [23] and lean startup [24]. Therefore, to avoid missing relevant characteristics, these approaches were added.

3.2 Characteristics of a CI Organization

A total of 40 books and papers was found and studied; see the list at the end of the paper, resulting in 736 characteristics of a CI organization. These characteristics were coded, double characteristics were removed, and similar characteristics were merged. The characteristics were iteratively combined into a model to come to an accessible overview. The first combination was about the value chain, followed by characteristics about process control. Other characteristics formed a group regarding management, and some characteristics described the role of supporting (staff) processes. Obviously, several characteristics were about improving. A large group of characteristics remained, which could be applied to all former groups. These characteristics were concerned with the people-side of a CI organization: leadership, people, and culture. To come to a useable model, finally, the characteristics were combined into 25 main characteristics, describing the eight elements of a CI organization. Table 1 shows the elements of a CI organization and their main characteristics.

Table 1. Main characteristics CI organization per element.

Element	Main characteristics
Value stream process	Use standardization and visualization
	Customer focus
	Stable processes with built-in quality leading to zero defects
	Flexible, smoothly flowing processes
	Long term cooperation with suppliers, based on mutual trust and improvement
Process control	For a limited number of strategy-based indicators, results are visible and transparent
	The organization is process-based; processes are controlled by teams using startup meetings
	External variations are actively smoothed, and countermeasures are taken to ensure a smooth-running process
	To stabilize the processes standards are clear and sustained, problems are prevented from happening
Management	Processes are clearly defined, with clear responsibilities (including improvement) for process owners and teams
	Long term focus in decision-making, partnerships, and people development
	Management is involved in the gemba and involves everyone in vision, goals, and improvement activities
Support processes	Support systems take a minimum effort and support a process-based organization
	The main focus of supporting units is on improving the process-based way of working
Improving	Everyone is involved in improvement activities in multi-disciplinary teams
	By developing people and processes, a better world is created for all the stakeholders by reducing waste and variation and adding more value
	Continuous learning of successes and failures by ongoing improvement activities, based on PDCA
	Control mechanisms, like indicators and standards, are the basis for improving
	Improvements are stepwise, or leapfrog, based on facts gathered at the gemba
Leadership	Leaders are an example; they demand the following of the standards and give autonomy to improve
	Leaders learn, experiment, and improve, persistent and critical, challenging, encouraging and supporting their people
People	People on all levels are leaders by being involved, persistent, creative, open-minded, progressive, proactive, and having self-discipline
	Ongoing development of people and motivating to use standards to reach goals
Culture	People work together in a safe working environment where they are involved, not blamed; they freely share knowledge and ideas
	People learn by continuous development, continuous reflection on their work and experimenting, mistakes are allowed, and detecting them is praised

4   Discussion

CI was defined as a never-ending, systematic, and cyclic approach that is happening all the time, striving for perfection by adding more value and eliminating waste, where everyone is involved in improving products, services, or processes in all the systems of an organization, thus coming to sustained improvements regarding effectiveness and efficiency, e.g., quality, speed, flexibility, cost, and sustainability. Based on this definition, the improvement approaches were selected, leading to the characteristics of a CI organization. Combining the results of this study resulted in eight elements with 25 main characteristics, see Table 1.

This paper aimed to help academics and practitioners with a clear definition of CI. Starting with the academics, an exact concept of CI helps in their empirical testing and academic discussions. Without such a concept studying and describing the implementation of CI and determining the quality of such implementations is difficult. The CI model can be used for that. More specifically, it helps academics studying the development toward the technical side and shift away from the people-orientated side of CI [20]. The CI model has three elements regarding people and can prevent this development, for instance, by distinguishing between management and leadership. Management and leadership are frequently used interchangeably; however, they are not the same [25]. In summary, management is system-oriented and aiming at control, while leadership is people-oriented and aiming at change [25–27]. They are connected, though, and dividing the two within an organization will generate problems [25]. Therefore, today's organizations need both leaders and managers [26]. Thus, the CI model supports studying the differences and similarities between management and leadership in CI organizations.

For practitioners, the model helps to know where they are heading, thus maintaining focus in their organizational transformation and creating or evaluating so-called maturity models. A clear idea of CI will help them create a dynamic assessment system, which evaluates and improves the used maturity model and its related checklists continuously [28]. Besides, the model may help practitioners to have a clear view of the meaning of CI. A question arising quite often in our work as practitioners is: Are not all organizations improving continuously? They probably are somehow, but the definition of CI makes a clear distinction between continuous (all the time) and continual (not all the time) improving.

Besides, the model helps practitioners to get a clear vision of their organization. An organization's value stream is the starting point of the CI model; all other processes are meant to support its smooth operation with minimal effort. That means minimum registration and inspections, and perhaps even the authority of value stream employees to accept or reject suggestions made by support departments. The study findings show that research on supporting processes seems to be lacking, yet in many organizations support departments seem to be planning and controlling, e.g., by writing policy documents, procedures, and work instructions. Therefore, to get a clear vision of their CI organization and have a proper implementation, the role and weight of the various processes and departments should be discussed, like the importance of support departments related to the value stream.

This study has some limitations. First, the literature on CI and related concepts is vast, and the selected literature in this study is relatively small. Though measures were taken to ensure the most relevant literature was incorporated, perhaps more characteristics exist, or characteristics need modification. For example, from our own practitioners' experience, clear customer-supplier-agreements between sub-processes or shifts in a process lacked in the literature. Though they are part of having a smooth process, they were not mentioned. The same applies to psychological ownership, which goes beyond 'all employees are involved'.

Second, this study assumes that one model of a CI organization fits all organizations worldwide. In contrast, two different variants of the CI term 'kaizen' can be recognized, a Japanese and a Western [29]. These variants assume that contextual factors can play a role in the CI model. This requires further investigation.

In order to improve the knowledge of a CI organization, practitioners and academics must join hands. Practitioners should use the CI model in choosing their CI path and report their findings. Academics should use these findings to improve the characteristics of a CI organization and the applicability of the CI model. The joined forces will contribute to better CI organizations capable of being successful in a rapidly changing era [30].

References

1. Gupta, S., Sharma, M., Sunder, V.: Lean services: a systematic review. Int. J. Product. Perform. Manag. **65**(8), 1025–1056 (2016)
2. Bhuiyan, N., Baghel, A.: An overview of continuous improvement: from the past to the present. Manag. Decis. **43**(5), 761–771 (2005)
3. Jurburg, D., Viles, E., Tanco, M., Mateo, R.: What motivates employees to participate in continuous improvement activities? Total Qual. Manag. **28**(13), 1469–1488 (2017)
4. Sanchez, L., Blanco, B.: Three decades of continuous improvement. Total Qual. Manag. **25**(9), 986–1001 (2014)
5. Shah, R., Ward, P.: Defining and developing measures of Lean production. J. Oper. Manag. **25**(4), 785–805 (2007)
6. Harvey, J.: Match the change vehicle and method to the job. Qual. Prog. **37**(1), 41–48 (2004)
7. Singh, J., Singh, H.: Continuous improvement philosophy – literature review and directions. Benchmark. Int. J. **22**(1), 75–119 (2015)
8. Zollo, M., Winter, S.G.: Deliberate learning and the evolution of dynamic capabilities. Harv. Bus. Rev. **13**(3), 339–351 (2002)
9. Kanbanize. https://kanbanize.com/lean-management/improvement/what-is-continuous-improvement. Accessed 17 Jan 2021
10. ASQ. https://asq.org/quality-resources/continuous-improvement. Accessed 17 Jan 2021
11. Getvetter. https://www.getvetter.com/posts/129-define-continuous-improvement-8-experts-definitions. Accessed 17 Jan 2021
12. Tallyfy. https://tallyfy.com/guides/continuous-improvement. Accessed 17 Jan 2021
13. ProductPlan. https://www.productplan.com/glossary/continuous-improvement. Accessed 17 Jan 2021
14. Triaster. https://blog.triaster.co.uk/blog/what-is-continuous-improvement. Accessed 17 Jan 2021
15. Viima. https://www.viima.com/blog/continuous-improvement. Accessed 17 Jan 2021

16. Carnerud, D., Jaca, C., Bäckström, I.: Kaizen and continuous improvement – trends and patterns over 30 years. TQM J. **30**, 371–390 (2018)
17. Lee, B., Jo, H.: The mutation of the Toyota production system: adapting the TPS at Hyundai motor company. Int. J. Prod. Res. **45**(16), 3665–3679 (2007)
18. Alhuraish, I., Robledo, C., Kobi, A.: A comparative exploration of lean manufacturing and six sigma in terms of their critical success factors. J. Clean. Prod. **164**, 325–337 (2017)
19. Prasanna, M., Vinodh, S.: Lean six sigma in SMEs: an exploration through literature review. J. Eng. Design Technol. **11**(3), 224–250 (2013)
20. Netland, T.: Exploring the phenomenon of company-specific production systems: one-best-way or own-best-way? Int. J. Prod. Res. **51**(4), 1084–1097 (2013)
21. Imai, M.: Kaizen: Key to Japan's Competitive Success. McGraw-Hill, New York (1986)
22. Shirose, K.: TPM New Implementation Program in Fabrication and Assembly Industries, 7th edn. Japan Institute of Plant Maintenance, Tokyo (2007)
23. Suri, R.: It's About Time. Taylor and Francis Group LLC, New York (2010)
24. Ries, E.: The Lean Startup, 1st edn. Penguin Books Ltd., London (2011)
25. Vasilescu, M.: Management versus leadership: a key theoretical distinction. Ann. Constantin Brânusi University **6**, 170–176 (2018)
26. Toor, S., Ofori, G.: Leadership versus management: how they are different and why. Leadersh. Manag. Eng. **8**(2), 61–71 (2008)
27. Oltean, A.: Lider versus manager, management versus leadership. Ann. Constantin Brânusi University **5**, 167–172 (2016)
28. Maasouman, M.A., Demirli, K.: Development of a lean maturity model for operational level planning. Int. J. Adv. Manuf. Technol. **83**(5–8), 1171–1188 (2015). https://doi.org/10.1007/s00170-015-7513-4
29. Suárez-Barraza, M.F., Ramis-Pujol, J., Kerbache, L.: Thoughts on kaizen and its evolution: three different perspectives and guiding principles. Int. J. Lean Six Sigma **2**(4), 288–308 (2011)
30. Oakland, J.: From quality to excellence in the 21st century. Total Qual. Manag. Bus. Excell. **16**(8–9), 1053–1060 (2005)

References Study of CI Characteristics (in Alphabetic Order)

31. Aalders, J., van den Broek, C., Mathot, J., Obers, P., Rensen, E.: Total Productive Maintenance. Kluwer Bedrijfsinformatie BV, Deventer (1995)
32. van Assen, M.: Handboek Lean Management. Academic Service, Den Haag (2014)
33. Bessant, J., Caffyn, S., Gilbert, J., Harding, R., Webb, S.: Rediscovering continuous improvement. Technovation **14**(1), 17–29 (1994)
34. van der Bij, H.: Kwaliteitsmanagement in beweging. Kluwer, Deventer (2014)
35. Blom, S.: Monozukuri, doen met aandacht. Blom Consultancy BV, Aarle-Rixtel (2012)
36. van Bokhoven, M.: TPM. MarkonTarget, Haarsteeg (2010)
37. Bong, S., Chong, E.: A framework for implementing quick response manufacturing system in the job shop environment. Sci. Int. (Lahore) **26**(5), 1779–1783 (2014)
38. Dean, J.W., Bowen, D.E.: Management theory and total quality: Improving research and practice through theory development. Acad. Man. Rev. **19**(3), 392–418 (1994)
39. Deming, W.E.: Out of Crisis. MIT Press, Cambridge (1986)
40. Dombrowski, U., Mielke, T.: Lean leadership - 15 rules for sustainable lean implementation. Procedia CIRP **17**, 565–570 (2014)
41. van Dun, D.: Improving Lean Team Performance: Leadership and Workfloor team Dynamics. University of Twente, Enschede (2015)

42. Evans, J.R., Lindsay, W.M.: The Management and Control of Quality, 3rd edn. West Publishing Company, New York (1996)
43. Feigenbaum, A.: Total Quality Control. McGraw-Hill, New York (1966)
44. Fernandes, N., Carmo, S.: Generic POLCA. Int. J. Prod. Econ. **104**, 74–84 (2006)
45. Goldrath, E.: The Goal, 3rd edn. North River Press, Great Barrington (2014)
46. Hammet, P.: The Philosophy of TQM. University of Michigan (2000)
47. Hartmann, E.H.: Successfully Installing TPM in a Non-Japanese Plant. TPM Press Inc., Pittsburgh (1992)
48. Imai, M.: Gemba Kaizen. McGraw-Hill, New York (2012)
49. Imai, M., Biekmann, L.: Gemba Kaizen: De toepassing van Kaizen op de werkvloer. Kluwer Bedrijfsinformatie, Deventer (1997)
50. Jenner, R.A.: Dissipative enterprises, chaos, and the principles of lean organizations. Omega **26**(3), 397–407 (1998)
51. Liker, J.K.: The Toyota Way. McGraw-Hill, New York (2004)
52. Mann, D.: Creating a Lean Culture. Productivity Press, New York (2010)
53. Maskell, B.: Making the Numbers Count, 2nd edn. Productivity Press, New York (2009)
54. Nakajima, S.: TPM Development Program. Productivity Press, Cambridge (1989)
55. Ohno, T.: Toyota Production System. CRC Press, Boca Raton (1988)
56. Pyzdek, T., Keller, P.: Six Sigma Handbook, 4th edn. McGraw-Hill, London (2014)
57. Rother, M., Shook, J.: Learning to See. Lean Enterprise Institute, Brookline (1999)
58. Satsomboon, W., Pruetipibultham, O.: Creating an organizational culture of innovation: case studies of Japanese multinational companies in Thailand. Hum. Resour. Dev. Int. **17**(1), 110–120 (2014)
59. Shook, J.: Managing to Learn. Lean Enterprise Institute, Cambridge (2008)
60. van Solingen, R.: De kracht van scrum, 2nd edn. Pearson B.V, Amsterdam (2016)
61. Suri, R.: Quick Response Manufacturing. A Companywide Approach to Reducing Lead Times. Taylor and Francis Group LLC, New York (1998)
62. Theissens, H.: Climbing the mountain, 1st edn. LSSA, Amsterdam (2015)
63. Timans, W., van Ahaus, K., Solingen, R., Kumar, M., Antony, J.: Implementation of continuous improvement based on lean six sigma in small- and medium-sized enterprises. Total Qual. Manag. Bus. Excell. **27**(3–4), 309–324 (2014)
64. Weggeman, M.: Het gaat om de gezamenlijke ambitie. Nederlands Tijdschrift Voor Evid. Based Pract. **14**(1), 21–23 (2016)
65. Willmott, P.: TPM The Western Way. Butterworth-Heinemann Ltd., Oxford (1994)
66. Womack, J.P., Jones, D.T.: Lean Thinking. Lean Management Instituut, Driebergen (2011)

Lean Coaching and Mentoring

Developing Middle Managers with Gemba Training

Eivind Reke[1(✉)] and Nadja Böhlmann[2,3]

[1] SINTEF Manufacturing, Raufoss, Norway
eivind.reke@sintef.no
[2] Teknos, Alzenau, Germany
[3] Teknos, Helsinki, Finland

Abstract. Much has been written and researched on the role of top management in Lean transformations. However, even with top management commitment companies can struggle to keep up kaizen activities once the so-called implementation period has passed. Little has been made of the role of the middle managers. Often overlooked and underappreciated, the development of middle managers leadership skills through gemba based training and hansei (self-reflection), might be the missing link in successful and sustainable lean transformations. Based on a literature review on the subject and case studies of three different companies, we present how middle managers can become an integral part in sustainable lean transformations by taking the role of trainers, and leading on-the-job development of both technical-, improvement- and teamwork skills. This should be done by teaching Toyota Production System (TPS) on the gemba, creating space for Hansei. The training should be carried out in a train-the-trainer system where each manager is responsible for the training and development of their direct reports.

Keywords: Learning lean · Middle-managers · On-the-job training · Training within industry · Gemba

1 Introduction

As the founder of Kaizen Institute and author of [1], Maasaki Imai once said; *"There are three most important requirements if you want to be successful and embrace kaizen and lean. The first of course is the Top Management commitment, second is the Top Management Commitment, third is Top Management commitment."* [2]. Indeed the commitment of top management was found to be one of five critical success factors in a study of lean implementations across two multinational companies in two different industries [3]. However, [4] found that middle managers are often overlooked as the resources needed for implementation and training is directed to front-line staff and senior management. Indeed, we suggest that the involvement of middle managers through hands-on-training, are vital if a company is to sustain a lean culture over time.

Middle managers are key to every change program [5]. However, it is not always clear if senior- or middle managers are aware of this. Including middle managers in lean training sounds like a logical recommendation. Even so, they often find

© IFIP International Federation for Information Processing 2021
Published by Springer Nature Switzerland AG 2021
D. J. Powell et al. (Eds.): ELEC 2021, IFIP AICT 610, pp. 271–277, 2021.
https://doi.org/10.1007/978-3-030-92934-3_27

themselves on the side-lines, swamped in everyday firefighting activities [1] and the pressure of delivery. Deliver now, and deal with the consequences later [6]. Yet, a key trait of a successful lean transformation is the inclusion of everyone in improving both the product itself and the process which makes the product. Middle managers play a crucial role in lean training and continuous improvement. Thus, they must be provided with the right training to develop the capabilities needed to thrive in a lean enterprise. They need to be developed towards their new role as trainers and improvers [7, 8].

2 Literature Review

According to [9] middle managers are the in-betweeners of traditional organizations, responsible for the implementation of strategic plans and to manage lower-level staff. Middle managers are both controlling and controlled, resistant and resisted, they are the doers to the top management thinkers. A major literature review [10] found that the role of middle managers changes depending on the contextual condition of a lean implementation. If lean is seen as a cost cutting program, middle managers will actively block the initiative. However, if lean is seen as integral management philosophy, middle managers tend to play a much more positive role in the lean implementation. Even though middle managers are more likely to succeed in a so-called soft lean implementation, there tends to be a tools focus in the coaching of middle managers in the typical consultancy led implementation [11].

[7] describe how Toyota still base its training of middle managers on the foundation of the Training Within Industry (TWI) program first developed to support the effective training of production supervisors during the second world war. TWI was rediscovered by the west as late as the early 2000s [1] and consists of three fundamental training programs, Job Instructions (JI), Job Methods (JM) and Job Relations (JR). A recent study of the impact on the organizations exposed to the program found that companies who had received training had experienced a substantial gain in both productivity and sales year on year. Furthermore, those who received the training mattered with JI key for middle managers [12]. It is no surprise then to find how [13] describe the rigorous, gemba[1]-based training that newly hired managers at Toyota go through to first, understand the job and second, improve the job together with front-line workers. Another example of the effectiveness of TWI was found in a study by [14] showed how TWI could be used to lower accidents in the UK construction industry. In fact, [15] suggest that TWI forms part of the roots of lean. A recent study of mentorship at Toyota Georgetown have also highlighted how American managers where systematically trained in both the thinking and practices of the Toyota Production System [8]. In addition, Toyota also emphasise leaders ability for self-reflection through a process called "Hansei" [16].

In contrast to the relatively little literature found on the middle-managers role as management support lean and how to train them [17], much more effort has been put into describing lean leadership traits. [18] goes so far as to describe lean leadership as

---

[1] Gemba meaning "actual place" (Source: Lean lexicon 5th edition, Lean Enterprise Institute).

the missing link, what is needed to change mindsets and behaviours in an organization. In a systematic review of the literature [19] found several attributes of lean leadership: 1) Creating and improvement culture, 2) self-development, and 3) the long term development of employees. Furthermore, in a multi-method study [20] found and validated no less than 16 lean leadership competencies. In an effort to demystify lean leadership, [21] summarized their research into six lean leadership practices: 1) Go and see, 2) Daily layered accountability, 3) Structured problem solving, 4) Continuous improvement, 5) Coaching, 6) Strategy Deployment.

3   Research Method

Our research is designed as an exploratory case study where we compare our findings in three different case companies located in Scandinavia with the findings of the literature review [22]. The data was collected from a combination of sources, including interviews and observations, as well as participatory action research. Our choice of research method is guided by the fact that even though prior research has studied the role of middle managers in lean implementations, there is little guidance for practitioners as to how they should train and develop middle managers to make them integral to the development of the continuous improvement culture needed to sustain a lean transformation. The case companies have in different ways addressed this problem. Case A cross-trained middle managers with operations by allowing them to study theory together inside working hours and practicing the theory within their own setting. The theory study was conducted in the form of book clubs and centred around a Norwegian practitioners' lean book called *"lean ledelse for lærende organisasjoner"* [23]. Case B developed their own version of middle management training with a focus on learning the basics of the Toyota Production System on the shop floor. Finally, Case C carried out a JI training program staying as true to the original material as possible.

4   Findings

All three case companies realized they needed to change the jobs of middle managers from firefighting, coordinating and controlling delivery to the development of people, improvement of workflow and quality, and sustaining healthy work environments. This also meant they had to change the way they involved and taught their middle managers lean, and how they developed their leadership skills. However, there where both differences and similarities to how they chose to go about the issue of training and development.

Case A – Training and Development Through Reflection and Practice
Case A is a typical Scandinavian SME that transformed its business during a 4-year period from 2008 to 2012. Doubling its turnover and achieving a profit margin of 10–15% in an industry where 3–5% is considered good. Based on the successful training of senior leadership through simply reading books on Lean together and reflecting them,

case company A decided to extend this approach to the rest of the organization, developing lean thinking skills in both operators and middle managers. A method referred to as the "book club" at case company A. According to one of our participants, after some initial problems, the first time they tried the "Ohno-circle" they forgot to tell the operators being observed and consequently *"got in a bit of a pickle with the union"*. However, once intentions where clarified, practicing the Ohno-circle of standing and observing an operation or machine over time became a regular event at their training sessions. After running the training over several years, the company did find that some middle-managers and some departments took to lean better than others. However, overall, they viewed the training program as a successful initiative. In addition, the improvements born out of the practical application of the theory, the book clubs also created a space for Hansei. Allowing operators and middle managers across functions to reflect deeply on the fabric and culture of the company and how they could support and develop it further.

Case B – Developing a Deeper Understanding of TPS and Muri, Mura, Muda

In many companies, lean often gets executed with staff teams working side by side with employees from the shopfloor right up to Management. Which was the case at company B, a large Scandinavian multinational operating in the automotive business. However, this led to a situation where the line management did not take responsibility for improvement or development. Therefore, case company B removed its lean staff and instead rotated line managers in trainer roles on a 6–12 month basis. Additionally, they invested heavily in a lean leadership training program in the form of a two week "classroom training" program that was almost 100% front-line "gemba" based training. Involving middle managers from high up in the organization down to team leaders. The focus of the training was to develop gemba awareness in leaders by teaching the foundations of TPS: just-in-time (make only what is needed when it's needed), Jidoka (Stop at defect and separate man and machine) and standardized work, as well as developing a deeper understanding of muri (strain or overburdening), mura (unevenness), muda (unnecessary effort or waste) [1]. In fact, a key component of the training was to understand the importance of removing muri before focusing on mura and muda, to improve the workers' conditions in gemba. The training was built up based on the real observations made while observing a previously chosen process and interviewing the operators involved in the process. In addition to this training program, the company have a dedicated training centre that allows both managers and operators to practice operational skills but also leadership and improvement skills, giving the student a safe environment for practice, reflection (hansei) and feedback. Changing the training from classroom to gemba gave the leaders an opportunity to change their understanding of their roles as managers. To summarize, case B trained their middle managers in lean by: 1) practicing and observing on the gemba, 2) offering a safe environment in a training centre where they can make observations, reflect and discuss 3) not having a staff of trainers and coaches but by challenging their line managers to take up the role as trainer and 4) by investing in people and, encouraging them to share the responsibility of lean.

Case C - Revisiting the Basics of Job Instructions

Case C is a large factory in the fish farming industry situated on the west coast of Norway. Part of a large multinational the factory is seen as the development ground for lean and Industry 4.0. In a bid to better understand standard work and to engage its middle managers in improvement the factory initiated a pilot based on the original TWI material. At the time of writing, the company has finished the first module of Job Instructions. The initial findings from this pilot suggest that the practical and hands-on nature of the TWI curriculum is still valid today. The training was carried out by strictly following the original JI training material. In the pilot around 25 people in three teams were trained in how to develop and teach standard work. Again, the results showed that the emphasis on understanding thoroughly value adding work had a positive impact on the middle managers. The emphasis had previously been on controlling and coordinating work, but through the JI training, they saw a need to change towards teaching and improving work. Furthermore, the teams discovered gaps in knowledge about the work (how and why work should be done in a certain way) among themselves and among operators. Additionally, they also saw opportunities for improvements and standardizations right away. Finally, we found that many of the middle managers found new motivation by attending the JI training.

5 Conclusion and Suggestions for Further Research

Even though the three different case companies took different approaches to the development of their middle managers, our findings suggest that there are some commonalities among them. Based on the literature review and the findings from the case studies we suggest companies develop their middle managers based on three fundamental principles: 1) Teach the TPS basics on the gemba, 2) make space for Hansei, and 3) Train the trainer.

Teach the TPS Basics on the Gemba

All three cases described above show that going back to basics is a key in order to continuously learn about their value creating processes, how to observe and how to improve them together with the workforce. We suggest that no matter how many Lego simulations are invented, the real learning can only happen on gemba. The three chosen examples show evidence to the fact that learning takes time and that companies should always go back to the basics of TPS. As good as classroom trainings and simulations might be, lean can only truly be learnt by practicing, observing, and improving over and over again.

Make Space for Hansei

Equally important, our findings suggest, is to make room for reflection. Hansei in lean terms. One of the fourteen principles developed by [16] to describe the Toyota management philosophy, engaging in deep reflection allowed middle managers in all of the case companies to digest and discuss what they had learned on the gemba and reflect on the theory of lean and what it meant to them together with others in the company.

Train the Trainer

To develop lean leaders, one must develop lean trainers. However, training for the sake of reaching your Key Performance Indicators (KPI) of a certain number of people having received a "belt" in something is not much use without practical deployment. Our findings suggest that managers taking the role of trainer is a sustainable way of shifting towards a continuous learning organization. As such, middle managers should be viewed as trainers with the responsibility of developing their own staff. Additionally, leaders themselves discover improvement potentials, gaps or even bad working conditions and have the capability to improve with their team immediately.

These findings have implications for practitioners and researchers alike. As our literature review showed, even though much has been written about what defines lean leadership, little effort has been made to understand the role of training and developing middle managers on the gemba. More effort is needed on the part of researchers to understand how organizations can effectively and systematically develop lean capabilities. The case companies mentioned in this paper should be considered to be revisited in order to get more information on their further development and their current status. Same as for the success rate of their Lean programs involving the middle managers. For practitioners, our findings point towards some guiding principles on how organizations can develop middle managers by teaching the basics of TPS, creating space for reflection and challenge the same managers to take the role of trainers.

Acknowledgement. The authors would like to acknowledge the support of the Norwegian research council for the research project "Circulær".

References

1. Imai, M.: Gemba Kaizen, 2nd edn. McGraw Hill Professional, New York (2012)
2. G Academy: Maasaki Imai on Gemba, Kaizen and Lean. Youtube (2019)
3. Netland, T.H.: Critical success factors for implementing lean production: the effect of contingencies. Int. J. Prod. Res. **54**, 2433–2448 (2016). https://doi.org/10.1080/00207543.2015.1096976
4. Holmemo, M.D.-Q., Ingvaldsen, J.A.: Bypassing the dinosaurs? Total Qual. Manag. Bus. Excell. **27**, 1332–1345 (2016)
5. Rich, N., Bateman, N., Esain, A., Massey, L., Samuel, D.: Lean Evolution. Cambridge University Press, Cambridge (2006)
6. Ballé, M., Chartier, N., Coignet, P., Olivencia, S., Powell, D.J., Reke, E.: The Lean Sensei. Go, See, Challenge. The Lean Enterprise Institute Inc., Boston (2019)
7. Liker, J.K., Mayer, D.P.: Toyota Talent. McGraw Hill Education, New York (2007)
8. Leuchel, S.R.: Sensei Secrets: Mentoring at Toyota Georgetown: A Qualitative Study of the Sensei-Protégé relationship at Toyota. Align Kaizen Publishing (2020)
9. Harding, N., Lee, H., Ford, J.: Who is 'the middle manager'? Human Relat. **67**, 1213–1237 (2014). https://doi.org/10.1177/0018726713516654
10. Hermkens, F., Dolmans, S., Romme, G.: The role of middle managers in becoming lean: a systematic review and synthesis of the literature. J. Econ. Manag. Trade **20**, 1–17 (2017). https://doi.org/10.9734/jemt/2017/38100

11. Rolfsen, M.: Lean thinking : Outside-in, bottom-up ? Total Qual. Manag. Bus. Excell. **29**, 1–21 (2016). https://doi.org/10.1080/14783363.2016.1171705
12. Bianchi, N., Giorcelli, M.: Not all management training is created equal: evidence from the training within industry program. SSRN Electron. J. (2019). https://doi.org/10.2139/ssrn.3457878
13. Spear, S.J.: Learning to lead at Toyota. Harv. Bus. Rev. **82**, 78–86+151 (2004)
14. Misiurek, K., Misiurek, B.: Methodology of improving occupational safety in the construction industry on the basis of the TWI program. Saf. Sci. **92**, 225–231 (2017). https://doi.org/10.1016/j.ssci.2016.10.017
15. Huntzinger, J.: The roots of lean training within industry: the origin of Japanese management and Kaizen. Target **18**, 6–19 (2002)
16. Liker, J.K.: Toyota Way: 14 Management Principles from the World's Greatest Manufacturer, 1st edn. McGraw Hill Education, London (2004)
17. Marodin, G.A., Saurin, T.A.: Implementing lean production systems : research areas and opportunities for future studies. 7543 (2013). https://doi.org/10.1080/00207543.2013.826831
18. Mann, D.: The missing link: lean leadership. Front. Health Serv. Manag. **26**, 15–26 (2009). https://doi.org/10.1097/01974520-200907000-00003
19. Aij, K.H., Teunissen, M.: Lean leadership attributes: a systematic review of the literature. J. Heal. Organ. Manag. **31**, 713–729 (2017). https://doi.org/10.1108/JHOM-12-2016-0245
20. Seidel, A., Saurin, T.A., Marodin, G.A., Ribeiro, J.L.D.: Lean leadership competencies: a multi-method study. Manag. Decis. **55**, 2163–2180 (2017). https://doi.org/10.1108/MD-01-2017-0045
21. Netland, T.H., Powell, D.J., Hines, P.: Demystifying lean leadership. Int. J. Lean Six Sigma **11**, 12–19 (2019). https://doi.org/10.1108/IJLSS-07-2019-0076/full/html
22. Eisenhardt, K.M.: Building theories from case study research. Acad. Manag. Rev. **14**, 532–550 (1989). https://doi.org/10.5465/amr.1989.4308385
23. Wig, B.B.: Lean ledelse for lærende organisasjoner, 2nd edn. Gyldendal, Oslo (2013)

Towards Effective Lean QRM Yellow-Belt Training Programs: A Longitudinal Analysis

Mitchell van Roij[1(✉)], Martin Linde[1], and Wilfred Knol[1,2]

[1] Han University of Applied Science, Arnhem 6826CC, The Netherlands
mitchell.vanroij@han.nl
[2] Institute for Management Research, Radboud University,
Nijmegen, The Netherlands

Abstract. This paper studies the effect of Lean QRM Yellow-belt training programs on employee's continuous improvement (CI) behavior. Training employees is still the most common approach organizations follow to implement Lean production. However, such training programs do not always have the desired effect. To understand why Lean training programs, such as the Lean QRM Yellow-belt training may or may not lead to expected results, this study draws on a process model to conduct a longitudinal analysis. The results indicate that the Lean QRM Yellow-belt training program positively influences the patterning or shared understanding of lean operating routines, but does not (yet) influence the enactment of lean operating or CI routines. The interaction between enacting and patterning lean operating and CI routines however is necessary to implement lean production and thus influences the potential success of its sustainable implementation. These results pose practical implications on the content and didactical form of teaching in Lean QRM Yellow-belt training programs.

Keywords: Routines · Lean · Yellow-belt · Training programs

1 Introduction

Lean production is a business philosophy focused on shortening lead times by removing waste and focusing on value-added processes. Lean production consists of lean operating routines (actual day-to-day activities) and continuous improvement (CI) routines (improvements on lean operating routines). Lean production is implemented successfully when it fosters a learning culture in organizations with the focus on continually improving all aspects of the organization. On the long run, Lean production supports companies to achieve incremental growth. For these reasons, Lean production has become a key business strategy that many companies are attempting to implement. In order to implement Lean production, organizations take different approaches, such as hiring consultants and training employees. However, while some companies successfully implement Lean production, others fail to achieve the expected results [1].

Training employees in Lean production is still one of the key strategies of organizations to implement Lean production [2]. However, most Lean training programs

D. J. Powell et al. (Eds.): ELEC 2021, IFIP AICT 610, pp. 278–288, 2021.
https://doi.org/10.1007/978-3-030-92934-3_28

don't properly contribute to successful long term Lean production implementation. For example, Bhasin's [3] research showed that a lack of adequate training is one of the main reasons why organizations fail to implement Lean production on the long term.

This paper takes a longitudinal approach and draws on the routine dynamics literature to explain why Lean training programs may not always have the expected result.

Routine dynamics literature distinguishes between enacting and patterning routines [4]. Patterning links to the shared understanding or the 'ostensive' aspect of a routine, while the enactment of a routine relates to the performative aspect [5]. In line with Knol et al. [6] this paper argues that if employees enact CI routines, it helps them to pattern and enact lean operating routines and that in turn, the enactment of lean operating routines helps them to pattern and enact in CI routines. Thus, it is the cyclical process of enacting and patterning lean operating and CI routines that help employees and thus organizations to successfully implement Lean production. However, most Lean training programs, such as our Lean QRM Yellow-belt training solely focus on the patterning aspect of routine development.

According to previous research on Lean training programs and Lean production implementation, training employees has a long-term effect on Lean production implementations [7]. Douglas, Douglas and Ochieng [7] state in their quantitative study that Lean training programs do have an effect on successful Lean production implementation, but they only focus on employees' understanding of Lean operating tools, that is patterning lean operating routines and related artefacts, to strengthen their claim. Though understanding Lean operating tools in itself is an important aspect for successful Lean production implementation, understanding CI tools and actually using these tools in CI activities is neglected as a possible explaining factor.

Thus, previous research mainly focusses on Lean training to pattern lean operating routines, whilst patterning CI routines and enacting both lean operating and CI routines are neglected. This study aims to close this gap by studying the influence of a "best-practice" Lean QRM Yellow-belt training on both enacting and patterning lean operating and CI routines and thus routine development. Therefore this paper poses the following research question:

"To which extent does our Lean QRM Yellow-belt training develop Lean Operating & Continuous Improvement Routines?"

2   Literature

2.1   Process Model

Knol et al. [6] state in their research that rather than management and expert driven activities, team leader and employee enacted CI routines helped to pattern and enact their lean operating routines. In turn, if team leader and employee enacted lean operating routines helped to pattern and enact their CI routines. This cyclical process of enacting and patterning lean operating and CI routines to successfully implement Lean production is shown in Fig. 1. This figure is further explained below.

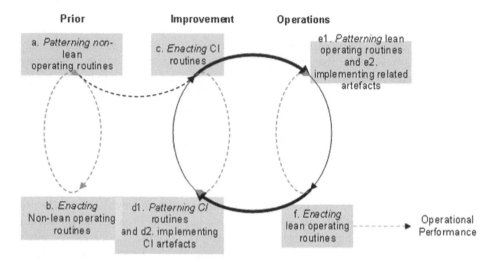

Fig. 1. The cyclical process of enactment and patterning lean operating and CI routines (Source: Knol et al. 2020)

Continuous Improvement Routines (a, d1 & d2 in Fig. 1) are, according to Knol et al. [6] aimed at continuously improving business processes, such as Ohno's Genchi Genbutsu or 'go and see the gemba' principle [8]. Rother and Shook's [9] 'Learning to see' approach, Rother's [10] Toyota Kata, CI [11], striving to continuously improve products and processes, solving problems in small groups, employee involvement, Kaizen activities, brown paper sessions, A3 & 8D-problem-solving reports, A3think-ing, PDCA, 5-whys, perfection, performance boards/ performance sessions and TQM. CI routine development may or may not be visible on the basis of related physical artifacts, such as performance boards, A3 and 8D problem-solving reports, and intangible artifacts, such as Deming's [12] Plan-Do-Study-Act circle and 5-whys. Employee knowledge about these subject can be improved by: CI meetings, engaging employees in enacting CI routines, discussions to clearly explain the need, challenging employees to perform CI routines, using cycles of CI activities and templates such as 8D or A3 problem-solving reports, continuously providing employees with coaching, time and resources [8]. Enacting (c) concerns employees actually conducting improvement activities using previously mentioned concepts in daily practice, or the ability of managers to realize lasting involvement of employees in CI. Patterning (d1 & d2) concerns the extent to which employees collectively have a thorough understanding of previous concepts [6].

Lean Operating Routines (e1, e2 & f in Fig. 1) relate, according to Knol et al. [6] mostly to hard lean practices (the tools). This concerns concepts such as: value stream, flow, takt time, Little's Law, Just-in-time deliveries (JIT), supplier integration, pull, Kanban (squares or containers of signals for production control), two-binsystems, 5S (like shadow boards) and setup routines (such as doing quicker product changeovers), lines and cells. Lean operating routines may or may not be visible on the basis of related physical artifacts, such as Kanban cards used to limit work-inprogress, 5S artifacts (such as shadow boards) and setup routines (such as doing quicker product changeovers), if also intangible artefacts, such as formulas and techniques like takt time and Little's Law (that help to determine the number of Kanban cards in the system). Opportunities to build up employee knowledge of these subjects are lean games, excursions (internal and external), training (Lean Yellow/Green/Black-belt training), involvement in improvement projects and coaching. Patterning lean operating routines (e1 & e2) concerns the extent to which employees have a thorough shared understanding of previously mentioned concepts. Enacting lean operating routines (f) concerns employees actually using previously mentioned concepts in practice [6].

2.2   Intervention - Lean QRM Yellow-Belt Training

In line with the distinction made by Knol et al. [6], an intervention to train team leaders and employees in Lean production best consists of the following four aspects. In addition we describe (behind the four aspects) the related parts from the given Lean QRM Yellow-belt training. The numbers in brackets indicate the order in which these parts were taught during our training.

Enacting Continuous Improvement Routines. Playing the 1st round of a lean sticklebrick game (2), playing the 2nd round of a lean stickle-brick game (jointly thinking up (and implementing) countermeasures in advance and reflecting afterwards on the improvements achieved or not) (6) and playing the 3rd & 4th round of a lean sticklebrick game (jointly devising (and implementing) countermeasures in advance and reflecting afterwards on the improvements achieved or not and the extent to which attention was paid to all Lean production principles) (11).

Patterning Lean Operating Routines. Explaining a summary of Lean QRM and CI (with emphasis on difference between source efficiency versus flow efficiency and therefore difference between focus on costs versus focus on lead time and quality (QDC) and difference between Lean (LVHV) and QRM (HVLV) and the 5 Lean production principles (LPs) (3), explanation about customer value & waste (LP1) (4), explanation about identify the value stream (LP2) with attention to tools such as the VSM, the swimlane and the Manufacturing Critical Timepath (MCT of Suri) combined with an explanation of the Response Time Spiral of Suri (5), summary of the first day (7), explanation of flow (LP3) with attention to the effect of batches on lead time and work in progress work (Little's law), influence of changeover times (SMED), takt time, line balancing, utilization rate versus lead time and the impact of variability on lead time (8), explanation of difference between push and pull (LP4) with attention to Kanban and two-bin systems (9) and explanation about 5S, standards and visual management (10).

Enacting Lean Operating Routines. By playing the 1st round of a lean stickle-brick game (2), making a homework assignment of at least 2 photos of waste at their own workplace (LP1) (4), preparing a homework assignment MCT for a product in which they are involved (5), playing the 2nd round of a lean stickle-brick game (jointly thinking up (and implement) countermeasures beforehand and reflect afterwards on the improvements achieved or not) (6), playing a short game regarding 5S, standards and the importance of visual management (10) and playing the 3rd & 4th round of a lean stickle-brick game (jointly devising (and implementing) countermeasures in advance and reflecting afterwards on the improvements achieved or not and the extent to which attention was paid to all lean principles) (11).

Patterning Continuous Improvement Routines. Explanation 'Why to improve?' (1), explanation about customer value & waste (LP1) (4), summary of the first day (7) and explanation about 'striving for perfection' (LP5) with attention to the importance of quality, the hidden factory, Jidoka, Andon, Mistake proofing & Poka Joke and the KATA-model (10).

2.3 Hypotheses

In line with Knol et al. [6] we formulated the following hypotheses:

H1a: 'The Lean QRM Yellow-belt training intervention has a positive effect on the enactment of CI routines of employees at Firm x'.

H2a: 'The Lean QRM Yellow-belt training intervention has a positive effect on the development of patterning Lean operating routines of employees at Firm x'.

H2b: 'The Lean QRM Yellow-belt training intervention has a positive effect on the enactment of Lean operating routines of employees at Firm x'.

H1b: 'The Lean QRM Yellow-belt training intervention has a positive effect on patterning CI routines of employees at Firm x'.

3 Method

3.1 Research Strategy

This research is based on a deductive, pre-post interventional research design, since theoretical concepts are derived from existing literature and the data regarding Lean operating- and CI routines were collected in two points in time, with the Lean QRM Yellow-belt training as intervention. Measuring differences in only two points in time is usually a poor method of longitudinal analysis, however both the information- and time difference between the first and the second measurement are fairly uniform for all subjects, making the effect of only two measurements negligible [13]. Pre-post interventional study designs have the strength of temporality to suggest that the outcome is impacted by the intervention, however, pre-post intervention designs do not have control over elements that are changing at the same time that the intervention is implemented [14].

3.2   Case Description

The subject of this research concerns the Lean QRM Yellow-belt training as given to one of the partners of our HAN Lean QRM Centre in the Netherlands. This partner firm is market leader in the production of front linkage systems and manufacturer of track systems for tractors. They first got to know Lean production in 2014 and in 2016 the first 55 (out of approximately 350) employees were trained at Yellow-belt level, followed by approximately 6 employees at Green-belt level. In 2019, the firm aimed to have at least 25% of its employees trained at the Yellow-belt level to improve their knowledge level of Lean production, QRM and CI. In the fall of 2019 and spring of 2020, 88 employees (in 7 groups) were subsequently trained in a 2-day Lean QRM Yellow-belt (with a week in between).

3.3   Data Quality

In order to test the hypotheses of this study, quantitative data is gathered through an exploratory questionnaire. The questionnaire is derived from existing, widely accepted literature and constructs are operationalized accordingly. Questions regarding to Lean operating routines are derived from the study of Spear and Bowen [15] on Toyota employees' behavior. Items related to CI routines are based on eight different key constructs identified by Bessant et al. [16]. Whereas Bessant et al. [16] use thirty-five items to measure the full extent of CI behavior, the seven items mostly related to understanding and enacting CI are used in this paper. Additionally, all items were carefully translated from English to Dutch by two independent individuals who used "back and forward translation". In order to ensure the research quality, the questionnaire was also checked by experts in the field of Lean production in order to control for inconsistencies and vagueness of concepts. The suggestions and feedback in turn were used to improve the questionnaire. All data was collected on an individual level.

3.4   Data Analysis

For the purpose of the research, the questionnaires (t1 & t2) were sent to all 88 participants of the Lean QRM Yellow-belt training. In total 49 questionnaires were filled in for both t1 and t2, generating a response rate of 58%. After checking for missing values, 17 pairs were deleted listwise, leaving only 32 pairs applicable for comparison. Additionally, we checked for outliers and normal distributed data. Hence, no extreme outliers were found and the data is fairly normally distributed. However, due to constraints, boxplots are used to compare the mean scores of the pre and post-tests.

4   Results

In order to determine whether the Lean QRM Yellow-belt training program affects the development of Lean operating and CI routines, this study looks at the mean differences between the pre- and post-intervention test. Table 1 shows the descriptive statistics of both the pre- and post-test related to both patterning and enactment of Lean operating and CI routines. According to the process model presented by Knol et al. [6], the development of routines starts with the enactment of CI routines. This studies hypothesis states that after the training program, employees of the firm would score higher on the enactment of CI routines. The descriptive statistics (Table 1) however show that the mean score of enacting in CI routines after the training program slightly decreased.

Table 1. Descriptive statistics

	Min.	Max.	Mean	Std. deviation
(Pre) Enacting CI Routines	3.83	8.00	5.95	1.18
(Post) Enacting CI Routines	2.33	7.67	5.91	1.19
(Pre) Patterning Lean Routines	2.00	8.00	5.23	1.34
(Post) Patterning Lean routines	4.50	8.00	6.42	1.00
(Pre) Enacting Lean Routines	4.00	7.67	6.01	0.770
(Post) Enacting Lean Routines	4.33	7.67	6.22	0.874
(Pre) Patterning CI Routines	3.50	8.00	6.05	1.04
(Post) Patterning CI Routines	4.50	8.00	6.57	0.90

The difference in both mean score and standard deviation is negligible, indicating that after the training, employees on average did not enact more in CI routines. The last two boxplots on the right illustrate the differences of CI enactment in the pre- and post-test. Figure 1 illustrates the differences in mean scores for the first two steps of the process model.

The descriptive statistics do show that patterning Lean operating routines is increased by 1.18 and that the standard deviation is slightly decreased by .330, implying that on average the respondents have better understanding of the Lean operating routines. The first two boxplots of Fig. 3 illustrate the difference in scores for patterning Lean Operating Routines. Whereas the first boxplot shows the preliminary results of patterning Lean Operating Routines, boxplot two illustrates the results of the post-test. Additionally, the descriptive statistics indicate that on average the scores of enactment in Lean operating routines is marginally increased, but that the standard deviation also is increased (see also the last two boxplots of Fig. 2). This would imply that there is no clear evidence that participants enact more in Lean operating routines after the intervention.

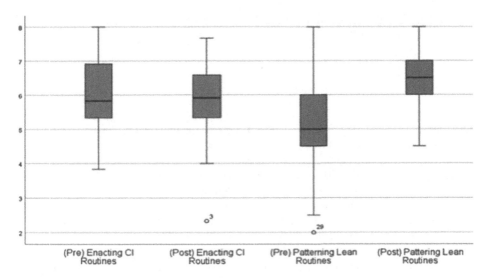

Fig. 2. Boxplots Differences in first two steps process model: Red: Lean Operating Routines, Blue: CI routines

The table above also provides information between the differences related to the patterning CI routines. The patterning of CI routines is on average increased with .467, and the standard deviation is decreased, which implies that the respondents slightly understood the CI routines better after the intervention (see also the first two boxplots of Fig. 3.

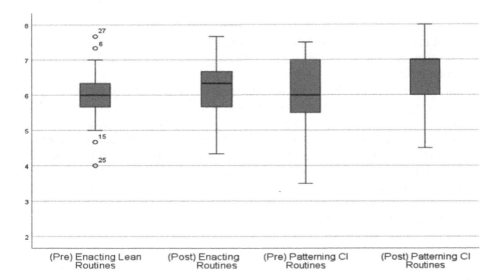

Fig. 3. Boxplot Differences in Mean scores. Red: Lean Operating Routines, Blue: CI Routines

On balance, this has led to the following overview of confirmed or rejected hypothesis (see Table 2):

Table 2. Overview of hypotheses

Construct	Hypothesis	Result
Enacting CI routines	H1a. The Lean QRM Yellow-belt training intervention has a positive effect on the enactment of CI routines of employees at Firm x	Rejected
Patterning lean operating routines	H2a. The Lean QRM Yellow-belt training intervention has a positive effect on patterning Lean operating routines of employees at Firm x	Confirmed
Enacting lean operating routines	H2b. The Lean QRM Yellow-belt training intervention has a positive effect on the enactment of Lean operating routines of employees at Firm x	Rejected
Patterning CI routines	The Lean QRM Yellow-belt training intervention has a positive effect on patterning CI routines of employees at Firm x	Confirmed

5 Conclusion

This study posed the following research question: "To which extent does our Lean QRM Yellow-belt training develop Lean Operating and CI routines?". Since lean operating routines originate from enacting CI routines, we studied what the effects of the training were on the development of enacting and patterning Lean Operating and CI routines.

The results of this deductive pre-post interventional exploratory research on the effects of the HAN Lean QRM Yellow-belt training on Lean operating and CI routine development are clear. Whereas Knol et al. [6] imply that the cyclical development of new routines starts with the enactment of CI routines, these results show that the current Yellow-Belt training, does not influence the enactment of CI routines by employees at this firm (H1a). The understanding and thus the patterning aspect of the lean operating routines (H2a) as well as the CI routines (H1b) is increased due to the two-day HAN Lean QRM Yellow-belt training program. In contrast, the enactment of the lean operating routines (H2b), shows no significant growth. Seen from the perspective of Knol et al. [6], these results imply that the HAN Lean QRM Yellow-belt training does not make a complete contribution to sustained Lean production implementation since the training in itself is not yet enough to develop and enact new Lean operating and CI-routines. So far, the HAN Lean QRM Yellow-belt training only contributes significantly to the patterning of lean operating and CI routines.

6 Discussion

6.1 Limitations and Future Research

Although this study reveals interesting findings, it has several limitations. First of all, even though the questionnaire is based on recognized literature, it has not been checked for internal validity and reliability, for example by using a factor analysis, or Cronbach's Alpha. Follow-up research should therefore pay attention to the content, validity and reliability of the questionnaire and develop it further. We used a pre- and post-measurement, over a period of three months. Developing routines can take longer, so the full effect of the training may not have been mapped. Follow-up research can therefore choose to schedule three measurement moments, over a longer period of time, in order to get an even better picture of the development of the routines. The current results show that in particular the enactment of both lean operating routines and CI routines lag behind. However, it is unclear what the effects of the Covid-19 outbreak are on the outcome of this study. Further research can replicate and complement this research.

Initially, we wanted to use a dependent samples t-test, in order to demonstrate the effects of the training with more certainty. We therefore checked for the requirements associated with such a test. For example, outliers, normal distribution of the data and a required sample size were examined. This study met two requirements, there were no outliers in the data, and the data was relatively normally distributed (the ShapiroWilk test was used for this), however, we did not have enough pairs to compare. We needed 45 pairs according to the G-power test, but in the end, we had 32 similar pairs in the data. To develop a firmer picture of the effects of Yellow-belt training on routine development, further research could therefore adhere to these guidelines.

6.2 Practical Implications

The results of this research are interesting for trainers of similar training courses such as the HAN Lean QRM Yellow-belt training. Explaining many concepts does lead to understanding or patterning of lean operating and CI routines, but the interventions deployed in the field of enactment, such as playing games (2, 6, 10 & 11) and giving homework (4 & 5), in this case turned out to have insufficient effect on a sustainable implementation of Lean production. On the one hand, it is therefore necessary to consider how the deployed interventions can lead to a higher return on the offer of enactment, and on the other hand, it must be considered which additional interventions can be used to achieve the same result. Participants in a Lean QRM Yellow-belt training can already during (or afterwards) the training think about how they can implement what they have learned during the training in their own practice, or even conduct a project during the training. In addition, management should also take responsibility for perpetuating the yields of the Lean QRM Yellow-belt in daily work practice.

6.3 Reflections

For us, these results have led to two clear improvement actions. First, we will from now on pay even more attention to the enacting of CI routines during our Lean QRM

Yellow-belt training, whereby we find it especially important to pay even more attention than before to the transfer of what has been learned in the course to the daily practice of the participants. Second have we developed a so-called 'CI Workshop', consisting of 5 afternoons (biweekly) in which the participants will work under our supervision on a joint improvement project (approximately 12 participants work on a total of 3 improvement projects) by means of the A3 problem-solving methodology. After these workshops, these participants are coached 5 more times or more by us in the further implementation of their improvement project and/or the start-up of a next improvement project. Future research will show what the effect of this will be on the enactment of lean operating and continuous operating routines. In addition, we will also extend this research to our HAN Lean QRM Green-belt training courses. Finally, in order to develop our understanding of CI- and Lean routines, it is necessary to improve the robustness of the questionnaire and amount of participants. So that we in turn can use statistical analysis to the fullest extent.

References

1. Secchi, R., Camuffo, A.: Lean implementation failures: the role of organizational ambidexterity. Int. J. Prod. Econ. **210**, 145–154 (2019)
2. Sisson, J., Elshennawy, A.: Achieving success with lean: an analysis of key factors in lean transformation at toyota and beyond. Int. J. Lean Six Sigma **6**, 263–280 (2015)
3. Bhasin, S.: An appropriate change strategy for lean success. Manag. Decis. **50**(3), 439–458 (2012)
4. Feldman, M., Pentland, B., D'Adderio, L., Lazaric, N.: Beyond routines as things: introduction to the special issue on routine dynamics. Organ. Sci. **27**(3), 505–513 (2016)
5. Feldman, M.S., Pentland, B.T.: Reconceptualizing organizational routines as a source of flexibility and change. Adm. Sci. Q. **48**(1), 94–118 (2003)
6. Knol, W., Lauche, K., Schouteten, R., Slomp, J.: Prototyping a routine building-intervention to develop towards a lean organization. Acad. Manag. Proc. **2020**(1), 11963 (2020)
7. Douglas, A., Douglas, J., Ochieng, J.: Lean six sigma implementation in East Africa: findings from a pilot study. TQM J. **27**(6), 772–780 (2015)
8. Liker, J.K.: The Toyota Way: 14 Management Principles from the World's Greatest Manufacturer. McGraw-Hill, Madison (2004)
9. Rother, M., Shook, J.: Learning to See. Lean Enterprise Institute, Brookline (1999)
10. Rother, M.: Toyota Kata, 2nd edn. Tata McGraw Hill, New Delhi (2010)
11. Bessant, J., Caffyn, S.: High involvement innovation through continuous improvement. Int. J. Technol. Manag. **14**(1), 7–28 (1997)
12. Deming, W.E.: Out of the Crisis (1986). http://www.amazon.com/Out-CrisisW-Edwards-Deming/dp/0262541157. Accessed 26 June 2015
13. Locascio, J.J., Atri, A.: An overview of longitudinal data analysis methods for neurological research. Dement. Geriatr. Cogn. Disorders Extra **1**(1), 330–357 (2011)
14. Thiese, M.S.: Observational and interventional study design types; an overview. Biochemia medica **24**(2), 199–210 (2014)
15. Spear, S., Bowen, H.K.: Decoding the DNA of the Toyota production system. Harv. Bus. Rev. **77**(5), 96–108 (1999)
16. Bessant, J., Caffyn, S., Gallagher, M.: An evolutionary model of continuous improvement behaviour. Technovation **21**(2), 67–77 (2001)

Skills and Knowledge Management

14 Steps Toward Lean Knowledge Management

Olve Maudal[1(✉)] and Torgeir Dingsøyr[2(✉)]

[1] Equinor, Oslo, Norway
olvm@equinor.com
[2] Department of Computer Science, Norwegian University of Science
and Technology, Trondheim, Norway
torgeir.dingsoyr@ntnu.no

Abstract. We present 14 learning principles supporting a lean approach
to knowledge management. These principles are discussed in relation to
research on competence development strategies. Further, we also relate
them to advice and lessons learned from the agile and lean software
development communities. The principles focus on learning as a team
activity, and learning focused on needs. They also suggest a perspective
where inventories of knowledge can be seen as waste, and where learning
is potentially a wasteful activity. We argue that the learning principles
are significant steps towards lean knowledge management.

Keywords: Lean knowledge management · Learning organization ·
Knowledge perspectives · Competence strategies · Organizational
learning · Software development · Knowledge flow · Collective learning

1 Introduction

Equinor is a multinational energy company with more than 21000 employees.
Lean thinking [10] and agile ways of working [4] are well established and have a
long history in the company.

In 2020, Equinor updated their IT competence strategy. A small, but significant part of this competence strategy was a list of 14 learning principles:

1. Information, knowledge and learning are corporate assets
2. Advanced learners and experts should lead
3. A critical mass of in-house knowledge is often enough
4. Do intermediate and advanced training first
5. Focus learning on current needs
6. Team up with learning buddies
7. Engage with professional networks
8. Establish knowledge bridges
9. Share your expertise - invite others to follow

© IFIP International Federation for Information Processing 2021
Published by Springer Nature Switzerland AG 2021
D. J. Powell et al. (Eds.): ELEC 2021, IFIP AICT 610, pp. 291–298, 2021.
https://doi.org/10.1007/978-3-030-92934-3_29

10. Diversity is a key to fast learning
11. Seek to build secondary and tertiary skills
12. Establish repeatable learning opportunities
13. Prefer collective learning and team building
14. Learn to unlearn

These learning principles generated a lot of curiosity and was very well received by the organization. Other parts of the company are now adapting similar principles into their own competence strategies [5].

In this paper we will first present an explanation of these learning principles. We then discuss the learning principles in relation to research on competence development strategies, lessons learned from agile ways of working and lean software development. Finally, we conclude on the significance of these learning principles as steps towards lean knowledge management.

2   Learning Principles Explained

Here are the learning principles explained, exactly as presented in a public article by Equinor [5]:

1. Information, knowledge and learning are corporate assets. New insights, competence and learning capacity gained during working hours is supposed to be made available to the rest of the organization. It shouldn't be used to make yourself or your team a more indispensable resource.

When a team is building competence in some area, they should also start thinking about how they can make new insights and abilities available to the whole organization by offering to teach, lead, mentor and help other teams. Encouraging teaching just as much as learning is key for a thriving knowledge organization.

2. Advanced learners and experts should lead. It is imperative that a knowledge organization prioritizes advanced learners and experts when it comes to learning opportunities. In return the advanced learners and experts are supposed to invite and lead – this includes identifying, initiating, asking for, and often organizing both internal and external learning opportunities for others.

3. A critical mass of in-house knowledge is often enough. If individuals and teams actively make their knowledge, competence and learning capacity available, there is a limit to how much other teams and individuals actually need to invest into learning things up front before tackling hard problems. A team can often assume that they can learn required things quickly from other teams when needed.

This enables an important discussion of what teams and individuals should not spend time learning more about, allowing them to focus their learning capacity towards something that is more useful for the organization.

4. Do intermediate and advanced training first. Do intermediate and advanced training first, later you may consider setting up introductory classes. Trying to build organizational competence by focusing on beginners does not work – advanced learners and experts will feel unprioritized and undervalued. A very common reaction is that your strongest knowledge workers and teams ignore the competence building activities, or sometimes start to work directly or indirectly against the learning effort.

5. Focus learning on current needs. When work is learning, and learning is work, it is imperative to focus most of your capacity on learning things your current need. These days, especially within IT, there is a surplus of available, effective and efficient ways for a team to learn new things quickly when needed.

In addition, you should rely on other teams and experts in your organization to be eager to support your immediate learning needs. Teams working in fast-learning organizations should be confident enough to delay many learning activities until they are strictly needed.

6. Team up with learning buddies. Learning new stuff and carry knowledge into an established organization is usually too difficult for an individual to do alone. Taking a course? Reading a book? Going to a conference? Not alone – find some learning buddies!

Learning activities should be coordinated among a collaborating group of knowledge workers with a common learning need, so that they learn approximately the same things at approximately the same time for approximately the same reason.

7. Engage with professional networks. Professional networks, also known as discipline networks and communities of practice, are very powerful and useful mechanisms for increasing the collective learning capacity of the organization.

Everyone should be encouraged to actively engage, contribute or start their own network of learners. The more the better! However, it is important to keep the networks open and inclusive, so they make their expertise available to others.

8. Establish knowledge bridges. When switching focus from one problem and solution domain to another, it is important that the organization allow knowledge workers to build an effective path to bring their existing knowledge, enthusiasm, competence, and learning capacity from the old into the new. Sometimes this requires extra investment in the old to enable a successful transition over to the new. Expecting advanced learners and experts within one domain to go back to "kindergarten" and relearn everything from scratch in a new domain is unnecessary, demotivating and counterproductive.

9. Share your expertise – invite others to follow. Sharing is caring – but it doesn't have to be daring! You can share through lunch talks, teaching courses, or organize learning events. It can also be as simple as updating some documents, engaging in online discussions, replying to emails, or just give a nod of recognition to someone that's trying to learn something.

However, some experts and advanced learners should be encouraged to do much more; give talks at conferences, publish articles, participate in external professional networks. For any organization that would like to learn fast, it is essential to take an active part in the industry dialogue.

10. Diversity is a key to fast learning. Diversity comes in many forms: gender, race, religion, sexual orientation, age, abilities, culture, education, background, mindset. Diversity is essential for building a healthy organization, but it also leads to better solutions as it brings different perspectives and counteracts monoculturalism. Collective learning benefits a lot by diversity – different perspectives and abilities accelerates the learning process.

11. Seek to build secondary and tertiary skills. A knowledge organization should encourage teams and professionals to build secondary and tertiary skills. Encouraging multiskilled knowledge workers and teams improves the communication lines inside the organization, and this will strengthen the collective intelligence significantly. To allow this to happen, advanced learners and experts must give space to others, and start to mentor, lead and teach with their primary skills.

12. Establish repeatable learning opportunities. If learning opportunities becomes a scarce resource, you may end up with the wrong people taking up the slots to learn the wrong things at the wrong time for the wrong reasons. To counter this, whenever organizing a potentially popular learning opportunity, make sure to repeat the same or similar opportunities later if it is a success.

Once this is a well-established practice, knowledge workers and teams will attend to learn the right things at the right time – and for the right reasons.

13. Prefer collective learning and team building. A team of knowledge workers that are used to learning things together are much better equipped to deal with upcoming challenges without requiring organized learning activities. When organizing a course like a team building event, students with different backgrounds and competence will start helping each other to achieve a collective result.

Soon, the attitude switches from "what can I benefit from this?" to "how can we learn to learn things together as a team?".

14. Learn to unlearn. An organizations knowledge capacity is limited - so it must learn to unlearn. One fascinating definition of intelligence says that it's "the ability to get rid of unnecessary information and knowledge". Scaling this up suggests that merely collecting information and creating large databases with knowledge is not optimal for an organizations ability to learn new things – it will actually slow you down. Some even say that learning to unlearn is the highest form of learning.

3   Discussions and Related Work

A fair share of employees in organizations like Equinor are knowledge workers. Typical for knowledge work is that it is mostly dominated by the challenge of learning new things fast and inventing unique solutions to non-trivial problems together with a team of other knowledge workers [11]. Once a task is properly understood the task is often completed. But, as the world is rapidly changing in unforeseen ways, combined with a continuous demand of new business needs, there is always a long line of new, hard problems waiting to be solved. There is a need to optimize the collective intelligence of the organization for reliably learning things fast, effectively, and efficiently.

The Equinor IT competence strategy and the learning principles attempts to address the challenge described above. The learning principles constitute a major departure from traditional knowledge management by taking a "knowledge as inventory" perspective (term inspired by [1]) to optimize for knowledge flow and increase the speed of learning.

In the following, we will try to link the learning principles up to competence development strategies and a lean mindset:

3.1 Competence Development Strategies

Most companies have strategies to develop employee competence. A study on learning strategies describe enterprise learning strategies as consisting of three parts: Learning systems and incentives, work design and organization of work, and competence development [2]. Such learning strategies can include policies, systems and practices used in development of personnel. These strategies used to consist of vocational training, education and certification but are increasingly informal and practice-based. Examples include "on-the-job and job-rotation training, online network knowledge exchange and communities of practice" [2]. Based on a study across 53 industries in 22 countries, Brandi and Iannone [2] lists several recommendations which include that competence development should focus on "soft skills" such as teamwork which could be embedded in core values, mission statements, workplace policies and practices. Further, Brandi and Ionne [2] recommend informal learning activities as part of the learning strategy, encourage peer-to-peer learning either self-initiated or through team-based work arrangements. Comparing the principles to findings reported in the study on competence development strategies [2], we note that the focus on learning as a team activity in research findings are present in several of the principles, such as 6 and 13. They recommend that learning should be triggered by work-related processes, industry standards, client needs and innovation objectives - gaps are identified when facing a problem, new requirement or from competitive benchmarks.

Another finding is that enterprises rely on employees to identify knowledge and skill gaps, and should foster activities "that uncover needs and learning potential through formal and informal evaluations". The most successful types of learning were identified as "short and demand-driven responses". The view on a competence strategy in that it should focus on current needs and be "demand-driven" is reflected in learning principle 5. They recommend to ease access to training and development, establish communities of practice, integrate problem identification and solving in work arrangements and to encourage "peer-to-peer, in-house learning activities by providing space for employee-led workshops, for example". Such ideas are captured in principles 7 and 9.

Another principle which is related to studies on organizational learning, is the principle 14 of unlearning. In the introduction to a special issue in "the learning organization" [6], the editors state that "unlearning, e.g. purposefully discarding knowledge that has been deemed obsolete, serves as an efficient way to initiate and facilitate successful learning and change in organizations". They argue that learning can be hampered as new knowledge compete with existing knowledge, and this can evoke cognitive dissonance and resistance in people. Unlearning then involves purposefully discarding elements such as assumptions, routines and procedures.

3.2 Lean Knowledge Management

When searching for literature on lean and knowledge management, we were only able find a few studies in the literature. A literature review on recent lean research found eight studies connecting lean practices with theories from organizational learning [3]. We have not seen studies we think addresses strategies for competence development in a "lean" fashion. The literature review [3] further suggests that future work on lean should include clarifying terminologi on lean concepts which could include "lean knowledge management" or "lean learning organization".

However, we find examples of studies which focus on lean as a "learning system" [10], where the authors argue that learning in corporate lean programs must be seen as a "process of deep thinking, reflection and improvement".

We argue that the learning principles summarize discussions in the software development community on agile methods [4] and lean software development [8]. The ideas in lean software development provides emphasis on learning for example through focus on speed of delivery, where fast delivery leads to feedback which gives more insight into product requirements. Learning principle 5 is about allowing knowledge to flow. Lean software development and agile methods, have a focus on teamwork, through organizing in self-managing teams (agile) and focusing on "team empowerment" (lean). Learning principle 11 is about amplifying learning, and to avoid delays, task switching, and handoffs.

Taiichi Ohno and The Toyota Production System further provides advice on reducing "waste" [7]. Inspired by how Mary and Tom Poppendieck translated ideas from lean manufacturing to lean software development [9], it is tempting to take this idea further and make an attempt to categorize waste also for knowledge management (Table 1).

Table 1. Initial attempt to categorize waste in knowledge management

Toyota production system	Software develpment	Knowledge management
Waste of overproduction	Extra features	Knowledge not needed
Waste of time on hand (waiting)	Delays	Time to learn
Waste in transportation	Handoffs	Documentation
Waste of processing itself	Relearning	Relearning
Waste of stock on hand (inventory)	Partially done work	Unapplied knowledge
Waste of movement	Task switching	Unfocused learning
Waste of making defective products	Defects	Incorrect knowledge

From this waste perspective there are many similarities between lean manufacturing and the 14 learning principles. Here are some examples: For knowledge-based work, waiting for others to help you can cause delays and it might be better to learn it yourself (principle 3 and 11). Focused learning on current needs is essential to increase learning capacity, and to avoid creating and maintaining knowledge not needed (principle 5, 12 and 14). Creating communities of practice and teaming up with learning buddies reduces the need for documentation and handoffs (principle 1, 6, 7 and 13). Allowing advanced learners and experts to invite and lead learning activites reduces the creation of incorrect knowledge (principle 2, 4 and 9). Bridging knowledge built in one context so that it can be used in another context might reduce relearning (principles 8). And finally, encouraging diversity will increase organizational learning capacity and reduce risk of incorrect knowledge (principle 10).

4   Conclusion and Further Work

We have presented 14 learning principles that we argue are significant steps towards a lean approach to knowledge management. These principles comply well with findings in studies on competence development strategies. We have further linked the principles to previous work on agile and lean software development. We hope these principles can inspire others who seek to renew their competence development strategy with a focus on speed of learning, learning as a team activity and learning focused on needs. The focus on needs seeks to avoid "wasteful" learning and creating inventories of knowledge.

The new learning principles were well received in Equinor. In the future we would like to learn more about how the principles are perceived to support speed of learning, focus on learning as a team activity and as a way to reduce wasteful learning. If we place the current work in the improvement cycle of Total Quality Management, we see the learning principles as in the "do" step and approaching a "check". Discussing the principles also with the lean community could lead to insight to decide on what to act on next.

While working on this paper we discovered and discussed several additional links between knowledge management, software development, and lean manufacturing. For example, if we consider modern agile software development as a method for codification of a collective learning process, then it becomes particularly interesting to compare lessons with recent work on lean as a learning system. Also, the "knowledge as inventory" perspective needs to be studied further. Finally, more work is needed to further define key characteristics of "lean knowledge management".

References

1. Alavi, M., Leidner, D.E.: Knowledge management and knowledge management systems: conceptual foundations and research issues. MIS Q. 107–136 (2001)
2. Brandi, U., Iannone, R.L.: Learning strategies for competence development in enterprises. Ind. Commer. Train. **49**, 1–5 (2017)
3. Danese, P., Manfè, V., Romano, P.: A systematic literature review on recent lean research: state-of-the-art and future directions. Int. J. Manag. Rev. **20**(2), 579–605 (2018)
4. Dingsøyr, T., Nerur, S., Balijepally, V., Moe, N.B.: A decade of agile methodologies: towards explaining agile software development. J. Sys. Soft. **85**, 1213–1221 (2012). https://doi.org/10.1016/j.jss.2012.02.033
5. Equinor: 14 steps toward a learning organization (2021). www.loop.equinor.com/en/stories/14-steps-toward-a-learning-organization.html. Accessed 01 June 2021
6. Nguyen, N., Grisold, T., Klammer, A.: Organizational unlearning: opportunities and interdisciplinary perspectives. Learn. Organ. **26**, 445–453 (2018)
7. Ohno, T.: Toyota production system: beyond large-scale production (1988)
8. Poppendieck, M., Poppendieck, T.: Lean Software Development: An Agile Toolkit. Agile Software Development Series, Pearson Education, London (2003)
9. Poppendieck, M., Poppendieck, T.: Implementing Lean Software Development: From Concept to Cash. A Kent Beck Signature Book, Addison-Wesley, Boston (2007)
10. Powell, D., Coughlan, P.: Corporate lean programs: practical insights and implications for learning and continuous improvement. Proc. CIRP **93**, 820–825 (2020)
11. Takeuchi, H., Nonaka, I.: The new new product development game. Harv. Bus. Rev. **64**, 137–146 (1986)

Taking the Playing Lean Experience Online
The Case of Using a Board Game to Teach Lean Startup Remotely

Bruno Pešec[1,2(✉)]

[1] Ashridge Executive Education, Berkhamsted, Hertfordshire, UK
[2] Playing Lean, Oslo, Norway
bruno@playinglean.com

Abstract. Using games and simulations to teach various lean methods and approaches enables the teacher to increase student's participation and identify each individual's challenges. Given the SARS-CoV-2 pandemic, most of the businesses and institution went into lockdown from early 2020, forcing the educators to deliver remote workshops. Playing Lean is a team-based board game in which players have to develop their business through iterative development and lean experimentation. The team that reaches early majority first—by crossing the proverbial chasm—wins. In this paper, the author explores how the game was adapted for teaching the lean startup method remotely.

Keywords: Lean startup · Innovation · Entrepreneurship · Gamification · Educational games · Game-based learning

1 Introduction

Lean educators are no strangers to using games and simulations for introducing various lean concepts. Although online and remote versions of such activities have existed ever since the days of modern internet, in 2020 they were the only safe option due to the SARS-CoV-2 pandemic. Taking existing games online isn't always as simple as developing an exact digital replica, especially when they rely on continuous player-to-player and player-to-facilitator interactions.

This paper presents key issues encountered when adapting Playing Lean, a board game for teaching Lean Startup, to an online experience. Described issues are: managing attention, storytelling, facilitator improvisation, communication, social learning, and technological aptitude. Succeeding them is a discussion of select counter-measures for each issue and their results.

Above case and discussion is preceded by brief introduction of relevant theory: the Lean Startup methodology, concept of gamification, and Playing Lean educational board game.

© IFIP International Federation for Information Processing 2021
Published by Springer Nature Switzerland AG 2021
D. J. Powell et al. (Eds.): ELEC 2021, IFIP AICT 610, pp. 299–307, 2021.
https://doi.org/10.1007/978-3-030-92934-3_30

2   Theoretical Background

2.1   Lean Startup

Lean Startup seeks to eliminate wasteful practices and increase value-creating practices during the product development phase by focusing on what customers truly want. It is a methodology for developing businesses and products, aiming to shorten product development cycles by adopting a combination of hypothesis-driven experimentation, iterative product releases, and validated learning. [2,13]

Lean thinking, business model design, customer development, and agile engineering are the core components of Lean Startup body of knowledge. Eric Ries [13] outlined following five underlying principles:

Entrepreneurs are everywhere This principle is about democratization of entrepreneurial activity, both in our communities as well as our organisations. Education or economic background should not dictate if someone is or isn't allowed to engage in entrepreneurship.

Entrepreneurship is management Just like Juran drew our attention to accounting practices, and how we can apply them to manage quality [6,10,15], so has Ries demonstrated how to apply a diverse set of practices from lean thinking to entrepreneurial ventures. As we once moved away from "hoping" for quality and taking rework as granted, now we are moving towards seeing entrepreneurship as something we can get better at, increase our odds of success, and reduce how much time, money, and human potential is wasted.

Validated learning Following on the previous point, Lean Startup, just like its predecessor, emphasizes decision making informed by data. Entrepreneurs and innovators ought to continuously test their assumptions, collect data, and adjust as they go. To borrow lean jargon, their *gemba* is where the customer is, since they are the ones who will be the ultimate judges what value is.

Build–Measure–Learn Taking a hint from the Plan–Do–Check–Act cycle, Lean Startup offers a Build–Measure–Learn loop. It acts as a framework for generating validated learning. Although executed as written, it is planned backwards. We begin by outlining what is the learning goal, then figure out what kind of data we need to collect, and then work out what is the smallest thing we have to build in order to learn what we wish to learn.

Innovation accounting New ventures can take more than three years to show returns. Since existing accounting practices focus on recording expenses and revenue, we are missing a set of practices that would allow us to measure the monetary value of learning and making better decisions when adopting a Lean Startup approach. Innovation accounting is still a nascent field, with an ample opportunity for further development.

2.2   Gamification

Gamification is a *"process of enhancing a service with affordances for gameful experiences in order to support user's overall value creation"* [4, 19]. By introducing game elements in a non-game environment, it makes the learning more enjoyable [5,7]. In [11], authors emphasize how using games as teaching tools benefits the teaching-learning process, especially when it comes to:

- introducing difficult concepts;
- developing problem-solving and decision-making skills;
- promoting an active participation of the student;
- increased interest among students;
- developing each student's talents, which welcomes students at different learning levels; and
- helping the teacher identify each student's difficulties.

Although gamification has positive effects on the learning [3,7,8], games and simulations are most effective when run by a facilitator who is knowledgeable and experienced about the topic being taught, and possesses good communication and storytelling skills [1].

2.3 Playing Lean

Playing Lean is a board game for teaching Lean Startup methodology, deploying a number of gamification practices: storytelling, social learning, motivation and reward structures, competition, and use of facilitator (game master).[1] Intended learning outcome of the game is improved understanding of the following 11 Lean Startup concepts:

- the Lean Startup methodology as a whole,
- the Build-Measure-Learn cycle;
- pivoting,
- "get out of the building",
- fast iteration,
- minimum viable product,
- innovation accounting,
- technical debt,
- problem-solution fit,
- product-market fit, and
- scalability and timing.

Playing Lean is a turn based game where students are divided into competing teams, each representing a fictional startup. They are all competing in the same industry, and the first team to reach 100 000 customers, wins. Each team starts with three employees, which can be assigned to three different activities. The game is divided into two phases:

---

[1] Detailed coverage of how Playing Lean gamifies teaching Lean Startup is presented in [12].

Planning Teams discuss what tasks will they assign to each employee: develop a product, conduct an experiment, attempt a sale, or attend a training. Conversations during this phase are quite lively, with students discussing questions such as *"How do we proceed? What strategy do we want to pursue? Do we want to satisfy all the customers or look for the quickest path to "victory"? How do we handle competition? Do we want to make the best product or do we want to invest time to find out what the customers want?"* Planning phase is usually time-boxed to 120 s, in order to keep the brisk pace and facilitate relevant discussions.

Execution Facilitator informs the teams of the outcomes of their activities. Product development is the only guaranteed activity—i.e., an employee will always succeed developing a single feature—whilst the other three are not. If teams are making an uninformed sales decision, they are risking disappointing the customer who won't give them another chance. Experimentation includes an element of randomness, meaning that the amount of collected customer insight varies.

The only way for teams to advance in the game is by selling their product successfully. Following the four stages of Lean Startup [9], all teams begin at the *Business Modelling* stage, working their way through *Problem/Solution fit* and *Product/Market fit* stages, before finally reaching *Scaling* stage. There is only one winner—the team that reaches the last stage first. Each experiment introduces one of the Lean Startup principles, tools or methods. It has a title, a short description and a "result" (how much has the team learned about a specific customer). To maximize learning, the facilitator should tell a story related to the card, sharing their experience and explaining real-life examples of how the tool or method can be applied.

Once the game is over, it's beneficial to hold a retrospective session in order to reinforce the learning. Recommended format is to mix-up the students into dyads, and ask them to discuss following questions:

- Why did your team win/lose?
- What was your overall strategy?
- Did your company need to pivot?
- Were you affected by technical debt?

By providing above frame for retrospection, the facilitator helps students contextualize and enrich their experience by including different—and sometimes opposing—viewpoints. Dyadic discussions are especially helpful for avoiding the "halo" effect [14], where the team that won believes all their actions must have been right because they won.

3   Case: Issues with Taking Playing Lean Online

Playing Lean has been designed for use in an intense workshop setting with everybody sitting at the table, and facilitator standing on one of the sides. It was never intended to be an online experience, but the SARS-CoV-2 pandemic changed that. Following subsections outline some of the issues encountered when re-creating the Playing Lean experience in a remote setup.

3.1   Facilitation Issues

Attention. Facilitator acts as both the game master and teacher, i.e., they have to balance playing with learning. Commanding attention is critical for facilitator's ability to shift focus from one to the other. In a physical workshop, facilitator can use their voice, body posture, and instruments to quickly get attention. Further, since students are all physically close, it's unlikely that any one of them will use their phone or laptop to perform other activities. In online workshops students are a click away from distraction.

Storytelling. Although each Playing Lean experiment card provides a story, it is the facilitator who delivers it. Aforementioned delivery isn't limited just to speech, but can also include using presentation slides, flip-charts, print-outs, and other audio-visual aids. All students have a unified experience of the story told by the facilitator, while that isn't always true for the online workshops. In the latter case, the facilitator is rarely in control of what the student sees and hears.

Improvisation. Tabletop board games with game masters embrace "rulings over rules" ethos, meaning that the facilitator can adjust rules on-the-fly, in order to create a better learning experience for the students. This improvisation is critical for the facilitator's ability to adjust the game to guide the students (and themselves) in achieving desired learning outcomes. Quickly adjusting the board game elements, team composition, or how planning and execution phases are conducted is nearly effortless in a physical workshop. When conducting online workshops, the scope of improvisation is heavily impacted by two factors: facilitator's skill with technology used, as well as the limitations of that same technology.

3.2   Collaboration Issues

Communication. As outlined earlier, students' conversations are important for both moving the game forwards, as well as the learning that comes from discussing different options and decisions. During a standard in-person workshop, the students would be sitting close to each other, and wouldn't be far from other teams as well. Everybody can hear everybody, and facilitator can jump-in as needed. It's possible to understand individuals, even if multiple people are speaking. In online workshops that is nearly impossible. Standard solution of putting people in separate "breakout" rooms only partially addresses the problem, whilst reducing the effect of social learning.

Social Learning. During the workshop, social learning is manifested by players interacting in multiple ways: helping each other understand the rules or reward mechanisms, arguing for different strategies, explaining to each other various business topics and Lean Startup concepts, and having game-related banter. Taking the workshop online dampens the effects of social learning because it becomes limited to team discussions, which happen in separate virtual rooms. That way the bigger group (i.e., all students) misses out on insights of other teams.

Technological Aptitude. Educators use Playing Lean to teach Lean Startup. They should spend as little time as possible to explain the game rules. During a physical workshop, students only need to use the board game, pen, and paper. In the online workshop each student must setup their microphone, camera, headphones, screen, video-conferencing tool, white-board tool, chat tool, and whatever other tool the facilitator might have selected. Given how many combinations of devices and operating systems there are, troubleshooting might take too much time.

4   Discussion of Select Counter-Measures for Each Issue and Their Results

Counter-measures for each issue are outlined in Table 1.[2] Guiding principles for designing them were to: (1) have as little interference between the students and the facilitator as possible, (2) attempt to transform issues into learning lessons themselves, and (3) keep them as simple as possible.

Some counter-measures had more effect than the others. Having specific roles for each student has worked particularly well. Each team had to name their *Head of Product*, *Head of Experimentation*, and *Head of Sales*. Each role corresponds to one of the core game activities, namely product development, experimentation, and sales. Each team also had to nominate their CEO, who would have to make all quick decisions the facilitator asks for. By doing this, each student had agency to speak up, and wasn't surprised when called out for a specific decision.

Using breakout rooms and additional guiding questions during reflection worked well to reduce the negative impact of the online format on social learning, although it was far from flawless. For example, since the facilitator does not join the breakout discussions, teams cannot ask quick rules-related questions (e.g. are they allowed to assign employees to specific action, can they attempt a sale to a particular customer, and so on). They have to make an educated guess and then bear the consequences. The author has addressed such cases through improvised storytelling, framing them as situations where the start-up is unsure about how the regulators will react, and they still have to decide if they will act

---

[2] Dr. Paige Wilcoxson and Priya Dasgupta-Yeung [16–18] offer several accessible texts on how to adapt workshop curriculum to the virtual classroom, grounded in their expertise with instructional design.

Table 1. Issues and their respective counter-measures

Issue	Counter-measures
Attention	Set the expectations in advance: no mobile phones, close all browser and windows tabs unrelated to the workshop, respect the time schedule. Introduce specific roles with clear accountabilities for each team member
Storytelling	Limit the storytelling to speech, use more vivid and less abstract examples, speak slower and more eligible, make sure to have attention of everyone before sharing
Improvisation	Gather student profiles and desired learning outcomes in advance and adjust the workshop design as necessary. Disseminate all student materials in advance, and have all relevant URLs listed in one document
Communication	Use video-conferencing solution (e.g. Zoom) for voice and white-board solution (e.g. Miro or Mural) as virtual classroom. Be explicit where communication is supposed to happen. Prepare a dashboard for each team. Ask how did students communicate with each other
Social learning	Use breakout rooms to facilitate social learning within teams, modify the reflection session to include discussion on each other's strategies and actions. Ask how did teams learn from each other
Technological aptitude	Select the least amount of technologies needed to deliver the workshop and achieve desired learning outcomes. Prioritise technologies with wide adoption and perceived ease of use. Include use of these technologies in the workshop agenda

and hope for the best, or if they will spend additional time to find out what is or isn't allowed. Whenever such improvisations are introduced, it's important to be consistent and apply them equally to each team, otherwise students might feel they are being judged unfairly.

Another thing that worked quite well was integrating instructions on how to use the selected technologies into the workshop itself. In this case, the author opted to use Zoom for video and chat, and Mural for the gameplay board. During the workshop, students had to perform various simple exercises that had double purpose: to move the workshop forward, while simultaneously improving their proficiency with the selected digital tool. For example, students were put into breakout rooms of three, and had to discuss their background, proficiency with the Lean Startup method, and what do they expect from the workshop. By doing so, they both bonded and got to experience the "breakout room" process. Another example was writing their name and a game role on a digital sticky note and moving it to their team corner. Again, in doing so they got slightly better at using Mural, as well as understanding the game rules.

Defining communication channels upfront produced desired results. Following the KISS dictum, we limited ourselves to voice during group and team discussions, and chat for 1-to-1 quick exchanges. Although Zoom's rudimentary chat features leave a lot to be desired, the author hasn't noticed any significant retardation to the information flow. That's most likely due to the fact that the bulk of strategic discussions happen during the breakout sessions, whilst chat is only used for quick agreements when one of the players is called out to make a decision (e.g. "how many employees will you send to this training?"). There weren't that many challenges with selected technologies, which can probably be attributed to the fact that all participants have been using them (and other, similar tools) throughout 2020.

5 Conclusion

This paper briefly introduced: (1) Lean Startup, a methodology for developing businesses and product by adopting a combination of hypothesis-driven experimentation, iterative product releases, and validated learning, (2) gamification, a *"process of enhancing a service with affordances for gameful experiences in order to support user's overall value creation"* [4, 19], and (3) Playing Lean, a board game for teaching Lean Startup. That was followed by a discussion of six key issues when using Playing Lean in an online and remote context, as well as results of attempted counter-measures. Although author's initial experience indicates that conducting remote Playing Lean workshops yields similar learning outcomes as the original, in-person variant, more research needs to be done in order to capture and verify exact effects.

References

1. Badurdeen, F., Marksberry, P., Hall, A., Gregory, B.: Teaching lean manufacturing with simulations and games: a survey and future directions. Simul. Gaming **41**(4), 465–486 (2010). https://doi.org/10.1177/1046878109334331
2. Eisenmann, T.R., Ries, E., Dillard, S.: Hypothesis-driven entrepreneurship: the Lean startup. Harv. Bus. School Entrep. Manag. Case (812–095) (2011)
3. Hays, R.T.: The effectiveness of instructional games: a literature review and discussion. Technical report 2005-004, Naval Air Warfare Center Training Systems Division (2005)
4. Huotari, K., Hamari, J.: Defining gamification: a service marketing perspective. In: Proceeding of the 16th International Academic MindTrek Conference, pp. 17–22 (2012)
5. Jakubowski, M.: Gamification in business and education - project of gamified course for university students. Dev. Bus. Simul. Exp. Learn. **41** (2014)
6. Juran, J.M., Godfrey, A.B. (eds.): Juran's Quality Handbook, 5th edn. McGraw Hill, New York (1999)
7. Kapp, K.M.: The Gamification of Learning and Instruction: Game-Based Methods and Strategies for Training and Education. Pfeiffer (2012)

8. Ke, F.: A qualitative meta-analysis of computer games as learning tools. In: Management Association, I.R. (ed.) Gaming and Simulations: Concepts, Methodologies, Tools and Applications, pp. 1619–1665. IGI Global (2011)
9. Maurya, A.: Running Lean: Iterate from Plan A to a Plan That Works. The Lean Series, 2nd edn. O'Reilly, Newton (2012)
10. Phillips-Donaldson, D.: 100 years of Juran. Qual. Prog. **37**(5), 25–39 (2004)
11. Ramos, A.G., Lopes, M.P., Avila, P.S.: Development of a platform for lean manufacturing simulation games. IEEE Rev. Iberoamericana Tecnol. Aprendizaje **8**(4), 184–190 (2013). https://doi.org/10.1109/RITA.2013.2284960
12. Rasmussen, T., Øxseth, S.: Playing Lean: gamification of the Lean startup methodology. Master's thesis, Norwegian University of Life Sciences, Ås/Norwegian University of Life Sciences (2016). http://hdl.handle.net/11250/2384042
13. Ries, E.: The Lean Startup: How Today's Entrepreneurs Use Continuous Innovation to Create Radically Successful Businesses. Crown Business (2011)
14. Rosenzweig, P.M.: The Halo Effect: ... And the Eight Other Business Delusions That Deceive Managers. re-issue edn. Simon & Schuster (2014)
15. Stephens, K.S., Juran, J.M. (eds.): Juran, Quality, and a Century of Improvement. Best on Quality, vol. 15. ASQ Quality Press (2005)
16. Wilcoxson, P., Dasgupta-Yeung, P.: Adapting your curriculum to the virtual classroom (2020). https://psychelxdesign.com/virtual-classroom/
17. Wilcoxson, P., Dasgupta-Yeung, P.: It's time to go virtual: workshops (2020). https://psychelxdesign.com/its-time-to-go-virtual-workshops/
18. Wilcoxson, P., Dasgupta-Yeung, P.: So you have to teach online... Now what? (2020). https://psychelxdesign.com/so-you-have-to-teach-online-now-what/

Bloom Taxonomy, Serious Games and Lean Learning: What Do These Topics Have in Common?

Gabriela R. Witeck(✉) [iD], Anabela C. Alves(✉) [iD],
and Mariana H. S. Bernardo(✉) [iD]

Centro ALGORITMI, Department of Production and Systems,
University of Minho, Guimarães, Portugal
anabela@dps.uminho.pt

Abstract. Lean Thinking principles and methods, in the context of highly valued production systems, are seen as the best practices and essential for competitiveness. Therefore, it requires educators, students, and employees well trained and prepared in Lean concepts to meet these demands. In this context, gamification is becoming a popular resource among educators who aim to train the principles of Lean Thinking. Among educators, Bloom's taxonomy is an objective-based assessment as it approaches a high level of detail when defining learning objectives. In the context of this paper, Bloom's Taxonomy encompasses the acquisition of knowledge, skills, and attitudes, to identify learning outcomes in serious games. This paper presents a literature review based on Bloom's Taxonomy, Serious Games and Lean learning. With this review, the authors intend to find evidence that Serious Games are suitable for Lean learning to reach the highest order level of Bloom's Taxonomy. Fourteen papers were identified in this review discussing the three topics. In just three of these papers, it was identified such evidence.

Keywords: Bloom taxonomy · Serious games · Lean learning

1 Introduction

In complex environment change, the companies need to adapt and transform themselves. Lean Thinking [1] has been a philosophy that help companies to seek more lean, flexible, sustainable and agile processes to face such complex scenarios [2, 3]. Nevertheless, there is "no Lean without learning", as referred by Powell and Reke [4]. Moreover, learning and knowledge dissemination are important processes in companies.

Lean Education, i.e., learning Lean, promote competencies that allowed to the apprentices to become able to improve Lean transformation in organizations [5, 6]. At a time when the industrial economy is being stronger allied to the knowledge economy, it is opportune to review the nature of practical education and the inherent skills developed [7]. Indeed, learning Lean implies effective and practical instructional methods such as gamification, hands-on, learning factories, role-play, simulations,

© IFIP International Federation for Information Processing 2021
Published by Springer Nature Switzerland AG 2021
D. J. Powell et al. (Eds.): ELEC 2021, IFIP AICT 610, pp. 308–316, 2021.
https://doi.org/10.1007/978-3-030-92934-3_31

among others [8]. In particular, some examples have demonstrated that simulation games are important instructional methods for learning Lean concepts [9–13]. Using these, learning is more practical, and objective and it is expected that using them will reduce the learning cycle time while add value to Lean teaching.

Gamification is the use of game elements in a non-game related context, with the aim of promoting learning behavior that enhances students' resources [14]. Applying game-based learning with its pedagogical benefits is a well-known instructional method recognized in Higher Education [9]. One of the main cognitive approaches to evaluate Serious Games (a second name for gamification) is Bloom's Revised Taxonomy [15]. Gamification can be a useful learning methodology for achieving all levels of Bloom's Taxonomy [16].

The increasing demand for knowledge within companies highlights the importance of measurable methods that respond to the gap between formal education and the workplace where is perceived [14]. The development and growing of a company can be determined by the ability and capacity of it and its individuals to be successfully involved in learning processes [7]. In this way, there is a clear need to evaluate empirically or study the effectiveness of gamification in relation to the results to which it is committed [16].

This article presents a literature review about Serious Games, Bloom Taxonomy and Lean learning to identify evidence that Serious Games allows effective learn Lean by allowing to highest order level of Bloom's Taxonomy.

This paper is structured in five sections. After the introduction of the objectives in this section, a brief background related to Bloom taxonomy, and game-based Lean is presented. The third section presents the research methodology. Results are presented in the fourth section. The last section, the fifth section presents the conclusions.

2   Background

2.1   Bloom's Taxonomy and Serious Games

Bloom's original taxonomy is a popular method among educators for objective-based assessment, as it allows for a high level of detail when defining learning objectives [15]. The three domains classified by Bloom [17] are: cognitive, affective, and psychomotor. The author also defines six different levels of cognitive skills: a) Knowledge; b) Comprehension; c) Application; d) Analysis; e) Synthesis; and f) Evaluation. Later it was revised by Krathwohl [18] as: 1) Remember; b) Understand; c) Apply; d) Analyze; e) Evaluate and f) Create. These are important levels for the achievement of better student learning and critical thinking results [7, 15].

The Serious Games, as a personalized learning approach to a real context, has a capacity to improve the quality of the educational system, offering opportunities to maximize the students' potential [19]. Given the current commitment of educators, Serious Games has been a positive choice for the development of Bloom's Revised Taxonomy as it can promote reaching the highest levels of learning [11]. Haring et al.

[15] considered that Bloom's Revised Taxonomy is one of the most appropriate cognitive tools for the evaluation of Serious Games. Furthermore, Nisula and Pekkola [20] concluded that the use of Serious Games based on Bloom's Taxonomy prepares professionals for complex and multidisciplinary work environments in a constantly changing world.

2.2 Game-Based Lean

Based on the fundamentals of the Toyota Production System [21], the Lean philosophy can be described as a set of principles that aim to reduce waste, add value and maximize the results of an organization through a systematic process of continuous improvement [9, 22–24]. Lean goes beyond a set of tools and methods. It means going deeper into a learning cycle, experiencing improvements, reflecting, and internalizing your insights, resulting in lean improvement lessons [6, 24].

For the global and sustainable evolution of Lean, fundamental values such as building a culture of continuous improvement, people development, management for learning, and problem solving are expected to be maintained [25]. Lean practices must promote, in addition to greater customer satisfaction, an ideal environment for learning: without guilt, fewer errors, availability of information, growth, and respect for workers [26, 27].

Lean learning can also support the company's ability to sustain innovation and knowledge linked to improvements in quality, productivity, and operational efficiency. This is because an organization that learns to learn in a Lean format expands the capacity to continuously create the established results [28]. In this way, models based on mechanisms of association, collaboration, sharing and mutual learning stand out for the development of a Lean learning environment [29].

In a context where current and holistic production systems are valued, the principles and methods of Lean Production are currently seen as the best practices and essential for competitiveness [28, 30]. Therefore, there is a requirement for students well trained and prepared in Lean concepts to meet these new demands [9, 31]. Since many companies suspect that the new generation of professionals is not prepared in terms of knowledge and skills appropriate to the needs of the industry [26], it is noted the importance of institutions to adapt Lean for curriculum and use alternative methods of teaching to connect to real contexts [9, 32]. For Lean educators, the consensus is the meaningful approach to the practical components [33].

This consensus is the consistent approach to the practical components of teaching [8, 13, 33]. Simulation games are currently a popular resource among educators who aim to train workers or students on the principles of Lean Production [33]. The use of game elements in non-game contexts can be defined by Gamification, or also Serious Games [14]. It is a plus pedagogical technique used to enrich the students' experience [16].

Game-Based Lean is designated to use Serious Games as a good alternative to teach Lean while provide new and important skills to professionals. Skills such as critical thinking, teamwork, communication, responsibility, motivation, and global learning are results achieved by Serious Lean Games, once that allow students to make decisions and solve problems based on the application of Lean concepts [13]. Alves et al. [5] provided some examples of the use of Serious Games, and considered a successful tool for the dissemination of Lean knowledge [10], as well as an important active teaching method applied to Lean Learning [8].

3   Research Methodology

For the research it was used a literature review that followed two phases. The first phase, identification, consists of finding the articles according to a topic, carried out in two different search databases: Scopus and Web of Science (WOS). These were selected as they have an ample coverage [34]. The search in the database of SCOPUS and WOS was carried out using the following search strings: ALL ("Bloom's Taxonomy" AND "Gamification" AND "Lean"), ALL ("Bloom's taxonomy" AND "Serious Game*" AND "Lean") and limited to the period from 2010 to 2021. This period was chosen as most publications about serious game assisted education appeared after 2009 [35]. Also, the search included articles, conferences, books, and book chapters. This search resulted in a total 32 articles.

The second phase consisted of filtering and excluding duplicated articles, using the Excel software, following the criteria explained next. Repeated studies within the databases were eliminated, studies published in WOS were maintained and those that were in Scopus were subtracted. To do this, all the files downloaded from the databases were grouped and duplicated files within were removed. In total, 22 articles resulted from this phase.

The analysis of the 22 articles allowed to discard eight papers applying some criteria: not having free access to full paper and considering only articles that are case studies approaching Bloom Taxonomy, Serious Games and Lean learning within universities or companies. With this, the authors want to answer to the question in the title of the paper: *Bloom Taxonomy, Serious Games and Lean learning: What do these topics have in common?* This means that the final selection resulted in 14 papers, which were deeply analyzed and made-up Sect. 4 of this study.

4   Results and Discussion

The 14 papers identified that have the keywords selected are in the Fig. 1. The deep analysis of all papers referred the three domains of Bloom Taxonomy, namely, cognitive, affective, and psychomotor. Some examples were the presence of the words problem-solving and confidence.

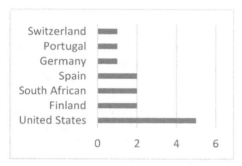

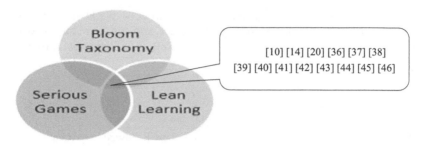

Fig. 1. Papers obtained from Scopus search.

Figure 2 and Fig. 3 shows respectively the countries and year from where the papers are. United States is the country with more publications (five papers), and considering the period of 2010 to 2021, the year of 2019 was the year with the highest number of publications. It is possible that papers discussing the three topics only appear in 2016.

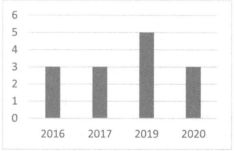

Fig. 2. Number of papers by Country.   **Fig. 3.** Number of papers by year.

Table 1 presents the hierarchical levels applied in the Serious Games in the 14 papers selected. Through the content analysis of the papers, the authors identified elements indicating all the hierarchical levels in only three papers [14, 39, 46]. This confirmation was based on the "Create" level keywords (e.g., formulate, build, invent, compose, develop, etc.) found as a description and conclusions of the games activities made by the apprentices. For instance, Souza et al. [10] and Crocco et al. [36] did not achieve the Create level. This means that, during or after the game, nothing was said about solutions created by the apprentices as an outcome of the game.

The Serious Games activity described in the papers [37, 41, 42, 44, 45] did not achieve the "Analyze" level. The keywords for this level (e.g., classify, break down, categorize, criticize, simplify, associate, etc.) were not found. However, Ponticorvo et al. [42] confirmed, while playing the game, the user expressed herself through behavioral acts that involved her body, or other forms of interactions. And this same author also referred that the educational aspect involved allowed users to foster transversal skills through innovative approaches to teaching, learning, and assessment.

Table 1. Bloom Taxonomy hierarchical levels applied in the serious games or simulations.

Author	Remember	Understand	Apply	Analyze	Evaluate	Create
[10]	X	X	X	X	X	–
[14]	X	X	X	X	X	X
[20]	X	X	X	X	–	–
[36]	X	X	X	X	X	–
[37]	X	X	X	–	–	–
[38]	X	X	X	X	–	–
[39]	X	X	X	X	X	X
[40]	X	X	X	X	–	–
[41]	X	X	X	–	–	–
[42]	X	X	X	–	–	–
[43]	X	X	X	X	–	–
[44]	X	X	X	–	–	–
[45]	X	X	X	–	–	–
[46]	X	X	X	X	X	X

Although the sample is small and the results are limited, from this research, the authors inferred that Serious Games could help to achieve the highest order thinking level of Bloom Taxonomy. The three topics discussed are complementary in their usefulness to the Lean apprentices and their mentors.

5   Conclusions

This paper contributes to the scientific understanding of how Bloom Taxonomy, Gamification and Lean learning are connected within engineering contexts. The literature review performed returned 14 papers that discuss the three concepts. This means that the three have something in common. Serious Games are practical instructional educational tools useful for people to learn Lean that allowed them to reach higher order Bloom Taxonomy levels. Lean learning is guided by practical and objective elements, often instructed by active learning methods.

Furthermore, Game-Based Lean when reaching the levels of Bloom Taxonomy, it is expected the achieve of better student learning and critical thinking results and reduction of the learning cycle time and added value to Lean learning.

The limitations of the study consist in its qualitative nature and its reliance on a literature review that may raise some issues regarding the generalizability of results. This lack of generalizability contributes to the small sample obtained and the subjective interpretation. Still, the present study should encourage future analyses, expanding the implementation steps of Serious Games to enhance Lean learning in university and industrial environments.

In future work, it will be meaningful to understand the real gains of the Serious Games, i.e., how it contributes to adding value to the companies.

Acknowledgements. This work has been supported by FCT – Fundação para a Ciência e Tecnologia within the R&D Units Project Scope: UIDB/00319/2020.

References

1. Womack, J.P., Jones, D.T.: Lean Thinking: Banish Waste and Create Wealth in your Corporation. Free Press, New York (1996)
2. Soliman, M., Saurin, T.A., Anzanello, M.J.: The impacts of lean production on the complexity of socio-technical systems. Int. J. Prod. Econ. **197**, 342–357 (2018). https://doi.org/10.1016/j.ijpe.2018.01.024
3. Saurin, T.A., Rooke, J., Koskela, L.: A complex systems theory perspective of lean production. Int. J. Prod. Res. **51**(19), 5824–5838 (2013). https://doi.org/10.1080/00207543.2013.796420
4. Powell, D., Reke, E.: No lean without learning: rethinking lean production as a learning system. In: Ameri, F., Stecke, K.E., von Cieminski, G., Kiritsis, D. (eds.) APMS 2019. IAICT, vol. 566, pp. 62–68. Springer, Cham (2019). https://doi.org/10.1007/978-3-030-30000-5_8
5. Carvalho Alves, A., Flumerfelt, S., Kahlen, F.-J. (eds.): Lean Education: An Overview of Current Issues. Springer, Cham (2017). https://doi.org/10.1007/978-3-319-45830-4
6. Flumerfelt, S., Kahlen, F.-J., Alves, A.C., Siriban-Manalang, A.B.: Lean Engineering Education: Driving Content and Competency Mastery. ASME Press, New York (2015)
7. Boles, W., Beck, H., Hargreaves, D.: Deploying bloom's taxonomy in a work integrated learning environment. In: Proceedings of the 2005 ASEE/AaeE 4th Global Colloquium on Engineering Education, pp. 1–10 (2005)
8. Alves, A.C., Flumerfelt, S., Moreira, F., Leão, C.P.: Effective tools to learn lean thinking and gather together academic and practice communities. In: Education and Globalization, vol. 5, p. V005T06A009 (2017). https://doi.org/10.1115/IMECE2017-71339
9. Herrera, R.F., Sanz, M.A., Montalbán-Domingo, L., García-Segura, T., Pellicer, E.: Impact of game-based learning on understanding lean construction principles. Sustainability **11**(19), 5294 (2019). https://doi.org/10.3390/su11195294
10. Sousa, R.M., Stadnicka, D., Dinis-Carvalho, J., Ratnayake, V.I.R.M.C.: Gamification based lean knowledge dissemination: a case study. In: Proceedings of the 2016 IEEE IEEM (2016)
11. Marley, K.A.: Eye on the gemba: using student-created videos and the revised bloom's taxonomy to teach lean management. J. Educ. Bus. **89**(6), 310–316 (2014). https://doi.org/10.1080/08832323.2014.903888
12. Sousa, R.M., Alves, A.C., Moreira, F., Dinis-Carvalho, J.: Lean games and hands-on approaches as learning tools for students and professionals. In: 7th International Conference on Production Research - Americas (2014)
13. Witeck, G., Alves, A.C.: Developing lean competencies through serious games. In: PAEE/ALE'2020, International Conference on Active Learning in Engineering Education, 12th International Symposium on Project Approaches in Engineering Education (PAEE) and 17th Active Learning in Engineering Education Workshop (ALE), pp. 179–186 (2020)
14. Henning, M., Hagedorn-Hansen, D., Von Leipzig, K.: Metacognitive learning: skills development through gamification at the Stellenbosch learning factory as a case study. South African J. Ind. Eng. **28**(3), 105–112 (2017). https://doi.org/10.7166/28-3-1845
15. Haring, P., Warmelink, H., Valente, M., Roth, C.: Using the revised bloom taxonomy to analyze psychotherapeutic games. Int. J. Comput. Games Technol. **2018**, 1–9 (2018). https://doi.org/10.1155/2018/8784750

16. Kenny, G., Lyons, R., Lynn, T.: Don't make the player, make the game: exploring the potential of gamification in is education. In: AMCIS 2017 - Americas Conference on Information Systems a Traditional Innovation, vol. 2017-Augus, no. August (2017)
17. Bloom, B.S.: Taxonomy of Educational Objectives. David McKay Company Inc, New York (1956)
18. Krathwohl, D.: A revision of Bloom's taxonomy: an overview. Theory Pract. **41**(4), 212–218 (2002)
19. Sukajaya, N., Purnama, K.E., Purnomo, M.H.: Intelligent classification of learner's cognitive domain using Bayes net, Naïve Bayes, and J48 utilizing Bloom's taxonomy-based serious game Artigo Lean_bloom. core.ac.uk (2015)
20. Nisula, K., Pekkola, S.: ERP based business learning environment as a boundary infrastructure in business learning. Educ. Inf. Technol. **24**(4), 2547–2566 (2019). https://doi.org/10.1007/s10639-019-09889-0
21. Ohno, T.: Toyota production system: beyond large-scale production. Product. Press **3**, 152 (1988). https://doi.org/10.1108/eb054703
22. Messaoudene, Z.: Lean thinking as a learning strategy at the service of global development. In: Alves, A.C., Kahlen, F.-J., Flumerfelt, S., Siriban-Manalang, A.B. (eds.) Lean Engineering for Global Development, pp. 33–60. Springer, Cham (2019). https://doi.org/10.1007/978-3-030-13515-7_2
23. Mazur, L., McCreery, J., Rothenberg, L.: Facilitating lean learning and behaviors in hospitals during the early stages of lean implementation. EMJ - Eng. Manag. J. **24**(1), 11–22 (2012). https://doi.org/10.1080/10429247.2012.11431925
24. Singh, J.: The lean prescription for non-traditional adult learners. Qual. Assur. Educ. **27**(3), 347–359 (2019). https://doi.org/10.1108/QAE-09-2018-0100
25. Qing, H., Found, P., Williams, S., Mason, R.: Lean thinking and organisational learning: how can they facilitate each other? In: Chiarini, A., Found, P., Rich, N. (eds.) Understanding the Lean Enterprise, pp. 61–77. Springer, Cham (2016). https://doi.org/10.1007/978-3-319-19995-5_3
26. Sremcev, N., Lazarevic, M., Krainovic, B., Mandic, J., Medojevic, M.: Improving teaching and learning process by applying Lean thinking. Procedia Manuf. **17**, 595–602 (2018). https://doi.org/10.1016/j.promfg.2018.10.101
27. Takeuchi, H., Osono, E., Shimizu, N.: The contradictious that drive Toyota's success. Harv. Bus. Rev. 96–104 (2008)
28. Tortorella, G.L., Fogliatto, F.S.: Method for assessing human resources management practices and organisational learning factors in a company under lean manufacturing implementation. Int. J. Prod. Res. **52**(15), 4623–4645 (2014). https://doi.org/10.1080/00207543.2014.881577
29. Alves, A.C., Dinis-Carvalho, J., Sousa, R.M.: Lean production as promoter of thinkers to achieve companies' agility. Learn. Organ. **19**(3), 219–237 (2012). https://doi.org/10.1108/09696471211219930
30. Blöchl, S.J., Schneider, M.: Simulation game for intelligent production logistics – The PuLL® learning factory. Procedia CIRP **54**, 130–135 (2016). https://doi.org/10.1016/j.procir.2016.04.100
31. Alves, A.C., Leão, C.P., Sousa, R.M.: Lean education as a platform to close the academic and professional gap. In: International Symposium on Project Approaches in Engineering Education, vol. 8, pp. 17–19 (2018)
32. Alves, A.C., Leão, C.P., Uebe-Mansur, A.F., Kury, M.I.R.A.: The knowledge and importance of lean education based on academics' perspectives: an exploratory study. Prod. Plan. Control **32**(6), 497–510 (2021). https://doi.org/10.1080/09537287.2020.1742371

33. de Vin, L.J., Jacobsson, L., Odhe, J.: Game-based Lean Production training of university students and industrial employees. Procedia Manuf. **25**, 578–585 (2018). https://doi.org/10.1016/j.promfg.2018.06.098

34. Mongeon, P., Paul-Hus, A.: The journal coverage of web of science and scopus: a comparative analysis. Scientometrics **106**(1), 213–228 (2015). https://doi.org/10.1007/s11192-015-1765-5

35. Zhonggen, Y.: A meta-analysis of use of serious games in education over a decade. Int. J. Comput. Games Technol. **2019**, 1–8 (2019). https://doi.org/10.1155/2019/4797032

36. Crocco, F., Offenholley, K., Hernandez, C.: A proof-of-concept study of game-based learning in higher education. Simul. Gaming **47**(4), 403–422 (2016). https://doi.org/10.1177/1046878116632484

37. Urbano, L., Terán, H., Gómez, F., Solarte, M., Sepulveda, C., Meza, J.: Bibliographic review of the flipped classroom model in high school: a look from the technological tools. learntechlib.org (2020)

38. Dancz, C., Parrish, K., Bilec, M., Landis, A.: Assessment of students' mastery of construction management and engineering concepts through board game design. J. Prof. Issues Eng. Educ. Pract. **143**(4), 04017009 (2017). https://doi.org/10.1061/(ASCE)EI.1943-5541.0000340

39. Abele, E., et al.: Learning factories for future oriented research and education in manufacturing. CIRP Ann. - Manuf. Technol. **66**(2), 803–826 (2017). https://doi.org/10.1016/j.cirp.2017.05.005

40. Lopez, C.E.: An introduction to the CLICK approach: leveraging virtual reality to in-tegrate the industrial engineering curriculum (2019)

41. Makransky, G., Borre-Gude, S., Mayer, R.E.: Motivational and cognitive benefits of training in immersive virtual reality based on multiple assessments. J. Comput. Assist. Learn. **35**(6), 691–707 (2019). https://doi.org/10.1111/jcal.12375

42. Ponticorvo, M., Dell'aquila, E., Marocco, D., Miglino, O.: Situated psychological agents: a methodology for educational games. mdpi.com (2019). https://doi.org/10.3390/app9224887

43. Cerezo-Narváez, A., Córdoba-Roldán, A., Pastor-Fernández, A., Aguayo-González, F., Otero-Mateo, M., Ballesteros-Pérez, P.: Training competences in industrial risk prevention with lego® serious play®: a case study. mdpi.com (2019). https://doi.org/10.3390/safety5040081

44. Polso, K.M., Tuominen, H., Hellas, A., Ihantola, P.: Achievement goal orientation profiles and performance in a programming MOOC. In: Annual Conference on Innovation and Technology in Computer Science Education, ITiCSE, pp. 411–417 (2020). https://doi.org/10.1145/3341525.3387398

45. Severengiz, M., Seliger, G., Krüger, J.: Serious game on factory planning for higher education. Procedia Manuf. **43**, 239–246 (2020). https://doi.org/10.1016/j.promfg.2020.02.148

46. Jooste, J.L., et al.: Teaching maintenance plan development in a learning factory environment. Procedia Manuf. **45**, 379–385 (2020). https://doi.org/10.1016/j.promfg.2020.04.040

Productivity and Performance Improvement

Lean Monitoring: Boosting KPIs Processing Through Lean

Bassel Kassem[(⊠)], Federica Costa[(⊠)],
and Alberto Portioli Staudacher[(⊠)]

Department of Management Engineering, Politecnico di Milano,
20156 Milan, Italy
{bassel.kassem, federica.costa,
alberto.portioli}@polimi.it

Abstract. The wide diffusion of the Lean Thinking philosophy has led different industries to improvements by analyzing their processes and identifying KPIs to monitor their performance, resulting in more notable importance to the KPIs results than the generation process. Waste appears in different environments, and the KPI monitoring process is no exception. The Lean approach within this area boosts the monitoring process, providing real-time analysis, allowing a better response time in corrective actions. This paper exhibits a continuous improvement project that successfully led to a significant reduction in KPI generation processing time. It relies on introducing a KPI dashboard, a digital tool for lean production.

Keywords: Lean · KPI dashboard · Continuous improvement

1 Introduction

The levels of globalization established in the modern area have set new rules for the business world. Dynamism and competitiveness of the current environment have driven customers' needs for accuracy and quality in the continuously growing service sector [1], and not just in the manufacturing sector [2–5] leaving companies with small or no space for mistakes [6, 7].

Awareness of the status, feedbacks, and requirements of the customers represents a highly competitive advantage, as opposed to the older and traditionally seen scenario where problems are commonly identified in a later stage, resulting in limited and somewhat costly intervention options [8]. For this reason, after the undeniable success of Lean methodology in manufacturing and mass production activities [4, 9], the lean scope is spreading and reaching different industries and contexts [10, 11]. Notwithstanding the astonishing benefits of lean, many firms struggle to integrate, harmonize and sustain lean principles within the different lines of the organization [12], resulting in a slow pace of lean growth, and an unexploited range of opportunities and benefits that lean at strategic levels can provide [2, 13].

The purpose of this article is to illustrate how an Italian company's journey, through the application of a lean problem-solving methodology, could become closer to developing a just-in-time monitoring process. Through an empirical project, the

© IFIP International Federation for Information Processing 2021
Published by Springer Nature Switzerland AG 2021
D. J. Powell et al. (Eds.): ELEC 2021, IFIP AICT 610, pp. 319–325, 2021.
https://doi.org/10.1007/978-3-030-92934-3_32

applicability of lean in the context of KPI monitoring activities through the A3 problem-solving methodology is represented, leading the company for just-in-time responses from the KPIs using the KPI dashboard and improving the relationship with the customers.

2 The Company XYZ

XYZ is an Italian company with more than 70-years of leadership in the production of innovative solutions for the textile industry. As part of a multinational group, the company focuses on developing the sustainability of its processes and providing products with the lowest possible environmental impact, as well as additional services for customers. The seeking for innovation and sustainability leads to extensive research targeting a constant improvement on productivity and quality. Together with techno-logical excellence, the company highly relies on an extensive network of local agents, and sales services to bring the best experience to the customers. The presented paper addresses a project carried out in the offices of the company; the headquarter is based in the north of Italy with more than 300 employees.

3 The Case

Part of the business of the company is related to the sale of services linked to its products. A portion of these services requires a presence on customers' sites, called Field Services (FS). These services encompass not only the activities on the field but some back-office activities as well. In the last years, FS has been facing changes introducing SAP and CRM Salesforce to support the management. And recently, the company was required to move to Salesforce Lightning which is considered to start in the following months of the beginning of this project. The objective of the company is to provide high-quality products and services.

FS directly affects relationships with customers and consequently their overall experience with the company. Therefore, it is of primary importance the ability of the company to monitor its FS activities.

3.1 The Problem

XYZ uses a traditional feedback and control cycle in the FS process. The problem perceived and under investigation has been a long time that is necessary to obtain the KPIs needed to monitor the Field Service Process. This is negatively impacting the possibility to maintain an updated overview of the current situation, making them losing control. The company is currently monitoring 6 KPIs; by shortening this time and the effort that is necessary to have the KPIs, it would be possible to increase the frequency of calculation that, currently, is once per quarter. The company, knowing that the process of formulating the KPIs takes more than four days in LT and 16 days in total workload continues postponing the creation until few days before the end of the quarter. Each week has its urgent tasks, and this eventually leads to the postponement and therefore became the new norm. However, this causes delays in understanding the situation and therefore the ability to uncover underlined problems and issues as customers started complaining and showed their concerns regarding some delays in the service and other peculiarities that we opted to omit due to privacy reasons. Surely, even with the current situation, it was possible to increase the frequency of KPI process formulation, but the concern was to spend the same duration each time. Therefore, the company asked the help of the Lean Excellence center at Politecnico di Milano to streamline the process, reduce its duration, and therefore allow for the increase of frequency. Increasing the frequency of calculation would allow the company's management to design timely and, therefore, more effective corrective actions, in case of arising problems, resulting in better overall performances.

3.2 Current Situation

For understanding the current situation, two different measures were considered: the lead time (LT) and the workload (WL). The LT represents the time necessary to obtain the KPIs from the top management's request to the moment when the report is delivered. The WL considers the man-hours dedicated to the activities related to KPIs' computation, during the whole quarter. Even if the LT and the WL are correlated, by monitoring both parameters it would be possible to better highlight the improvements to the company. The computation of the KPIs relates to three macro-activities: data storage, pre-computation, and actual computation. These activities allowed the calculation of the LT and WL. The results for the current situation were LT of 4.1 working days and WL value of 15.8 working days (1 working day equal to 8 h).

3.3 Target Definition

The targets emerged from brainstorming sessions with the After-Sales department leading to a must-have goal of reducing the LT from 4.1 to 1.6 working days (60% reduction), and a reduction of WL from 15.8 to 9.5 working days (40% reduction). Additionally, a nice-to-have target is set as a reduction of LT from 4.1 to 0.8 working days (80% reduction), and a reduction of WL from 15.8 to 6.3 working days (60% reduction).

3.4 Root Cause Analysis

The Root Cause Analysis allowed understanding which was the main reasons that were triggering a long time to compute the KPIs of interest. The Ishikawa diagram, showed in Fig. 1, was the tool applied that gave structure to the symptoms, and the 5 Whys method allowed us to go further in defining the root causes. Afterward, a Priority Matrix showed in Table 1, was designed and, through a Pareto Analysis with the Marginal Increase method, the list of the A classes was drafted. Specifically, the A class root causes which have been tackled by the countermeasures were:

- Functionalities Dispersed in Many Platforms: the functionalities needed to run the F.S. process were dispersed in several softwares;
- Excel Files and Salesforce Cases Updated Occasionally: the software that was used to support the F.S. process was often not up to date;
- High Complexity of Excel Files: Excel files that were used for KPIs computation were born for other purposes, resulting in high underlined complexity.

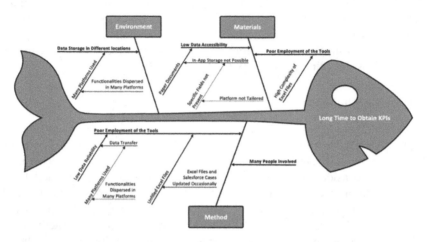

Fig. 1. Ishikawa diagram integrated with 5 WHYs

Table 1. Prioritization matrix of the root causes

Criteria/Root Causes	Data Inaccuracy	Data Incompleteness	Difficulty to Act on Data	Priority Number	Classes
Functionalities Dispersed in Many Platforms	3	9	3	5,0	A
Excel Files and Salesforce Cases Updated Occasionally	9	1	3	4,3	A
High Complexity of Excel Files	3	1	9	4,3	A
Platform not Tailored	1	3	3	2,3	B
Many People Involved	3	1	1	1,7	C

3.5 Countermeasures

Once the root causes have been defined, the next focus was the proposal of counter-measures to address mainly the Class A root causes and to be implemented. Precisely, the countermeasures applied have been:

- KPI dashboard: The team created a digitally-enabled KPI dashboard. Through exploiting all the possible functionalities provided by the implementation of the new platform, the system can store all data needed for the KPIs calculation;
- Even if the reporting functionalities offered by Salesforce Lightning are very effective, usually the last steps still need to be performed in an Excel spreadsheet. Therefore, for most of the KPIs, simplified Excel files were designed to execute those final activities;
- Salesforce Reports: A new reporting system Salesforce Lightning that can be used to pre-aggregate data present in the platform;
- Good Practice for Constant Case Closure: Office personnel every week should check all opened cases on F.S. Lightning and close the ones which are finished;
- Dashboard to Monitor Open Cases: it is a visual tool to have a complete overview of the case interventions that are open at the time of the control.

To have a better understanding of the countermeasures, an effort-benefit analysis was carried out. It's important to underline that the countermeasures do not require additional costs for the company, since they are a consequence of complete exploitation of the Salesforce Lighting platform, usage of Excel, and good practices (Table 2).

Table 2. Evaluation table for countermeasures

Impact/Effort	LOW	MEDIUM	HIGH
HIGH			
MEDIUM		Good Practice for Case Closure	KPI dashboard
LOW		SF reports; Open cases Dashboard	

For the application of the countermeasures, a simulation was run to assess the results that can be obtained after the Go-Live phase. Different people belonging to the After Sales department went through all the different phases of the Data Storage and KPIs computation, applying the countermeasures. Afterward, the average time for each person (to perform each activity) and the average number of interventions per quarter (executed in 2019) were considered to obtain the final results. The whole simulation was executed in a test environment of F.S. Lightning, previously populated with specific data by the team.

3.6   Results Monitoring

Based on the numbers obtained after the simulations performed, the achievements have been 96% reduction for LT (well beyond the Nice-to-Have target) and 48% reduction for WL (overcoming the Must-Have target, but not reaching the Nice-to-Have one, yet). These results will lead to a saving of 240 h per year from monitoring activities. To achieve the predetermined Nice-to-Have target for the WL, further training of the personnel may be necessary. Higher expertise can be reached through the daily usage of the new functionalities of F.S Lightning. Additionally, a brief survey was submitted to understand the overall personnel's perception of the new procedures. The new methodology has been positively perceived, obtaining 4.5 stars out of 5 on the overall satisfaction scale. The positive perception of the methodology will certainly play a key role in the implementation of F.S. Lightning and the application of the whole procedure to KPIs computation after the Go-Live phase.

In the closing phase, some activities were performed to help to achieve the standardization of the work allowing the improvements to last in the long term. Tutorial videos and a PDF guide have been made by the team to support the personnel in performing the activities. Tutorial videos will help the people by providing a complete explanation about how to insert data in F.S. Lightning and the PDF guide will precisely show the new procedure of KPIs calculation, step by step.

4 Conclusion

The practicality of lean in monitoring activities is still in discussion. In this sense, the present article intends to put under focus the applicability of the lean framework within this area and to provide a contribution in the field of knowledge through this case study.

The company has reached an impressive reduction in lead time through the application of lean principles in this improvement project. The A3 problem-solving tool is proven to be an effective lean methodology that would grasp the problem in its entirety to be able to solve it. In addition, applying such methodology and viewing its success allowed the company to welcome it and apply it in all of their successive projects. The outcomes of this successful story were shared with the top management and the personnel. Moreover, an environment of satisfaction was perceived by the direct actors on the monitoring activities, due to the countermeasures implemented and the guidelines for the sustainability of these measures.

Furthermore, we introduced the KPI dashboard, a lean digital tool inside their operations. We showed the successful implementation and the usefulness of a simple digital tool to make the process even leaner by reducing non-value-added activities.

Hence, this paper can be of great help to managers and area leaders to visualize how Lean approach can be successfully implemented in their organization, and to understand the advantages and synergies that this could bring for the relationship with the customers. For the academicians and students, this article will bring an optic of applications of lean methodology outside the common productive environment.

References

1. Torri, M., Kundu, K., Frecassetti, S., Rossini, M.: Implementation of lean in IT SME company: an Italian case. Int. J. Lean Six Sigma 12(5), 944–972 (2021)
2. Rossini, M., Portioli-Staudacher, A., Cifone, F.D., Costa, F., Esposito, F., Kassem, B.: Lean and sustainable continuous improvement: assessment of people potential contribution. In: Rossi, M., Rossini, M., Terzi, S. (eds.) ELEC 2019. LNNS, vol. 122, pp. 283–290. Springer, Cham (2020). https://doi.org/10.1007/978-3-030-41429-0_28

3. Kassem, B., Costa, F., Portioli-Staudacher, A.: JIT implementation in manufacturing: the case of giacomini SPA. In: Rossi, M., Rossini, M., Terzi, S. (eds.) ELEC 2019. LNNS, vol. 122, pp. 273–281. Springer, Cham (2020). https://doi.org/10.1007/978-3-030-41429-0_27
4. Rossini, M., Audino, F., Costa, F., Cifone, F.D., Kundu, K., Portioli-Staudacher, A.: Extending lean frontiers: a kaizen case study in an Italian MTO manufacturing company. Int. J. Adv. Manuf. Technol. **104**(5–8), 1869–1888 (2019). https://doi.org/10.1007/s00170-019-03990-x
5. Rossini, M., Cifone, F.D., Kassem, B., Costa, F., Portioli-Staudacher, A.: Being lean: how to shape digital transformation in the manufacturing sector. J. Manuf. Technol. Manag. **32**, 239–259 (2021)
6. Kumar, M., Vaishya, R., Parag: Real-time monitoring system to lean manufacturing. Procedia Manuf. **20**, 135–140 (2018)
7. Portioli-Staudacher, A., Costa, F., Thürer, M.: The use of labour flexibility for output control in workload controlled flow shops: a simulation analysis. Int. J. Ind. Eng. Comput. **11**, 429–442 (2020)
8. Dallasega, P., Rauch, E., Frosolini, M.: A lean approach for real-time planning and monitoring in engineer-to-order construction projects. Buildings **8**, 38 (2018)
9. Kundu, K., Rossini, M., Portioli-Staudacher, A.: Analysing the impact of uncertainty reduction on WLC methods in MTO flow shops. Prod. Manuf. Res. **6**, 328–344 (2018)
10. Melara, J., Lima, R.M., Souza, T.: Lean office: a systematic literature review. In: 3rd International Joint Conference ICIEOM-ADINGOR-IISE-AIM-ASEM - New Global Perspectives on Industrial Engineering and Management, pp. 319–326 (2017)
11. Costa, F., Portioli-Staudacher, A.: On the way of a factory 4.0: the lean role in a real company project. In: Rossi, M., Rossini, M., Terzi, S. (eds.) ELEC 2019. LNNS, vol. 122, pp. 251–259. Springer, Cham (2020). https://doi.org/10.1007/978-3-030-41429-0_25
12. Dorval, M., Jobin, M.-H.: A conceptual model of Lean culture adoption in healthcare. Int. J. Product. Perform. Manag. (2021, in press)
13. Sousa, R.M., Dinis-Carvalho, J.: A game for process mapping in office and knowledge work. Prod. Plan. Control **32**, 463–472 (2021)

A Simulation-Based Performance Comparison Between Flow Shops and Job Shops

Christoph Roser[✉], Daniel Ballach, Bernd Langer,
and Claas-Christian Wuttke

Karlsruhe University of Applied Sciences, Moltkestrasse 30, Karlsruhe, Germany
christoph.roser@hochschule-karlsruhe.de

Abstract. It is well known in manufacturing that a flow shop usually outperforms a job shop for comparable products. This paper aims to understand the impact of routing variability on WIP inventory, lead time, and delivery performance. A perfect flow shop has no routing variability, with every part following exactly the same sequence of processes. A job shop, on the other hand, often has significant routing variability. In reality, there can be any degree of variability, from the perfect flow shop to a perfectly random routing in a job shop, or anything in between. This paper compares the performance of a perfect flow shop with a perfectly random job shop, aiming to keep all other factors influencing the performance as comparable as possible. The goal is to isolate the impact of routing variability on system performance.

Keywords: Flow shop · Job shop · Utilization · Delivery performance · Lead time

1 Introduction

The material flow in manufacturing can be arranged in two basic ways. In a flow shop, the processes are arranged in the sequence in which the material is processed. (Almost) all parts have the same sequence of processes. In a job shop, the processes are arranged by their function. The parts often have very different sequences of processes and hence very different material flows.

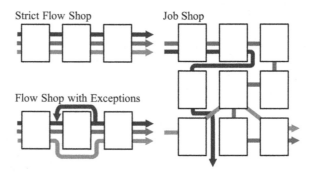

Fig. 1. Illustrative example of a strict flow shop, a flow shop with exceptions, and a job shop

D. J. Powell et al. (Eds.): ELEC 2021, IFIP AICT 610, pp. 326–332, 2021.
https://doi.org/10.1007/978-3-030-92934-3_33

These differences in material flow make the management of a job shop much more challenging and complex than the management of a flow shop. As a result, the performance of a comparable flow shop is usually vastly superior to the performance of a job shop. Figure 1 shows an illustrative example of the material flow in flow shops with or without exceptions and a job shop.

Much research has been done on the question of managing the material flow of a job shop. An exhaustive literature review would exceed the scope of this paper. An overview of such job shop scheduling approaches can be found at (Arisha, Young and Baradie 2001) or (van Hoorn 2018). Swarm intelligence was also a popular approach in research (Gao et al. 2019), albeit not in industry. Nowadays, artificial intelligence and genetic algorithms are often employed in research, albeit its use in industry is still infrequent. See (Amjad et al. 2018) for an overview. There is also research on flow shop scheduling (González-Neira, Montoya-Torres and Barrera 2017), albeit in a much smaller quantity as the problem is often less complex.

2 Relevance for Lean Manufacturing

One key concept in lean manufacturing is *flow*, often summarized as "*Flow where you can, pull where you can't*" (Roser 2021). The idea is to increase the share of time within the lead time where the part is improved and value is added to the part, and to reduce all non-value adding times, especially the waiting times. Flow shops are much easier to manage compared to job shops, which leads to much less fluctuations and variations in the material flow, which requires much less inventory to buffer said fluctuations, which in turn has much less inventory waiting. Overall, the flow is much better in a flow shop compared to a similar job shop.

Within this paper, we try to compare the behavior of a flow shop with the behavior of a job shop, keeping everything as similar as possible (number of processes, arrivals, cycle times, etc.), except for the routing of the parts. In the flow shop, all parts follow the same process sequence. In the job shop, all parts have a random sequence, but with the same number of process steps. Thus, we try to extract the effect of the process sequence variability onto the overall system performance. It would be difficult to create such a system in the real world, hence we used simulation software which allowed us a high level of control over all system parameters.

3 Simulated Systems

This paper that originated from a thesis (Ballach 2021) aims not to schedule but to compare the performance of flow shops and job shops. This comparison is for systems that are as similar as possible. The only desired difference was the routing of the parts through the system. For the flow shop, each part followed exactly the same process sequence.

Figure 2 shows the simulated flow shop system for 5 processes. There is an infinite supply and demand of parts. The processes all have exactly the same gamma distributed cycle time, with a buffer of capacity 10 between the processes. For consistency with the job shop, a buffer before the first process has also been added, even though this buffer after an infinite supply will always be full and has no impact on the productivity but increases only inventory and lead time.

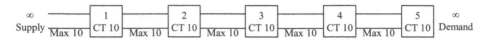

Fig. 2. Flow shop system with 5 processes

The job shop system for 5 processes is also shown in Fig. 3. Every process also has the same identical gamma distributed cycle time, with a buffer of capacity 10 in front of every process. The main difference is the routing, which is perfectly random. Every arriving part gets randomly assigned to one of the processes. After completion the part gets again randomly assigned. A counter keeps track of the number of iterations, and releases the part after a number of iterations that matches the number of processes.

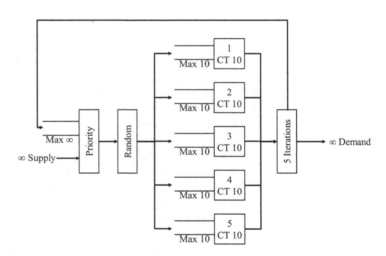

Fig. 3. Job shop system with 5 processes

Besides the iterating random material flow there is another difference, which was necessary to make the system feasible. Parts in iteration have priority of new parts arriving. A queue in front of the prioritization holds iterated parts that cannot yet be assigned due to the queue of the target process being full. Hence this additional queue with infinite capacity is necessary to make the system feasible, but will increase the inventory and lead time.

To compare the effect of the size of the system, we created similar flow shops and job shops not only for 5 processes but also for 3, 10, 25, and 50 processes to model larger and smaller systems. A system with a single process was also simulated, although in this case the flow shop and the job shop systems are identical. As it is standard with simulations, we determined a suitable warm-up period and results-collection period to eliminate the behavior during the ramp-up of the system.

4   Comparison Results

4.1   Utilization

The utilization of the flow shop and the job shop for different number of processes is shown in Fig. 4. The utilization of the flow shop was not significantly affected by the number of processes, and a system with 50 processes has almost the same utilization as a system with 3 processes.

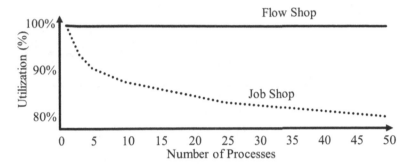

Fig. 4. Comparison of utilization of flow shop and job shop

This, however, was not true for the job shop system. Increasing the number of processes in the system increased the randomness of the routing and decreased the utilization. For the largest system with 50 processes, the utilization was reduced to 79.9%, compared with an almost perfect utilization of 99.6% for the flow shop.

4.2   Line Takt

The line takt in Fig. 5 shows a similar behavior. Since the line takt is measured as a time per part, its behavior is inverse to the utilization. The line takt is also the inverse of the throughput. All of the processes had a mean cycle time of 10 time units. With a near-perfect utilization the flow shop was able to complete a part almost every 10 time units. The job shop, however, suffered more and more as the number of processes increased. For the worst case with 50 processes, the job shop had a line takt of 1 part every 12.52 time units, where the flow shop completed a part every 10.04 time units.

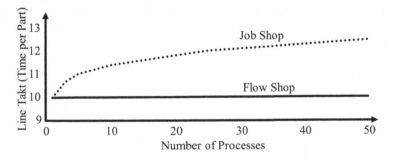

Fig. 5. Comparison of line takt of flow shop and job shop

4.3 Inventory

Figure 6 shows the inventory of the flow shop and the job shop for different numbers of processes. As the number of processes increase, the buffer capacity increases as does the inventory.

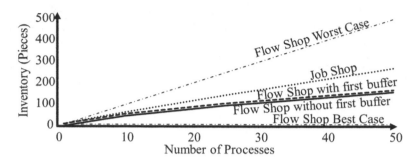

Fig. 6. Comparison of inventory of flow shop and job shop

As explained above for Fig. 2, the flow shop included a buffer in front of the first process for comparability, even though this buffer is always full and has no positive impact on the utilization or line takt, but increases inventory and lead time. Hence, in Fig. 6 we show the inventory in the flow shop both with and without the first buffer.

The job shop shows a very linear increase, although the inventory increases slower than the number of processes. For 3 processes a job shop has in average 7.8 parts per process, whereas for 50 processes the average is 5.4 parts per process.

For flow shops the increase is significantly less, regardless if we include the first buffer or not. However, this depends heavily on the cycle time of the different processes. In our example all processes had equal cycle times. Due to random fluctuations, this caused the buffers to become emptier the further downstream they were. If the cycle times were different, the system would behave different. If the last process would be the bottleneck, then the flow shop would have much more inventory than the job shop. If the first process in the flow shop would be the bottleneck, then the flow shop would never have more inventory than the first buffer and the process inventory. The best case and the worst case are also shown in Fig. 6. Please note, however, that in a worst-case situation with the bottleneck at the end, it would be possible to reduce buffers throughout the system without any impact on the utilization or line takt, but with a drastic improvement of the inventory and the lead time. Hence the extreme situations in the flow shop are not realistic examples.

4.4 Lead Time

The lead time is directly related to the inventory and the line takt. This relation is known as Little's law (Little 1961). Figure 7 shows the lead time for the flow shop and the job shop example. The flow shop extremes from Fig. 6 are not included, since the line takt of the flow shop system and comparable job shop systems would also be impacted by the cycle times, and hence the lead time would also show a more complex behavior. Yet, for the simulated systems the flow shop also has a significantly faster lead time than the job shop, regardless if we include the first buffer of the flow shop or not.

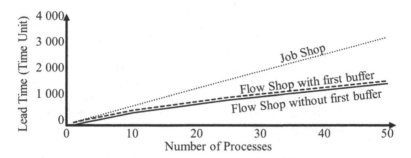

Fig. 7. Comparison of lead time of flow shop and job shop

5 Summary

Overall, the flow shop performs much better than the job shop with respect to utilization, line takt, inventory, and lead time. Hence, it makes a significant difference if you have the same process sequence for all parts (flow shop) or if they differ (job shop). The utilization and the line takt can easily be 20% worse purely based on the process sequence variability.

This is, however, for an idealized system where only the routing of the parts differs. In the real world there may be additional differences. For example, the flow shop routing may not be as perfectly random as in this simulation, which would improve the behavior of the job shop.

However, there are many more effects which make the job shop behavior worse. For example, in this simulation it was always clear which job goes where next, and there was always 100% capacity available at every process. In reality, however, a job shop has much more confusion regarding the next job for a process, the next process for a job, or the moving around of operators among different processes to reduce idle times of the operators.

Overall, a job shop is almost always more difficult to understand and to manage, and in reality there will be more losses due to searching, waiting, or gathering information or tools. These losses also enable more mistakes, which generate subsequent losses. A flow shop will have much fewer problems of this type. Hence, it will be beneficial in many cases if the job shop could be transformed into a flow shop, making it easier to *flow* material and hence making it a leaner system.

References

Amjad, M.K., et al.: Recent research trends in genetic algorithm based flexible job shop scheduling problems. Math. Probl. Eng. 2018 (2018). https://doi.org/10.1155/2018/9270802

Arisha, A., Young, P., El Baradie, M.: Job shop scheduling problem: an overview. In: Conference Papers (2001). https://doi.org/10.21427/D7WN5Q

Ballach, D.: Simulation und gegenüberstellung verrichtungs- und flussorientierter fertigungssysteme. Bachelor Thesis. Karlsruhe University of Applied Sciences, Karlsruhe, Germany (2021)

Gao, K., Cao, Z., Zhang, L., Chen, Z., Han, Y., Pan, Q.: A review on swarm intelligence and evolutionary algorithms for solving flexible job shop scheduling problems. IEEE/CAA J. Autom. Sin. **6**(4), 904–916 (2019). https://doi.org/10.1109/JAS.2019.1911540

González-Neira, E., Montoya-Torres, J., Barrera, D.: Flow-shop scheduling problem under uncertainties: review and trends. Int. J. Ind. Eng. Comput. **8**(4), 399–426 (2017)

Little, J.D.C.: A proof for the queuing formula: $L = \Lambda W$. Oper. Res. **9**(3), 383–387 (1961). https://doi.org/10.1287/opre.9.3.383

Roser, C.: All About Pull Production: Designing, Implementing, and Maintaining Kanban, CONWIP, and Other Pull Systems in Lean Production EBook. AllAboutLean.com, Offenbach, Germany (2021)

van Hoorn, J.J.: The current state of bounds on benchmark instances of the job-shop scheduling problem. J. Sched. **21**(1), 127–128 (2018). https://doi.org/10.1007/s10951-017-0547-8

A Framework and Qualitative Comparison on Different Approaches to Improve the Lean Skillset

Christoph Roser[(⊠)], Bernd Langer, and Claas-Christian Wuttke

Karlsruhe University of Applied Sciences, Moltkestrasse 30, Karlsruhe, Germany
christoph.roser@hochschule-karlsruhe.de

Abstract. Lean Manufacturing is considered to be one of the major frameworks in improving manufacturing performance. Yet, a majority of lean improvement projects fail to generate a benefit for the organization. The problem usually lies with an incomplete understanding of lean manufacturing by the personnel involved. Therefore, industry is very interested in improving lean skills. This paper gives a qualitative overview and comparison on how to improve the abilities of the personnel to do lean, as well as suggestions on how to combine these approaches for a well-rounded improvement of the ability to do lean. It shows opportunities both for corporations as well as individuals on how to improve lean performance.

Keywords: Lean Manufacturing · Operational excellence · Training · Lean skillset

1 Introduction

Lean manufacturing based on the Toyota Production System is one of the significant frameworks for manufacturing optimization (Womack 1990). It includes a multitude of different methods and philosophies, from eliminating waste, fluctuations, and overburden to improving quality. Lean was also either the basis or a significant inspiration for other process-performance-related frameworks like six sigma, Agile, and others, and is often related to topics like Industry 4.0 (Buer et al. 2020; Lai et al. 2019).

Lean can have significant positive effects on the performance of a production system (Fleischer and Liker 1997; Krafcik 1988; Shah and Ward 2003). However, doing lean is not easy. By some estimates, 70–90% of all lean projects fail or at least do not generate any measurable benefit (Ignizio 2009). This may be due to different reasons. Some papers look in more detail at the implementation techniques (Sundar, Balaji and Kumar 2014; Mostafa, Dumrak and Soltan 2013). Others look in more detail at the different tools of lean manufacturing (Pavnaskar, Gershenson and Jambekar 2003; Feld 2000). Again others look at the role of people for lean (Jeretin-Kopf et al. 2016).

© IFIP International Federation for Information Processing 2021
Published by Springer Nature Switzerland AG 2021
D. J. Powell et al. (Eds.): ELEC 2021, IFIP AICT 610, pp. 333–338, 2021.
https://doi.org/10.1007/978-3-030-92934-3_34

2 Problematization

This paper looks at the options for training individuals in lean manufacturing. How does one become a lean expert? How do you turn your employees into lean experts? There are different approaches to learning, and the success also depends on the personality of the learner (Romanelli, Bird and Ryan 2009). (Dinis-Carvalho 2021), besides giving an overview of the topic also looks at collaborations between academia and industry.

Based on our industry experience, the state of lean training leaves much to be desired. Much lean training in industry is based on classroom training, sometimes augmented with simulation games. As described further below this has the advantage of quickly and inexpensively introducing many people to lean concepts. Unfortunately, it lacks practical use, and is often not very effective (Negrão, Filho and Marodin 2016; Shrimali and Soni 2017).

Often, industry also falls back on certifications, often six sigma white/yellow/green/black/master black belts. While some of these certificates have an honest aim to teach lean, there are also many low cost online certificates of dubious quality available.

Hence, the state of lean training often leaves much to be desired, despite lean training being one of the key factors for a successful lean transformation (Netland 2015; Yamchello et al. 2014). In this paper, we will look in more detail at the different training approaches and how they can be combined.

3 Overview of Common Lean Learning Approaches

There are a number of common approaches to train lean, or for that matter almost any subject in industry. They are loosely sequenced by their effectiveness from least effective to most effective. Unfortunately, this is also roughly the inverse of the effort needed to train, in that the easiest approach is also the least effective one.

3.1 Theoretical Training

A popular approach in industry is theoretical training. This is most commonly an in-classroom training with a trainer, but it may also be a self-taught course using videos, written texts, and/or other materials. This approach is popular because it is comparatively inexpensive, and a large number of people can be trained simultaneously. However, it is also usually not very effective. The challenge is that the trainee may or may not be mentally engaged, and may even have misunderstood or missed parts.

3.2 Interactive Training and Simulations

A better approach is to motivate the learner to be engaged. In its easiest form, this could include exams or quizzes to test the learner's knowledge. However, it is difficult to narrow down such a complex topic as lean to a few quiz questions.

Simulations, games, and other interactive training models also require more engagement. This may also be called "gamification." Such simulations are a transition toward learning by doing, albeit a simulation environment is often an environment with limited complexity and geared toward easier problems.

3.3 Learning by Doing

Another approach to train lean is by simply doing it. This is also called on-the-job training (OJT). Doing lean projects helps people to become better in lean. This can be done by the trainee on its own, or in a small group. However, done incorrectly it can also lead to and reinforce bad behavior. Hence, it is advised to not rely solely on learning by doing, but instead to use this as part of a larger framework. Coaching especially is useful for the initial phase in a lean journey.

Key to a successful learning-by-doing approach, or lean in general is the well-known lean principle Plan, Do, Check, and Act (PDCA) (Matsuo and Nakahara 2013). The trainee must thoroughly verify if the lean implementation actually works (the Check part) and implement countermeasures if it does not (the Act part).

3.4 Being Coached

Very related to learning by doing is coaching. The trainee is actively coached, supported, and occasionally challenged by a mentor or coach. This approach is somewhat overlapping with learning by doing, where a group leader can also take the role of the coach. The coach should have a good knowledge of lean, and should be able to guide the trainee in the right direction. It is common in lean coaching to not give the answers to the trainees directly, but rather to ask the right questions and let the trainee find out the answers him/herself.

3.5 Teaching and Coaching of Others

The ultimate learning experience is to teach and coach others. Teaching others requires the coach to thoroughly analyze and understand the lean approach. Furthermore, the questions of the trainees challenge the coach to expand his horizon. Finally, the trainee may have different ideas or views, which also furthers the understanding of the coach. Effectively, for the coach this training and mentoring is also a form of learning by doing for the coach.

4 Comparison of Lean Learning Approaches

The approaches above all have different benefits. Mostly it is a trade-off between the quantity of knowledge gained on one hand, and the ease and cost thereof on the other hand. Table 1 gives you an overview thereof.

Table 1. Overview of the different aspects of lean training approaches

Approach	Number of trainees	Speed of training	Knowledge gain	Expense
Theoretical training	Many	Fast	Very low	Low
Interactive training and simulations	Many	Fast	Low	Low
Learning by doing	Few	Slow	Medium	Medium
Being coached	One	Medium	Good	High
Teaching and coaching of other	One	Very slow	Very good	Medium

5   Combination of Approaches

For a more effective way to learn lean manufacturing, it is advised to combine these approaches. Similar to obtaining a driver's license for a car, a combination of learning by doing and theoretical training will be the backbone of the early stages of the lean journey.

Figure 1 shows the use of different learning approaches throughout the lean journey. Learning by doing is, especially at the beginning, the key approach, supported by theoretical trainings and interactive simulations and trainings as needed. Often, the lean journey is started with a theoretical introductory training, albeit it may be better to start with learning by doing, and add specific theory as the trainee encounters problems in his practical work. In this way, the trainee is more actively engaged in the theory training, as he is actively looking for a possible solution to resolve the problem in his practical work.

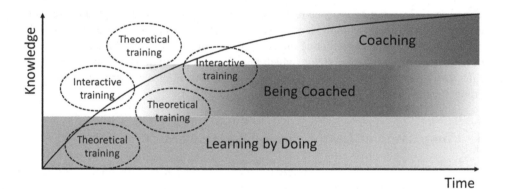

Fig. 1. Learning styles throughout the lean journey

The learning by doing never really ends, no matter how far you advance in your lean journey. However, as the trainee (or by then, the expert) reaches the coaching stage, the *learning by doing* shifts to *learning by having others do*.

Learning by doing is supported by a small group of learners with various levels of experience. As the trainee advances, the learning by doing is best supported by individual coaching, where a skilled expert guides not the group but coaches the individual learner. Hence, coaching and mentoring is usually limited to few participants due to the need for a qualified coach or mentor, albeit it gives a much deeper understanding and knowledge. The primary author was for many years a lean coach in industry, doing multiple six-month programs with two trainees per coach. These were very successful, albeit training a large number of people is usually economically infeasible. Hence this is usually reserved to train lean experts.

Finally, the coaching of others is only advisable for learners who have already progressed far along their lean journey. Yet, the coach will also learn while coaching others.

6 Summary

Overall, building knowledge in lean manufacturing is a longer process, depending on the desired level of expertise. Different approaches to transfer and build knowledge should be used at different stages of this lean journey. Our subjective impression is that many companies focus more on theoretical trainings and interactive trainings and simulations, which allow the training of many trainees simultaneously and inexpensively. However, we believe that such a theoretical training is best used only as a support for the actual problem-solving process in the learning-by-doing approach of understanding lean manufacturing.

References

Buer, S.-V., Semini, M., Strandhagen, J.O., Sgarbossa, F.: The complementary effect of lean manufacturing and digitalisation on operational performance. Int. J. Prod. Res. **59**, 1976–1992 (2020). https://doi.org/10.1080/00207543.2020.1790684

Dinis-Carvalho, J.: The role of lean training in lean implementation. Prod. Plan. Control **32**(6), 441–442 (2021). https://doi.org/10.1080/09537287.2020.1742376

Feld, W.M.: Lean Manufacturing: Tools, Techniques, and How to Use Them. CRC Press, Boca Raton (2000)

Fleischer, M., Liker, J.K.: Concurrent Engineering Effectiveness: Integrating Product Development Across Organizations, 1st edn. Carl Hanser Verlag GmbH & Co., Cincinnati (1997)

Ignizio, J.P.: Optimizing Factory Performance: Cost-Effective Ways to Achieve Significant and Sustainable Improvement (Kindle Edition), 1st edn. Mcgraw-Hill Professional, New York (2009)

Jeretin-Kopf, M., Wuttke, C.C., Haas, R., Wiesmüller, C.: Methods and tools for enabling employees to contribute to technological progress. Int. J. Eng. Pedagogy (IJEP) **6**(3), 45–52 (2016)

Krafcik, J.F.: Triumph of the lean production system. Sloan Manage. Rev. **30**, 41–52 (1988)

Lai, N.Y.G., Wong, K.H., Halim, D., Lu, J., Kang, H.S.: Industry 4.0 enhanced lean manufacturing. In: 2019 8th International Conference on Industrial Technology and Management (ICITM), pp. 206–11 (2019). https://doi.org/10.1109/ICITM.2019.8710669

Matsuo, M., Nakahara, J.: The effects of the PDCA cycle and OJT on workplace learning. Int. J. Hum. Res. Manage. **24**(1), 195–207 (2013). https://doi.org/10.1080/09585192.2012. 674961

Mostafa, S., Dumrak, J., Soltan, H.: A framework for lean manufacturing implementation. Prod. Manufact. Res. **1**(1), 44–64 (2013). https://doi.org/10.1080/21693277.2013.862159

Negrão, L., Filho, M., Marodin, G.: Lean practices and their effect on performance: a literature review. Prod. Plan. Control, 1–24 (2016). https://doi.org/10.1080/09537287.2016.1231853

Netland, T.: Critical success factors for implementing lean production: the effect of contingencies. Int. J. Prod. Res. **54**, 2433–2448 (2015). https://doi.org/10.1080/00207543.2015. 1096976

Pavnaskar, S.J., Gershenson, J.K., Jambekar, A.B.: Classification scheme for lean manufacturing tools. Int. J. Prod. Res. **41**(13), 3075–3090 (2003). https://doi.org/10.1080/0020754021000 049817

Romanelli, F., Bird, E., Ryan, M.: Learning styles: a review of theory, application, and best practices. Am. J. Pharm. Educ. **73**(1) (2009). https://www.ncbi.nlm.nih.gov/pmc/articles/ PMC2690881/

Shah, R., Ward, P.T.: Lean manufacturing: context, practice bundles, and performance. J. Oper. Manag. **21**(2), 129–149 (2003). https://doi.org/10.1016/S0272-6963(02)00108-0

Shrimali, A., Soni, V.: A Review on issues of lean manufacturing implementation by small and medium enterprises. Int. J. Mech. Prod. Eng. Res. Develop. **7**, 283–300 (2017). https://doi. org/10.24247/ijmperdjun201729

Sundar, R., Balaji, A.N., Kumar, R.M.S.: A review on lean manufacturing implementation techniques. Procedia Eng., "12th Glob. Congr. Manufact. Manage." GCMM - 2014, **97**, 1875–1885. https://doi.org/10.1016/j.proeng.2014.12.341

Womack, J.P.: The Machine That Changed the World: Based on the Massachusetts Institute of Technology 5-Million-Dollar 5-Year Study on the Future of the Automobile. Rawson Associates, New York (1990)

Yamchello, H., Samin, R., Tamjidyamcholo, A., Bareji, P., Beheshti, A.: A review of the critical success factors in the adoption of lean production system by small and medium sized enterprises. Appl. Mech. Mater. **564**, 627–31 (2014). https://doi.org/10.4028/www.scientific. net/AMM.564.627

Lean Warehousing: Enhancing Productivity Through Lean

Matteo Rossini[✉], Bassel Kassem, and Alberto Portioli-Staudacher

Politecnico Di Milano, Milan, Italy
matteo.rossini@polimi.it

Abstract. The environment dynamism of the modern era is fostering the spreading of lean in every industry. Followed by the demonstrated benefits of the lean principles in operational performances many companies have taken the leap. Nonetheless, beyond the manufacturing and mass production industries, the dissemination of lean has a long road ahead yet. Logistic and warehousing activities form part of this group, and authors agree about the limited contribution in the literature. For this reason and considering the necessity of the current era for constant optimization of the activities, increasing attention on warehousing is important. The purpose of this article is to illustrate an empirical performance of how lean is applied to the context of a logistic service provider distinguished by flexible solutions. The continuous improvement story exploits the A3 framework for building the problem solving.

Keywords: Lean · Continuous improvement · Warehousing

1 Introduction

Modern era has set complexity and competitiveness in the way to perform businesses. The dynamic environment leaded to the born and spread of what we know today as lean [1]. The set of concepts proved to provide several benefits marking the direction for testing lean principles' adaptability and expansion from the automobile industry to other industries as construction, textile, service, food, medical, etc. [2–4].

This new era has brought new challenges such as digitalization [5, 6], shorter product life cycle [7], shorter time-to-market impacting the logistics activities [8]. In order to meet customer needs, warehouses require to be constantly optimizing their activities by reducing inefficiencies and making them more reliable in terms of cost [9, 10]. Despite the spread of lean principles and the increasing attention in warehousing activities in the last years, the contributions in the literature are still limited [11]. Lean warehousing (LW), represented by the implementation of lean concepts in warehouses, is a quite new concept that must not be left behind [12].

The aim of this research is to give evidence of the creation of a leaner warehousing system through the LW, following the principles and tools of lean. The importance of this story relies on the how a leading company well consolidated can increase value with the implementation of LW approach by improving productivity. The methodology exploited is the case study. Case studies present a fundamental feature because they rely on a variety of sources of evidence that usually are not available for other

© IFIP International Federation for Information Processing 2021
Published by Springer Nature Switzerland AG 2021
D. J. Powell et al. (Eds.): ELEC 2021, IFIP AICT 610, pp. 339–347, 2021.
https://doi.org/10.1007/978-3-030-92934-3_35

explanatory method such as a history [13]. In fact, case studies can leverage information coming from documents, artefacts, interviews of people part of the study and
direct observation of the studied events [13]. Then, the case study is the method that
best fits the previously mentioned criteria and, hence the final aim of the research.

2   Case Study

2.1   Company Overview

The company MNB is one of the main logistics service providers both in Italy and
worldwide. It provides services for a wide range of customers across a variety of
businesses offering flexibility and suitable solutions for everyone. It is a global logistics
provider which has air freight services, ocean freight services, transport execution,
transport management, and contract logistics as its main services. The case of this paper
is based in North of Italy, developed in one of the more than 60 establishment in Italy.
The establishment serves to more than 20 customers.

2.2   Reasons for Actions

Among the main costs that MNB sustains, Handling Costs are the most significant
part. We find Material & Handling Equipment cost (11%) and Labour cost (89%). As
in all the warehouses of the company, the workforce is provided by the Cooperative,
which has a cost-per-hour contract defined and not negotiable. In the last two years,
there has been an increase of 9.5% (twice the increased in 2013–2017). This cost trend
cannot be reflected upon customers, so the Business Process Excellence (BPE) team
started the "Warehouse Process Improvement" initiative with the objective to improve
productivity. The warehouse under analysis is dedicated to customers of the tyre
industry. MNB serves two main customers in this warehouse, defined for privacy
reason as Customer1 and Customer 2.

2.3   Current State

Despite the two customers are immersed in the same industry, they run business in
different ways, as showed in Table 1. Furthermore, the team performed an analysis of
process mapping and the estimation of incidence of the Value-Added (VA), Non-
Value-Added (NVA), and Business-Value-Added (BVA) activities in the overall
process (Fig. 1).

Table 1. Customer analysis

Aspect	Customer 1	Customer 2
Business strategy	Manufacturer	Distributor
Products portfolio	Passenger (88%), Truck (7%), Moto, Industry, other tyres	Passenger tyres
Level of involvement and willingness to invest	Up to date; collaborative; new and innovative techniques and tools used; high level of awareness in warehouse process; Investment oriented	Traditional and old techniques used; No interest in warehouse operations improvement
WMS sharing	Shared database and WMS; Free access to process information	No WMS; No data sharing

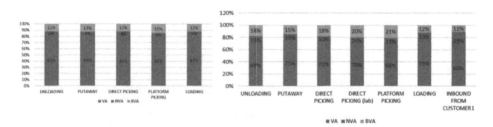

Fig. 1. VA / NVA analysis – customer 1 (left side) & customer 2 (right side)

The definition of the current situation highlighted the presence of problems that strongly affect productivity, which have been detailed and analysed in the root-cause phase.

2.4   Targets of the Project

Assisted by the internal analysis, the team defined the set of proper targets discussing with the company. The following targets have been fixed:

- Increase the overall warehouse productivity by 5% in terms of tons/hour moved
- Decrease performances gap between the two customers by 10%.

2.5   Root-Cause Analysis

After target definition, the root-cause analysis has been performed on each customer flow in order to identify gaps between current situation and the targets. The procedure of the root cause analysis was quite similar for the two customers' flows: the team interviewed operators involved in the processes and executed Gemba Walks. These activities have been resumed in an Ishikawa diagram, presented in Fig. 2 and Fig. 3.

In summary, despite the similarities in the process, the two systems present different issues and problems. Customer 1 flow presents a well-structure procedure; nonetheless, some opportunities can be provided. In Customer 2 flow, the lack of a defined procedure affects significantly in the BVA and NVA activities like wrapping with plastic film and check phase for all the picking typologies, and these have a great impact on the productivity of the system.

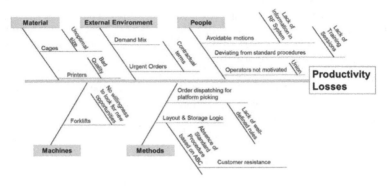

Fig. 2. Ishikawa diagram – customer 1

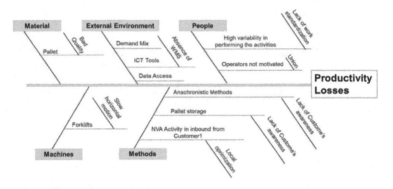

Fig. 3. Ishikawa diagram – customer 2

2.6 Countermeasure Proposals

Once the root-causes have been identified and prioritized, potential countermeasures have been proposed. Due to the high number of proposals a prioritization of the interventions resulted necessary. For this reason, the countermeasures proposed were assessed according two dimensions: the impact or benefits expected on the targets and the effort required for the interventions. In Table 2 and Table 3 the proposed countermeasures of Customer 1 and Customer 2 are presented, respectively.

Table 2. Proposed countermeasures, customer 1

Category	Description	Action area	Effort	Impact
Layout	Layout and storage allocation policy optimization based on ABC Analysis on qty picked Class A: estimate pc/SKU for Platform picking; Layout + MHE: saving area impact	Methods	3	3
Outbound	Using Conveyor to load truck	Methods	4	3
Outbound	Order the check list as the picking list so that the activity time needed can be lowered	Methods	1	1
Outbound	Planification methodology: definition of fixed rules for planning the platform picking activities (no truck). Introduce specific Call-off for each route	Methods	2	5
Inbound	Fix 2 gate for unloading and define rules to select the correct unloading area	Methods	2	1
MHE/5S	Introduction flag on the screen: show on the screen/barcode scanner if the in the next picking task, the cage is going to be emptied	People	3	3
MHE/5S	Use order picker instead of forklift for platform Picking (low pc/line): trade-off between up & down of cages and truck speed	Machines	1	2

Table 3. Proposed countermeasures, customer 2

Category	Description	Action area	Effort	Impact
Inbound	Change Inbound from Customer1 logic: Pick the needed material with order picker and prepare pallet for put away. Avoid double touch of goods	Methods	2	3
5S	Definition of Standard procedure for Loading Activity	People	1	4
Inbound	Fix 2 gate for unloading and define rules to select the correct unloading area	Methods	2	1
5S	Define the area of customer3 area with "signage" or iron protection to prevent road blocking	External Environment	1	1
5S	Definition of Standard procedure for Unloading Activity	People	1	4
5S	Definition of Standard procedure for Picking Activity	People	1	4
In/Outbound	Use cages instead of pallet. Definition of the right cage's size	Methods	5	3
In/Outbound	Introduce a Warehouse Management System	External Environment	3	4
Outbound	Use conveyor for loading trucks	Methods	3	3
In/Outbound	Use cages instead of pallet just for picking Definition of the right cage's size	Methods	2	3

2.7   Implementation of Countermeasures

The team estimated for each countermeasure a priority mark based on effort, costs, benefits and impact. Then, a structured implementation plan have been developed for the proposals with the highest mark values and in agreement with the company's project sponsor. Some of the countermeasures were evaluated through pilot test while others were assessed performing business cases using a simulation tool. The plan, for both customers, includes cycle time (CT) saving target, the starting data and due date for the implementation.

Focusing on the Customer 1, as shown in the Table 4, the countermeasures considered are presented with its respective % of CT saved expected. "The Flag on the screen" is a quick-win countermeasure aiming to avoid multiple motions for full-cage picking activities, with an immediately implementation and some training session to Cooperative's workers. Order planning & release required trial phase to assess the goodness of the solution and review of specific cut-offs of specific route to achieve the maximum benefits. In this sense, an ABC analysis among the different SKUs was performed to identify class A products and set new fixed positions. As regards the new layout and storage allocation strategy, a transition period for moving the cages and fix them in the new position is required, as well as other internal movements. Meanwhile, for the conveyor to load truck, a big increase on the productivity would be perceived; nonetheless. some further considerations are required based on the installation and operators' training.

Table 4. Implementation plan customer 1

Description	%CT saved	Start date	Due date
Layout and storage allocation policy optimization	7%	01/08/2019	31/07/2020
Using conveyor to load truck	37%	01/07/2019	31/07/2020
Orders planning & release	4.6%	01/07/2019	31/07/2020
Flag on the screen	0.4	01/07/2019	31/10/2019

For Customer 2, the countermeasures for implementation with the respective % of CT saved expected are shown in the Table 5. The standardization activities in the loading, unloading and picking, as well as changing the inbound from Customer1 have been implemented. A set of training sessions are required to set the pilot tests and assessments of the results. For the usage of cages for outbound and installation of automatic conveyor, the company must discuss in detail with the customer to reach an agreement leading to a long-term horizon countermeasure.

Table 5. Implementation plan customer 2

Description	%CT saved	Start date	Due date
Use cage instead of pallet for outbound (plt size)	5%	01/07/2019	31/07/2020
Using Conveyor to load truck	34%	01/07/2019	31/07/2020
WMS	18% + 12%	01/07/2019	01/07/2020
Inbound from Customer1	24%	01/07/2019	31/11/2020
Loading Standardization	19%	01/07/2019	31/11/2019
Unloading Standardization	23%	01/07/2019	31/11/2019
Picking Standardization	10%	01/07/2019	31/11/2019

2.8   Standardization and Future Steps

The company and the Cooperative approved countermeasures and set the need for definition of the new procedures as new standards. Then, for both customers, new work instructions have been written in coordination with BPE team, Cooperative shift leaders, and engineers. The new procedure allows to reduce the variability of performance providing positive effects on the warehouse productivity. The new work instructions have been attached on the warehouse walls, fully visible to be consulted at any moment. A second deliverable has been developed only for Customer 2. A feasibility assessment with productivity calculations has been developed to improve bargain power in the negotiation phase with the Customer 2.

The approved countermeasures will have high impact on company's performances, but opportunities for further improvements are still present. For the next steps, the team is already studying the implementation of a new update process that has as "zero defects". After the implementation of the new process, data would be gathered and analysed to intervene with corrective actions to solve possible issues. The steps should follow the monitoring Gantt Diagram showed in the Fig. 4.

The focus will mainly be on Customer 2 since it is the most critical one, a strong collaboration with the manager is needed to achieve common goals and reduce shared risks. A possible next goal could be increased the awareness of Customer2 manager in process improvement increasing the number of visits on site and brainstorming sessions to arise issues and develop new ideas. And for what concerns Customer1, the monitoring of the long- term countermeasure should be tackled and evaluated.

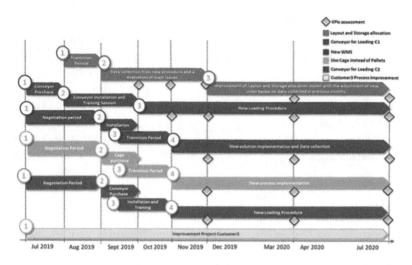

Fig. 4. Next steps – Gantt diagram

3   Conclusions

According to the best of authors knowledge, only few studies involve the implementation of lean in a logistic and warehousing context, which involves by nature in their core non-value adding activities. This article is an attempt to increase the empirical evidence of this topic, presenting a continuous improvement project that includes the application of the Lean framework in a real case study.

The company achieved significant increase in productivity through this improvement project. The findings of the project have been shared with the Top management of the company and the Cooperative shift leaders and engineers. As well the new procedures were published across the plant, and the feasibility assessment and investment for the long-term countermeasures of the warehouse of Customer 2 were delivered to the company.

Further, this paper can be helpful for managers to understand how Lean methodology can be implemented in their organisation and to incentive data-based decision making in all aspects of the business process. Finally, this case study will help the academicians and students to understand the practicality of Lean methodology adopted in logistic industry.

References

1. Azadegan, A., Patel, P.C., Zangoueinezhad, A., Linderman, K.: The effect of environmental complexity and environmental dynamism on lean practices. J. Oper. Manage. **31**, 193–212 (2013). https://doi.org/10.1016/j.jom.2013.03.002
2. Bhamu, J., Sangwan, K.S.: Lean manufacturing: literature review and research issues. https://doi.org/10.1108/IJOPM-08-2012-0315

3. Davim, J.P.: Research Advances in Industrial Engineering. Springer International Publishing, Cham (2015). https://doi.org/10.1007/978-3-319-17825-7
4. Liker, J.: The Toyota Way : 14 Management Principles from the World's Greatest Manufacturer. Mcgraw-hill (2004)
5. Rossini, M., Audino, F., Costa, F., Cifone, F.D., Kundu, K., Portioli-Staudacher, A.: Extending lean frontiers: a kaizen case study in an Italian MTO manufacturing company. Int. J. Adv. Manufact. Technol. **104**(5–8), 1869–1888 (2019). https://doi.org/10.1007/s00170-019-03990-x
6. Torri, M., Kundu, K., Frecassetti, S., Rossini, M.: Implementation of lean in IT SME company: an Italian case. Int. J. Lean Six Sigma (2021). https://doi.org/10.1108/ijlss-05-2020-0067
7. Kundu, K., Cifone, F., Costa, F., Portioli-Staudacher, A., Rossini, M.: An evaluation of preventive maintenance framework in an Italian manufacturing company. J. Qual. Maint. Eng. (2020). https://doi.org/10.1108/JQME-02-2020-0007
8. Rossini, M., Portioli, A.: Supply chain planning: a quantitative comparison between lean and info-sharing models. Prod. Manuf. Res. **6**, 264–283 (2018). https://doi.org/10.1080/21693277.2018.1509744
9. Dotoli, M., Epicoco, N., Falagario, M., Costantino, N.: A lean warehousing integrated approach: A case study. In: IEEE International Conference on Emerging Technologies and Factory Automation, ETFA (2013)
10. Dotoli, M., Epicoco, N., Falagario, M., Costantino, N., Turchiano, B.: An integrated approach for warehouse analysis and optimization: a case study. Comput. Ind. **70**, 56–69 (2015). https://doi.org/10.1016/j.compind.2014.12.004
11. Gopakumar, B., Sundaram, S., Wang, S., Koli, S., Srihari, K.: A simulation based approach for dock allocation in a food distribution center. In: Proceedings - Winter Simulation Conference, pp. 2750–2755 (2008)
12. Prasetyawan, Y., Ibrahim, N.G.: Warehouse improvement evaluation using lean warehousing approach and linear programming. In: IOP Conference Series: Materials Science and Engineering. Institute of Physics Publishing, p. 012033 (2020)
13. Yin, R.K.: Case Study Research: Design and Methods. Sage Publications Inc, Thousand Oaks, CA (2009)

New Perspectives on Lean

Lean Office in a Manufacturing Company

Federica Costa[(✉)], Bassel Kassem[(✉)],
and Alberto Portioli Staudacher[(✉)]

Politecnico di Milano, via lambruschini 4,b, Milan, Italy
{federica.costa,bassel.kassem,
alberto.portioli}@polimi.it

Abstract. Since its inception with Toyota, the lean approach has been a growing point of interest in various industries and organizations worldwide. However, the focus of attention, in terms of its application in the literature, has remained constant in improving the performance of production processes compared to improving that of offices. Instilling Lean as a strategic vision requires not only a focus on production but also other areas of the focal firm. This paper showcases a success story in a manufacturing company that has decided to opt for a lean office continuous improvement project through A3 problem solving tool to compenetrate the strategic vision of Lean Culture and to develop one first step for the creation of synergy with its production processes.

Keywords: Lean office · Lean culture · Continuous improvement

1 Introduction

Globalization has put increasing importance on adaptability as effect of environmental complexity and dynamism. The increase of competition generated requires companies to focus on the development of competitive advantage [1]. In this direction, The Toyota Production System (TPS), originated by Toyota Motor Corporation as a set of concepts, practices, and tools focused on identifying and reducing wastes, has proved to bring several benefits changing the way production is organized and managed [2, 3]. Later on, TPS was referred to as Lean Production, after the work published by Womack, Jones, and Roos [4].

Lean practices became recognized to gain a competitive advantage, causing the spread of the principles from manufacturing [5–8] to services [9] to supply chain [10] into companies' culture [11] even in companies willing to start their digital journey [12, 13]. Nonetheless, the application and studies of lean on offices (Lean Office) are a recent theme demonstrated by the small number of articles on the subject [3]. Lean Office relies on implementing lean manufacturing principles in offices and administrative processes to streamline information flow by reducing total cycle time [14].

Many manufacturing companies that strive to implement lean culture start focusing on the shopfloor appealing to the sense that it hides wastes. Nonetheless, waste is not exclusive to manufacturing processes, but it is also present in administrative areas and must not be forgotten. To have a true lean transformation and synergy through the company, the lean principles must be applied to all the operations in the organization, not only restricted to the shopfloor [3, 15, 16].

© IFIP International Federation for Information Processing 2021
Published by Springer Nature Switzerland AG 2021
D. J. Powell et al. (Eds.): ELEC 2021, IFIP AICT 610, pp. 351–356, 2021.
https://doi.org/10.1007/978-3-030-92934-3_36

The objective of this paper carried out through an empirical activity was to put in evidence the application of Lean Office in a manufacturing company that has started its lean transformation in 2000. The importance of this study relies on how a company with experience in the use of lean can still get advantages from applying the principles in administrative activities that work in alignment with the operational areas.

2 The Company

The company ABC, with several facilities all over the world, is a leader in developing and manufacturing products and services for primary and secondary distribution of natural gas. At the turn of the 20th and 21st century, the company undertook a radical change of strategy to face a financial crisis. This particular change of strategy led to the start implementation of the lean methodology. As of today, the company invests in training its employees about the lean culture by the hands of the Kaizen Promotion Office (KPO). KPO is dedicated full time to continuous improvement, organizing one Kaizen Week per month, in which they address specific problems involving the employees and a Japanese sensei. The present paper addresses a project carried out in one of the plants of the company, based in Milan. The plant counts around 400 employees and produces several equipment for the gas industry. Please note that abbreviations of the various rooms, locations and departments have been used with no referring to the original names for privacy reasons.

3 Research Methodology

This paper relies on the A3 problem-solving tool which has been used as a way for transforming organizations and starting their lean journey [17, 18]. It has been used by various cases in the literature [6, 7, 9].

3.1 Problem Background and Breakdown

The plant is expected to grow in the next years due to higher demand, thanks to the substitution of the old generation of gas meters. This growth of production will generate an increase in personnel. The effect will be perceived in the internal demand of support materials managed by the "Centro di Servizio" (CS). In the CS, the workload on the current four employees will increase: leading to a necessity for increasing efficiency in the support materials management system. The highest impact of the increasing demand will be an increase of the average inventory levels of four categories of materials (clothing, visual tools, personal protective equipment, and stationery), reduction of the service level, particularly lower on-time deliveries and higher customer lead time.

Four categories of products have been identified: Clothing, Visual Tools, Personal Protective Equipment (mainly Working Gloves and Safety Shoes), and Stationery, all managed by CS (Table 1).

Table 1. Categories of items

Category	Description
Clothing	Each worker (either employed by the temporary worker agency or by the company) should receive a subset of the clothing at the entrance in the company and the full set after one month. Nonetheless, the policy is not followed, and clothes are given once the team leaders of the different departments make the request on CS Portal
Visuals	The team leaders of the different departments make the request on CS Portal, CS workers print the request and deliver the quantity required
Personal Protective Equipment (PPE) heading	Each worker that enters the company must receive PPE depending on the tasks to be performed in the company. the team leaders of the different departments make the request on CS Portal, CS workers print the request, prepare the order and deliver the quantity required. Considering working gloves, CS stock point is managed with ROP (reorder point) model and Kanban, once the ROP is reached, the Kanban is detached, and CS workers make the order to the supplier. For the safety shoes, CS worker orders the shoes based on experience
Stationary materials	They are managed through a ROP model with Kanban (the Kanban states the reorder point and the quantity of reordering for each item). Internal customers make the request on CS Portal, CS workers print the request, prepare the order, and deliver the quantity required

3.2   Target Definition

Prioritization on the categories of products has been performed with parameters (Annual expenditure, actual space occupied, and operational importance) in agreement with the company. The analysis led to focus on two out of four groups. In this sense the targets were defined for "Clothing" and "PPE", as follows:

- Clothing: must-have target, reduction of 50% of Average inventory Level AIL $[m^3]$, corresponding to 4.25 equivalent closets (to free the RSU Room, that should not be used to stock closets). Nice-to-have target, reduction of 65% of AIL $[m^3]$.
- Working Gloves: must-have target, reduction of 50% of AIL [€] (to eliminate one of the two stock points). Nice-to-have target, reduction of 65% of AIL [€].
- Safety Shoes: must-have target, reduction of 25% of AIL [€] (considering the already existing and effective management system adopted by CS). Nice-to-have target, reduction of 40% of AIL [€] (to stock all safety shoes in one closet rather than 2, increasing saturation).

3.3   Root Cause Analysis

In the analysis of root causes, the Five Whys technique has been used for all the three types of products chosen. Uncovering multiple root causes, the method was repeated asking a different sequence of questions each time and involving many figures in the company for each level of the analysis. The root causes are presented for each type of product with the corresponding possible countermeasures.

4   Results

From potential countermeasures proposed in Table 2, five of these have been implemented (*) and the other 2 have been accepted to be implemented by CS workers in the future (**). Besides the quantitative improvements following showed, the satisfaction of the customer was expressed at the end of the project.

- Clothing: A reduction of 53% of the AIL [m^3] has been achieved, and according to an estimation, once the new supplier will start to work, a reduction of 65% of the AIL will be reached.
- Working Gloves: A reduction of 50% of the AIL [€] has been achieved.
- Safety Shoes: A reduction of 43% of the AIL [€] has been achieved. All safety shoes are now stocked in one closet.

Table 2. Countermeasures and tackled root causes

Cat.	*/ **	Countermeasure	Root cause tackled
Clothing	**	New management system using Periodic Review Model	- CS is not aware of the actual quantity and type of clothing - CS is not aware of the future demand - Team leaders are not responsible for managing to clothe
	**	New shipment agreement with the supplier	- The company is changing supplier - Supplier accepts only huge orders
	*	Revision of the clothing policy	- Clothing policy establishes the consumption of clothes
		Creation of a dedicated room with a dedicated resource for managing to clothe	- CSS is not aware of the actual quantity and type of clothing - Team leaders are not responsible for managing to clothe
Safety shoes	*	New management system using Kanban	- Reorder procedure is not up to date - Avoid out of stock
	*	New communication procedure between HR and CS	- CS receives the info from Randstad only the day before - Avoid out of stock
		New shipment agreement with the supplier	- Avoid out of stock
Working gloves	*	Reorganization of the stock point both in CS and Mechanical Department	- There are 2 independent stock points (CS and Mech Dept)
	*	Improvement of the visibility between CS and Mechanical Department	- CS is not aware of the stock point in Mech Dept

Additionally, follow-up actions were suggested as for the tackled categories of material (PPE and Clothing), they are mainly monitoring actions. In particular for clothing, CSS will have to monitor the evolution of the new shipment agreement with the supplier. For the other categories (stationery materials and visual tools), CS, in collaboration with KPO office, will have to further address the analysis.

5 Conclusion

Even considering the lean presence in the company's operations for 20 years, the lean office approach revealed to be essential to address and tackle the effects of a future increase of demand on the management of support materials. Thanks to this satisfactory project, the company has gained awareness of the impact of production changes on areas as the CS, aiming to improve the synergy. Through the lean office approach, and adopting the A3 methodology, CS was able to reduce their numbers of inventory levels. Additionally, thanks to this experience, the company is now more concerned about the lean office. Indeed, the pleasing results within the CS allow the workers to be more worried about the importance of their work in the whole organization and the positive impact that they can achieve with lean thinking, generating workers with higher commitment and motivation for their job.

This article contributes to showcasing a real case in which the lean concept is mutually applied in the production and in the offices of a manufacturing company that has been implementing lean for a while. Lean is a culture and not a simple concept, therefore embarking on a lean journey requires not only applying it in the area in which it is born, production [19] but in the entire organization. The spreading of the lean culture entails small continuous improvement projects, focused and aiming at specific targets that would allow measuring the performance and the success of the project. Managers could benefit from this research to effectively understand the simplicity and the usefulness of using A3 methodology as a preliminary step to instilling the lean culture beyond the production areas. Future research will try to further this topic by delineating the struggles and the peculiarities that a manufacturing company already applying lean in its production processes might face in applying lean in its offices as well.

References

1. Azadegan, A., Patel, P.C., Zangoueinezhad, A., Linderman, K.: The effect of environmental complexity and environmental dynamism on lean practices. J. Oper. Manag. **31**, 193–212 (2013). https://doi.org/10.1016/j.jom.2013.03.002
2. Ohno, T.: Toyota production system beyond large -scale production Taiichi Ohno (1979)
3. Sousa, R.M., Dinis-Carvalho, J.: A game for process mapping in office and knowledge work. Prod. Plan. Control **32**, 463–472 (2021). https://doi.org/10.1080/09537287.2020.1742374
4. Womack, J.P., Jones, D.T., Roos, D., Massachusetts Institute of Technology: The Machine That Changed the World: The Story of Lean Production. HarperCollins (1991)

5. Kassem, B., Costa, F., Portioli-Staudacher, A.: JIT implementation in manufacturing: the case of giacomini SPA. In: Rossi, M., Rossini, M., Terzi, S. (eds.) ELEC 2019. LNNS, vol. 122, pp. 273–281. Springer, Cham (2020). https://doi.org/10.1007/978-3-030-41429-0_27
6. Torri, M., Kundu, K., Frecassetti, S.: Rossini M implementation of lean in IT SME company: an Italian case. https://doi.org/10.1108/IJLSS-05-2020-0067
7. Rossini, M., Audino, F., Costa, F., Cifone, F.D., Kundu, K., Portioli-Staudacher, A.: Extending lean frontiers: a kaizen case study in an Italian MTO manufacturing company. Int. J. Adv. Manuf. Technol. **104**(5–8), 1869–1888 (2019). https://doi.org/10.1007/s00170-019-03990-x
8. Kundu, K., Rossini, M., Portioli-Staudacher, A.: Analysing the impact of uncertainty reduction on WLC methods in MTO flow shops. Prod. Manuf. Res. **6**, 328–344 (2018). https://doi.org/10.1080/21693277.2018.1509745
9. Torri, M., Kundu, K., Frecassetti, S., Rossini, M.: Implementation of lean in IT SME company: an Italian case. Int. J. Lean Six Sigma (2021). https://doi.org/10.1108/IJLSS-05-2020-0067
10. Rossini, M., Portioli, A.: Supply chain planning: a quantitative comparison between lean and info-sharing models. Prod. Manuf. Res. **6**, 264–283 (2018). https://doi.org/10.1080/21693277.2018.1509744
11. Rossini, M., Portioli-Staudacher, A., Cifone, F.D., Costa, F., Esposito, F., Kassem, B.: Lean and sustainable continuous improvement: assessment of people potential contribution. In: Rossi, M., Rossini, M., Terzi, S. (eds.) ELEC 2019. LNNS, vol. 122, pp. 283–290. Springer, Cham (2020). https://doi.org/10.1007/978-3-030-41429-0_28
12. Rossini, M., Cifone, F.D., Kassem, B., Costa, F., Portioli-Staudacher, A.: Being lean: how to shape digital transformation in the manufacturing sector. J. Manuf. Technol. Manag. **32**, 239–259 (2021). https://doi.org/10.1108/jmtm-12-2020-0467
13. Costa, F., Portioli-Staudacher, A.: On the way of a factory 4.0: the lean role in a real company project. In: Rossi, M., Rossini, M., Terzi, S. (eds.) ELEC 2019. LNNS, vol. 122, pp. 251–259. Springer, Cham (2020). https://doi.org/10.1007/978-3-030-41429-0_25
14. Wang, J.X.: Lean manufacturing: business bottom-line based (2010)
15. Chiarini, A.: Lean Organization: From the Tools of the Toyota Production System to Lean Office. Springer, Milan (2013). https://doi.org/10.1007/978-88-470-2510-3
16. Productivity Press Development Team: The Lean Office. Productivity Press (2005)
17. Sobek, D.K.: Understanding A3 Thinking. CRC Press (2008)
18. Sobek, D., Jimmerson, C.: A3 reports: tool for organizational transformation. In: 2006 IIE Annual Conference and Exhibition (2006)
19. Womack, J.P., Johnes, D.T., Roos, D.: The machine that changed the world (1990)

Lean Thinking Application in the Healthcare Sector

Federica Costa[(✉)], Bassel Kassem, and Alberto Portioli-Staudacher

Politecnico di Milano, Via Lambruschini 4b, 20156 Milan, Italy
federica.costa@polimi.it

Abstract. As years go by, companies operate in an ever-changing environment requiring strategic tools to provide effectiveness and efficiency. The Healthcare sector is not different, dealing with several challenges like increasing patient numbers and limited resources. Healthcare centers need to improve continuously while offering the best services to their patients. The lean thinking, which was initially a symbol of manufacturing process, has evolved over the years to reach the scope of service improvement. This paper exhibits a successful initiative of engaging lean thinking in the healthcare industry to develop more efficient and effective services.

Keywords: Lean healthcare · Italy · Continuous improvement

1 Introduction

Healthcare sector is facing the effects of globalization and major challenges that call for higher quality healthcare services [1]. In fact, the rapid growth of patients demand to higher quality expectations and fast responses collides with the limited available resources of the hospital, forcing the healthcare centers to find ways to innovate and improve utilizing scare resources with efficient and effectiveness [2]. Specially, during scenarios of crises such as the COVID-19 outbreak, which has put to the test the resilience of healthcare systems all over the world [3].

Ever since its origin in the manufacturing sector, Lean methodology has proved to bring great results in operational performances, which enhanced its dissemination throughout different industries in the manufacturing and service [4] context during the last three decades [5]. In the healthcare sector, the first appearance of Lean was in the UK and Australia in 2001; and twenty years later, only few healthcare organizations specially in the USA, have managed to maintain results with Lean approach [6, 7]. Meanwhile, in Italy, there are only few examples of successful cases in the seek for operational excellence [8, 9] in the healthcare sector.

The objective of this paper is to present and discuss an empirical successful case of continuous improvement performed in an Italian healthcare company. The importance of this case relies in how a company in the healthcare sector can improve its visibility on the process flow and provide more efficient and effective services in benefit of the patients.

© IFIP International Federation for Information Processing 2021
Published by Springer Nature Switzerland AG 2021
D. J. Powell et al. (Eds.): ELEC 2021, IFIP AICT 610, pp. 357–364, 2021.
https://doi.org/10.1007/978-3-030-92934-3_37

2   Company Overview

The company is a leading hospital located in the north of Italy, and one of the points of reference of its region. The company is a health company that provides hospital services, for acute and specialized needs, and territorial services, for outpatient care, as a supplement to the previous ones. The company incorporates in its management a wide range of territorial poles situated in different locations of its city. Amongst them, there is one pole, which provides child neuropsychiatry, psychiatry and some territorial services such as legal medicine. The improvement case is focused on this particular territorial center, in particular, on the administrative processes between the company and the territorial center. A correct functioning of them is necessary to allow a good management of the territorial pole.

Furthermore, since most of the services provided require a payment, a system to track all the activities and control the economic flows of the company is needed. Therefore, to collect and control data about the visits made, the informative system of the hospital is used. It contains the agendas of all the doctors and specialists. However, often patients are exempt (e.g., under 18 years old). In those cases, the economic income is provided by the Agenzia di Tutela della Salute (ATS) but, in any case, the registration of the activity is necessary.

3   The Problem and Current Situation

The main problem highlighted by the hospital was the inefficient management of the administrative processes characterized by few information and data. This prevented company's general management from efficiently monitoring and improving them. There were many problems that top managers were slightly aware of but had not yet been analyzed and improved.

The analysis of the current situation was based on the administrative processes, especially those regarding the first contact with the patient and the acceptance, in order to discover problems and find opportunities for improvement. Furthermore, the starting of the Covid-19 pandemic brought difficulties worldwide; in the conduction of the case at the hospital, it led to difficulties for the collection of data and delaying the start of the analysis. Nonetheless, once the data were available, the analysis was performed, identifying the most critical activities in the processes (for child neuropsychiatry, psychiatry, and territorial services) by considering those contributing to one or more of the 7 Lean Wastes. Those activities were then prioritized through a matrix considering two dimensions: Not-Value-Adding (NVA) time and severity of consequences. It was noticed that only the departments of child neuropsychiatry and psychiatry had activities with medium-high criticality, as presented in Fig. 1. Therefore, it was decided to develop an A3 framework on these two processes.

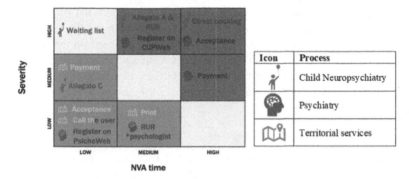

Fig. 1. Priority matrix of processes

4   Child Neuropsychiatry

Considering the prioritization made according to severity and NVA time, it was decided to focus on the reduction of lead time in the Child Neuropsychiatry for direct booking, acceptance and registration processes. As must-have target, it was defined to reduce the NVA time of the two activities with high priority ("Direct booking" and "Fill Allegato A and RUR") by 2.5 h/week. Meanwhile, for the nice-to-have target, it was decided to act on "Update waiting list" activity, acting on both overprocessing and transportation times in order to achieve a reduction of 0.4 h/week.

After setting the targets, a root cause analysis was performed initially on the basis of an Ishikawa diagram, later on analyzed through the 5WHYs method, as shown in the Fig. 2.

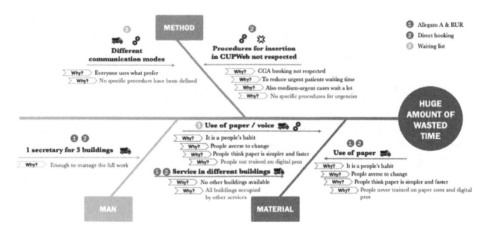

Fig. 2. Root cause analysis of child neuropsychiatry

From the analysis of the root-causes, the team developed a list of possible countermeasures, followed by a prioritization analysis, as shown in the Fig. 3 and Fig. 4, to identify the best ones to implement. The prioritization was based on the impact on the target and the effort in terms of both time and cost, with a scale up to 5. The final effort was calculated as a weighted average of time (60%) and cost (40%). And the final grade between impact and effort was computed as 30% and 70% respectively.

MAIN CAUSE	ROOT CAUSE	COUNTERMEASURE	IMPACT	EFFORT TIME	EFFORT COST	FINAL EFFORT	GRADE	CLASS
Use of paper	*People have never been trained on paper disadvantages and digital benefits*	1. Explanatory session	4	4	4	4	4	
		2. Digitalize "Allegato A"	4	5	5	5	4,7	
		3. Use digital prescription	5	5	5	5	5	
		4. Use mail as method of communication	2,5	5	5	5	4,25	
		5. Create info exchange function in CUPWeb	5	2	2	2	2,9	
Different buildings	*All buildings occupied by other services*	6. Reorganize Ippocrate to centralize neuropsychiatry in 1 building	2,5	1	1	1	1,45	
1 sec x 3 build	*Enough to manage the work*	7. Add 1 secretary for each building	2,5	2	2	2	2,15	
Procedures for CUPWeb not respected	*No specific procedures for urgencies*	8. Define a specific procedure for urgent cases	2,5	4	5	4,4	3,83	

Fig. 3. Countermeasures for must-have target of child neuropsychiatry

MAIN CAUSE	ROOT CAUSE	COUNTERMEASURE	IMPACT	EFFORT TIME	EFFORT COST	FINAL EFFORT	GRADE	CLASS
Different communication modes	*No specific procedure has been defined*	1. Define a standard procedure for communicating new patients to specialists	2,5	5	5	5	3,75	
Use of paper / voice	*People never trained on paper cons and digital pros*	2. Explanatory session	4	4	4	4	4	
		3. Use Excel for all the waiting lists	2,5	4	5	4,4	3,45	
		4. Add waiting list function to CUPWeb	5	2	2	2	3,5	

Fig. 4. Countermeasures for nice-to-have target of child neuropsychiatry

Due to pandemic situation, it was not possible to implement the countermeasures identified in the short term. In this sense, a set of KPIs has been developed to provide the company all the tools and procedures to implement and monitoring the countermeasures, as shown in the Fig. 5 and Fig. 6.

Must-To-Have

COUNTERMEASURE	KPI NAME	KPI FORMULA	THRESHOLD	TIME SPAN	MEASURING FREQUENCY	OWNER	SOURCE OF DATA	IMPACT ON TARGET
Explanatory session	% of acceptance towards change	# positive answers / Total questions	< 60% ⬤○○ 60-80% ○⬤○ > 80% ○○⬤	1 month	Once a week	NPI primary	Administrative and Doctors	↑ staff accord
Digitalize "Allegato A" and RUR Mail as method of communication	Digitalization Rate	# digital communications / bookings for next visits	< 50% ⬤○○ 50-80% ○⬤○ > 80% ○○⬤	1 month	Once a week	NPI primary	Administrative and Doctors	Transport time ↓ (Waiting time)
Define a specific procedure for urgent cases	% regular booking for first visit	# CCA accepted patients / # first visits	< 50% ⬤○○ 50-80% ○⬤○ > 80% ○○⬤	1 month	Once a week	NPI primary	CUP, Administrative and Doctors	Overproc. time Transport time ↓ (Waiting time)

Fig. 5. KPIs for must-have target of child neuropsychiatry

Nice-To-Have

COUNTERMEASURE	KPI NAME	KPI FORMULA	THRESHOLD	TIME SPAN	MEASURING FREQUENCY	OWNER	SOURCE OF DATA	IMPACT ON TARGET
Define standard procedure for communicating new patients to specialists	% paper & voice used	# (paper + voice) commun / # new patients	> 30% ⬤○○ 30-10% ○⬤○ < 10% ○○⬤	1 month	Once a week	NPI primary	Doctors	Transport time ↓ (Waiting time)
Use Excel for all the waiting lists	% excel used	# Excel communications / # new patients	< 70% ⬤○○ 70-90% ○⬤○ > 90% ○○⬤	1 month	Once a week	NPI primary	Doctors	Transport time ↓ (Waiting time)

Fig. 6. KPIs for nice-to-have target of child neuropsychiatry

5 Psychiatry

From the prioritization made in Fig. 1, it was decided to focus on Psychiatry as a must-have target on the reduction of the 75% of the acceptance and payment time, which in the same time will bring benefit in customer satisfaction. Meanwhile the nice-to-have target was based on the problem of communication flow between the CPS (Centro Psicosociale) and the administrative offices. Therefore, the nice-to-have target was set as a reduction of 75% in the lost or postponed visits.

A specific root cause analysis was performed on the basis of the Ishikawa diagram and the 5WHYs method for each of the two targets, as shown in the Fig. 7 and Fig. 8. The analysis of the causes helped the brainstorming of proposed countermeasures, and applying the same methodology previously used for the "Child Neuropsychiatry", a prioritization analysis was performed, as shown in the Fig. 9 and Fig. 10.

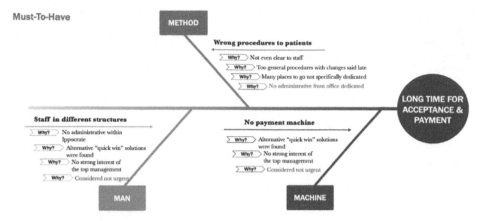

Fig. 7. Root cause analysis of psychiatry for must-have target

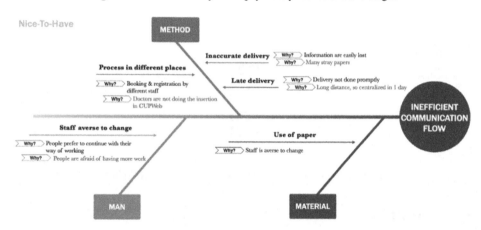

Fig. 8. Root cause analysis of psychiatry for nice-to-have target

MAIN CAUSE	ROOT CAUSE	COUNTERMEASURE	IMPACT	EFFORT TIME	EFFORT COST	FINAL EFFORT	GRADE	CLASS
Staff in different structures	*Considered not urgent*	1. Meeting with top managers	5	5	5	5	5	
		2. Share an existent administrative of Ippocrate	2	4	4	4	3,4	
		3. Add 1 administrative for psychiatry	4	4	3	3,6	3,72	
Wrong procedures to patients	*No administrative front office dedicated*	4. Create a front office in the CPS (pav 25)	5	2	2	2	2,9	
		5. Create a front office in pavilion 1	5	4	4	4	4,3	
No payment machine	*Considered not urgent because few cases require payment*	6. Meeting with top managers	5	5	5	5	5	
		7. Buy a new POS	4	4	4	4	4	
		8. Share the POS of territorial services	1	1	1	1	1	
		9. Add a Cash Machine	4	2	2	2	2,6	

Fig. 9. Countermeasures for must-have target of psychiatry

MAIN CAUSE	ROOT CAUSE	COUNTERMEASURE	IMPACT	EFFORT TIME	EFFORT COST	FINAL EFFORT	GRADE	CLASS
Process in different places	*Doctors are not doing the insertion in CUP*	1. Have the doctors doing the insertion of 28/SAN in CUPWeb	4	4	4	4	4	
Staff averse to change	*People are afraid of having more work*	2. Explanatory session	5	4	4	4	4,3	
Inaccurate delivery	*Many stray papers*	3. Past papers in a unique folder	0	5	5	5	3,5	
		4. Use digital format (Allegato A and electronic prescription)	2	3	4	3,4	2,98	
Late delivery	*Centralized in 1 day*	5. Improve the number of deliveries	2	3	2	2,6	2,42	
		6. Send via digital	2	3	4	3,4	2,98	
		7. Create info exchange function in CUPWeb	2	1	1	1	1,3	

Fig. 10. Countermeasures for nice-to-have target of psychiatry

The situation of the pandemic affected as well this part of the project, and it was not possible to implement the countermeasures defined. KPIs have been defined to monitor the countermeasures in the future, Fig. 11 and Fig. 12.

Must-To-Have

COUNTERMEASURE	KPI NAME	KPI FORMULA	THRESHOLD	TIME SPAN	MEASURING FREQUENCY	OWNER	SOURCE OF DATA	IMPACT ON TARGET
Meeting with top managers	*% of acceptance towards change*	$\dfrac{\text{# positive answers}}{\text{total questions}}$	< 60% ●○○ 60-80% ●●○ > 80% ●●●	1 day	Once	Top Management	Complex Structure Director	↑ Top mgmt. engagement
Add 1 administrative for psychiatry / Create a front office in pavilion 1	*% front office utilization*	$\dfrac{\text{# accept patients in pav 1}}{\text{total 28 SAN visits}}$	< 70% ●○○ 70-95% ●●○ > 95% ●●●	1 month	Once a week	Administrative and Doctors	Complex Structure Director	↓ Acceptance & Payment time
Buy a new POS	*% new POS utilization*	$\dfrac{\text{# paid visits with new POS}}{\text{total paid visits}}$	< 50% ●○○ 50-80% ●●○ > 80% ●●●	1 month	Once a week	Administrative and Doctors	Complex Structure Director	↓ Payment time

Fig. 11. KPIs for must-have target of psychiatry

Nice-To-Have

COUNTERMEASURE	KPI NAME	KPI FORMULA	THRESHOLD	TIME SPAN	MEASURING FREQUENCY	OWNER	SOURCE OF DATA	IMPACT ON TARGET
Explanatory session	*% of acceptance towards change*	$\dfrac{\text{# positive answers}}{\text{total questions}}$	< 60% ●○○ 60-80% ●●○ > 80% ●●●	1 month	Once a week	Psychiatry primary	Administrative and Doctors	↑ Staff accord
Have the doctors doing the insertion of 28/SAN in CUPWeb	*% of Registration on CUP*	$\dfrac{\text{# registered 28 SAN visits}}{\text{total 28 SAN visits}}$	< 60% ●○○ 60-80% ●●○ > 80% ●●●	1 month	Once a week	Psychiatry primary	Administrative and Doctors	↓ Lost visits

Fig. 12. KPIs for nice-to-have target of psychiatry

6 Conclusions

Despite the intense effort and level of resilience required during the data collection phase and the periods of uncertainty due to the pandemic, the case shows the applicability of the lean thinking in the healthcare industry, with the aim to reduce lead times and provide a better service to patients. The data collection phase necessary to depict the current situation involved employee at the hospital and the top management increased the attention on it, providing visibility on the severity of the inefficiencies as well as an increased commitment to change and improve. This application case allowed to engage the staff and actors in the Lean thinking approach, and to generate awareness on the importance of their collaboration for the hospital's Lean program. The increase of commitment reduced the resistance to change and the spread of Lean thinking in the company.

References

1. Al-Zuheri, A., Vlachos, I., Amer, Y.: Application of lean six sigma to reduce patient waiting time: literature review. Int. J. Qual. Res. **15**, 241–258 (2021). https://doi.org/10.24874/IJQR15.01-14
2. Cifone, F.D., Portioli-Staudacher, A., Silla, A.: Lean healthcare: how to start the Lean journey. In: Rossi, M., Rossini, M., Terzi, S. (eds.) ELEC 2019. LNNS, vol. 122, pp. 321–329. Springer, Cham (2020). https://doi.org/10.1007/978-3-030-41429-0_32
3. Prado-Prado, J.C., Fernández-González, A.J., Mosteiro-Añón, M., García-Arca, J.: Increasing competitiveness through the implementation of lean management in healthcare. Int. J. Environ. Res. Public Health **17**, 1–26 (2020). https://doi.org/10.3390/ijerph17144981
4. Torri, M., Kundu, K., Frecassetti, S., Rossini, M.: Implementation of lean in IT SME company: an Italian case. Int. J. Lean Six Sigma (2021). https://doi.org/10.1108/IJLSS-05-2020-0067
5. Costa, F., Portioli-Staudacher, A.: On the way of a factory 4.0: the Lean role in a real company project. In: Rossi, M., Rossini, M., Terzi, S. (eds.) ELEC 2019. LNNS, vol. 122, pp. 251–259. Springer, Cham (2020). https://doi.org/10.1007/978-3-030-41429-0_25
6. Dorval, M., Jobin, M.-H.: A conceptual model of Lean culture adoption in healthcare. Int. J. Prod. Perform. Manag. ahead-of-print (2021). https://doi.org/10.1108/ijppm-06-2020-0345
7. Borges, G.A., Tortorella, G., Rossini, M., Portioli-Staudacher, A.: Lean implementation in healthcare supply chain: a scoping review. J. Health Organ. Manag. **33**, 304–322 (2019)
8. Chiarini, A., Bracci, E.: Implementing Lean Six Sigma in healthcare: issues from Italy. Public Money Manag. **33**, 361–368 (2013). https://doi.org/10.1080/09540962.2013.817126
9. Portioli-Staudacher, A.: Lean healthcare. An experience in Italy. In: Koch, T. (ed.) APMS 2006. ITIFIP, vol. 257, pp. 485–492. Springer, Boston, MA (2008). https://doi.org/10.1007/978-0-387-77249-3_50

Lessons Learned from Toyota Kata Implementation in the Norwegian Construction Industry

Eirin Lodgaard[1]([⊠]), Maria Flavia Mogos[1], and Natalia Iakymenko[2]

[1] SINTEF Manufacturing, Enggata 40, 2830 Raufoss, Norway
eirin.lodgaard@sintef.no
[2] SINTEF Community, Strindveien 4, 7034 Trondheim, Norway

Abstract. Successfully completed construction projects with regards to quality, time consumed, and cost are the main concern focus of the construction industry. Thus, there is a growing attention to continuous improvement programs dealing with the streamlining of material and information flows and the minimization of waste in all processes at a construction site. Unfortunately, many companies are struggling with this, and the continuous improvement programs fails. Drawing on an action research case study involving a leading Norwegian construction company and their key suppliers, this study aims to identify enablers and inhibitors influencing the implementation process of a specific continuous improvement method – Toyota Kata. The aim was to learn from the implementation process to avoid pitfalls and to know what to solve, increasing the likelihood of success. The case is a housebuilding project where the case companies collaborate on common improvement tasks. Main enablers identified is the importance of having a common area for collaboration between actors in a value chain to solve common problems and application of a systematically approach ensuring learning and lasting improvement. The main inhibitors are lack of culture for CI and Kata is not a part of a management system such as Lean Construction, enabling to involvement of the entire workforce in the CI process. The results of this study can help managers and practitioners to address the identified enablers and inhibitors, paving the way for successful implementation of Kata.

Keywords: Continuous improvement · Toyota Kata · Lean construction · Enablers · Inhibitors · Action research

1 Introduction

The intense competitive situation is forcing the construction industry to continuously improve production performance enabling to sustain their competitiveness. The construction industry is characterized by high variability, cost/delay overruns poor performance with high level of production waste and unsafe working environment. To overcome this, implementation of Lean Construction and with continuous improvement (CI) philosophy at its core is seen as a valuable approach to create value and eliminate waste at the construction site [1]. CI includes application of a standardized

© IFIP International Federation for Information Processing 2021
Published by Springer Nature Switzerland AG 2021
D. J. Powell et al. (Eds.): ELEC 2021, IFIP AICT 610, pp. 365–372, 2021.
https://doi.org/10.1007/978-3-030-92934-3_38

problem-solving approach, targeting the elimination of waste and the improvement of quality in all the systems and processes of an organization.

Lean approaches such as CI, 5S, visual management, and the minimization of lot sizes and set-up times span across several industrial environments. However, approaches such as the takt and Kanban are best applied in environments with higher degree of repetitiveness, higher volumes, and rather stable customer demand [2]. For instance, the calculations of takt time are based on average sales rate. With relatively stable customer demand, this average will give a reasonable number to work with. But if demand varies each week, the sale rate figures will not be as meaningful. This is particularly the case of project-based industries such as the construction industry. Thus, the Lean Construction strategy was developed, based on the main lean principles in lean for the manufacturing sector [3], and having the CI philosophy at its core [4].

Even though CI has been known for decades, companies often achieve significant improvement only in the short run. One of the methods applied within the field CI is Toyota Kata (Kata) with the aim to achieve sustainable benefits through the development of a culture of CI and learning at all the organizational levels [5]. Kata combines systematic improvement and coaching methods, emphasizing the leaders' role as a facilitator and that interdisciplinary teams should move into an experimentation and learning zone when solving problems. It provides a standardized approach for managers and supervisors to support, guide and demand the improvement work.

Similar to Lean, Kata should be implemented, tested and adapted to the environment where it is applied. However, apart from very few studies [6], there is little research on the application of Toyota Kata in the construction industry [7]. To guide the investigation, we adopt the following research question: *What are the enablers and inhibitors of effective Toyota Kata implementation in the construction industry?*

2   Continuous Improvement

CI is a central principle regardless of the type of management system the company has chosen to apply [8], including the Lean Construction management system [1]. There are several CI definitions and one of the most common is: *"CI is a continuous stream of high-involvement, incremental changes in products and processes for enhanced business performance"* [9]. Consequently, CI is based on incremental improvement through the active participation of the company's entire workforce [10].

The implementation of CI as a part of a management system is not a straightforward process. Today, many companies often find their CI program in a fledging state and it is of high importance to know what aspects they need to consider when implementing CI with the entire workforce [10]. To avoid pitfalls, it is of high importance to understand the enablers and inhibitors of a successful implementation process. Due to their existence or absence in the organization, enablers and inhibitors can act as catalysts or on the contrary restrain of the implementation process.

One of the main reasons why many companies fail with the implementation of CI is the lack of top management commitment and involvement [10]. Unfortunately, all managers do not acknowledge the importance of their own commitment and involvement in succeeding with sustainable CI [11].

Another important factor is the independence of the contractors working at the construction site and of the supply chain actors [12]. This complicates the CI process and points the importance of having a management strategy to involve all the partners in the CI.

The knowledge and thorough understanding of CI is of high importance [13]. One of the pitfalls is too much focus on what CI methods to apply alongside the under-estimation of the human factors.

Motivated people who are actively involved in the CI in their daily work, is one of the key factors in succeeding with sustainable CI [14], as confirmed by recent studies [10, 11, 15]. A study by [15] shows the importance of teamwork and involvement as a common practice that is needed alongside knowledgeably about the approach and the issues to be improved by the team.

Another important factor is that the CI process is embedded in the culture, so that the systematic problem-solving comes naturally to the workforce. It means that the companies have to thoroughly reflect on how to organize for the integration of CI for the entire workforce [10].

By applying a CI method for problem solving and an effective learning process, companies can gradually reduce waste, the number of defects and the repair costs at the construction sites. Main benefits of applying a systematic approach include constantly inserting incremental improvements in the companies and their value chains, mini-mizing the amount of product defects and achieving better efficiency [16].

Important factors also include having knowledge about what available CI methods to apply and especially, having the required competence to apply the chosen method so that people can contribute in a proper way [11]. This is also perceived as a motivation factor [17].

A study by [18] emphasizes the need of tailoring the CI methods and developing a framework for the construction industry, as it has been done in the manufacturing industry. Another study from the construction industry has shown that certain CI methods from other industrial sectors have been found useful without having to adapt them, for instance the Deming's cycle [16].

3   Toyota Kata as a Method

According to Rother [5] leading people to implement specific Lean solutions like assembly cells, Six Sigma tools, Kanban, etc. will not make companies continuously improving and competitive. Rather, it is important to teach people to sense and understand the situation and react to in a way that moves the organization forward. This is what Kata is about – behavioral routines and habits that are practiced daily.

Kata refers to two linked behaviors: improvement Kata and coaching Kata [5]. By consistently practicing Kata routines, they become habitual for employees and the organization. The improvement Kata is a four-step routine for working toward a goal in a systematic way: (1) Understand the desired direction, (2) Grasp the current condition, (3) Set the next challenge, (4) Run small experiments (PDCA – Plan, Do, Check, Act). To ensure that improvement happens, and that people internalize the CI process, coaching Kata is used, where coaches teach people by guiding them in making improvements in processes.

Research on Kata implementation in construction is limited. Tillmann et al. [19] and Casten et al. [6] report on their efforts on Kata implementation in construction projects. However, Tillmann et al.'s report on Kata application is limited – the work stopped at the problem identification. Casten et al. stated that the main inhibitor to Kata implementation in construction is the "specificity of work area preparation". What that means is that in a manufacturing environment the production process flow is reliable, predictable, and understandable for everyone involved. This makes it relatively easy to evaluate and continuously improve the process. In construction, on the other hand, even a small change like the size of the crane instantly introduces variability into the operation. The authors state that even after years of effort to balance and control the workflow and implement lean practices, they find work area conditions the most challenging to stable and reliable workflow. In these conditions, the improvement Kata will likely be constrained by the conditions of the construction environment, according to the authors.

In addition to this inhibitor specific for the construction industry, Michals et al. [7] found following Kata implementation inhibitors: lack of direction and environmental preparedness (i.e. lack of solid understanding of target condition and current condition), lack of value stream mapping, and lack of coaching Kata meetings. Among the Kata enablers, the authors distinguished three: collaboration, learning and CI.

4   Research Design

This study is part of an ongoing national research project concerning performance improvement and operations at a construction site. The involved companies are the entrepreneur and three of the main suppliers in a housebuilding project in Norway. The purpose of this study was to explore the enablers and inhibitors of implementing a CI method – Toyota Kata. The aim of implementing Kata at all the involved companies was to strengthen the collaboration and learning process between them when addressing causes of inefficient flow in a joint housebuilding project.

According to these criteria, the action research as part of a case study investigation was identified as a suitable approach. The action research had a focus on developing knowledge in action and was a collaborative research between the partners in the housebuilding project and a Norwegian research institute. Data was generated through the researcher's active involvement and the process of inquiry during the application of the cyclical action research process - diagnosing, action planning, action taking, evaluation and identification of findings [20].

Five workshops, semi-structure interviews, formal and informal meetings were conducted during the research period of one and a half year. Field notes were taken through the entire research period including observations, experiences, results and reflections, which helped in understanding the Kata implementation process and the behavior of the involved personnel.

5   A Case of Toyota Kata Implementation

This section will briefly present the case companies involved in this study and the Kata implementation process during the action research. The research involves actors in a housebuilding project including following case companies: the entrepreneur, supplier of steel-reinforcement, supplier of concrete, and the supplier of formwork. All companies have a standardized system for internal deviation handling and customer complaints. In addition, the supplier of steel-reinforcement has prior experiences with Kata. A senior adviser from the research institute had the role as the Kata coach in the implementation process at the case companies. The role of the coach was to assist the involved personnel from the case companies with advice and guidance on the implementation.

A one-day workshop was organized where representatives from all companies received training on the Kata methodology and the principles of behavioral routines and habits that are practiced daily. Thereafter, a Kata game was conducted with the workshop participants, so they could understand the importance of being led by a coach, and of approaching a goal in a systematic way through the plan-do-check-act cycle. One of the important goals of the game was to learn that failed attempts also provide important insight which can help you reach the goal.

The next step was performing an extended value stream mapping (VSM) on a predefined housebuilding project facilitated by the coach. A mandate for this process was prepared including the delimitation of the VSM to a concrete process that involved all the case companies – the 'in situ' (at the construction site) casting process. The aim of the VSM was to grasp the current condition (step 2 in the Improvement Kata) - to create a common understanding for all the case companies of types of production wastes at the construction site. Four VSM workshops were conducted with the case companies and the results were presented at a subsequent workshop that resulted in an agreement on two specific problem areas. This study presents results up to the step 4 in the Improvement Kata - run small experiments and learn from the step-by-step process through the PDCA cycle.

6   Results and Discussion

Right from the beginning of this study, the case companies were planning to go from a culture based on a firefighting approach to problems, to a proactive improvement approach based on continuous learning from experiments and mistakes. This is a good starting point, and an important enabler of a successful application of a CI method.

The aim for the extended VSM was to create a common overview over the situation for the casting process including identification of problems area creating an inefficient flow. This was seen as a good starting point regards to the step 2) grasp the current situation and step 3) set the next challenge. This statement is in line with existing literature when implementing Kata [7]. It emerged in the discussions while working with the VSM that there was a common opinion that a large number of deviations occur every day at the construction site and adjusted in a reactive manner instead of proactive with identification and elimination of root causes – a firefighting approach.

Another important factor revealed during the VSM activities is that the suppliers indirectly sustain the firefighting approach by correcting defects created by the customer in order to please the customer. Many of the suppliers' problem areas were related to: i) the customer changes the plan close to the delivery date, and ii) the lack of a smooth process for the delivery of goods at the construction site. One of the suppliers reported that the customer had changed the plans 12 times during the same day and claimed that this was the entrepreneur's routine approach. Consequently, the supplier's internal production process had to undergo several changes, to adapt to this fire-fighting approach. It appears that the project partners please the customer at the expenses of their internal processes and original production plans. Surprisingly, only few of the entrepreneur's identified problem areas addressed the suppliers. This discovery was surprising to everyone and strengthen their opinion about a strong customer focus affecting unnecessary problems for the suppliers.

It was also revealed that the implementation of the CI method was not self-driven. The manager's commitment and effective leadership are crucial to implementing and performing CI as reported in existing literature [7, 10]. One of the manager's re-marks indicates that he was unaware of this: "Contact me if our dedicated team does not show up for this. If so, I will ask them to do it". The implementation of CI has to be integrated into the management system, to involve the entire workforce and generate lasting improvements [10]. The concrete supplier feared their personnel's lack of trust in the Kata implementation due to their failed attempt to implement Lean Construction as a management approach in the past.

An effective Kata implementation requires both knowledge about the method and the skills to perform Kata in the correct way. Although Kata teaching and training was provided during a workshop, an appropriate Kata competency also requires learning by interaction and experiences. The participants from the case companies were not used to systematic problem-solving methods and related learning process. Nevertheless, they saw the necessity to apply a systematic approach in order to change the firefighting culture. Furthermore, it appeared that they appreciated the coach helping with advice and guidance on Kata implementation. This is in line with [10]. Coaching is particularly useful at achieving that the participants become actively involved.

Lastly, the empirical findings also include the importance of having a common meeting place where the customer and the suppliers collaborate on streamlining the flow at construction sites. A common meeting place, our workshops, encourages the participants to spend time on making lasting improvements for both the customer and the suppliers.

7   Concluding Remarks and Further Work

The objective of this study was to identify enablers and inhibitors of effective Kata implementation. To this end, the authors conducted an action research study in collaboration with value chain actors in a housebuilding project.

The first conclusion drawn for this ongoing study of Kata as a CI method, reveals that the focus on CI ought to be increased in the future as a means to stay competitive in the housebuilding sector. Undoubtedly, implementing an unknown method as Kata it is not a straightforward process, especially when the organization lack a culture for CI and a firefighting approach is a common working mode at the construction site.

One of the enablers identified in this study was the fact that the common meeting area created by the Kata implementation process motivated the value chain actors to address common problem areas, enabling lasting improvement. A coach assisted the involved participants with encouragement and guidance on applying the unknown method. To change from a firefighting approach to a systematic approach the value chain actors are dependent on each other, and a common CI method that is used collaboratively can be an enabler.

Unfortunately, an inhibitor is the lack of a management system integrating Kata as a CI method into the organizations and their value chains, with only a limited possibility of achieving this after the end of the research project. This conclusion is in line with [10]'s study, which highlights that the management commitment and involvement is an essential factor when embarking on CI.

This study supports earlier research literature, but also extends the knowledge about influencing factors. However, our findings cannot be generalized. The results from action research should be regarded as indicative only, since they reflect the experience of the people involved in the implementation process. Therefore, future research should compare the findings in this research with findings from other studies of Kata implementations in the construction industry, to increase their external validity.

Acknowledgement. The current research was funded by the case companies and Norwegian Research Council, through the research project "Increased competitiveness for in-situ casting of concrete constructions".

References

1. Bajjou, M.S., Chafi, A.: The potential effectiveness of lean construction principles in reducing construction waste: an input-output model. J. Mech. Eng. Sci. **12**(4), 4141–4160 (2018)
2. Buer, S.-V.: Investigating the relationship between Lean manufacturing and industry 4.0. Doctoral thesis. Norwegian University of Science and Technology (2020)
3. Koskela, L., Ferrantelli, A., Niiranen, J., Pikas, E., Dave, B.: Epistemolgical explanation of lean construction. J. Constr. Eng. Manag. **145**(2), 04018131 (2019)
4. Netland, T.H., Powell, D.J.: The Routledge Companion to Lean Management. Taylor & Francis (2016)
5. Rother, M.: Toyota Kata: Managing People for Improvement, Adaptiveness and Superior Results. McGrawHill, New York (2009)
6. Casten, M.H., Plattenberger, J., Barley, J.M., Grier, C.: Construction Kata: adapting Toyota Kat to a Lean construction project production system. In: 21st Annual Conference of the International Group for Lean Construction, IGLC (2013)
7. Michals, E., Forcellini, F.A., Fumagali, A.E.C.: Opportunities and barriers in the use of Toyota Kata: a bibliographic analysis. Gestao Prod. Oper. Sist. **14**(5), 262–285 (2019)
8. Dahlgaard-Park, S.M.: The quality movement - where are you going? Total Qual. Manag. Bus. Excell. **22**(5), 493–513 (2011)

9. Ljungström, M., Klefsjö, B.: Implementation obstacles for a workdevelopment-oriented TQM strategy. Total Qual. Manag. **13**(5), 621–634 (2002)
10. Garcia-Sabater, J.J., Marin-Garcia, J.A.: Can we still talk about continuous improvement? Rethinking enablers and inhibitors for successful implementation. Int. J. Technol. Manag. **55** (1), 28–42 (2011)
11. Lodgaard, E., Ingvaldsen, J.A., Aschehoug, S., Gamme, I.: Barriers to continuous improvement: Perceptions of top managers, middle managers and workers. Proc. CIRP **41**, 1119–1124 (2016)
12. Bajjou, M.S., Chafi, A., En-nadi, A.: A comparative study between lean construction and the traditional production system. Int. J. Eng. Res. Afr. **29**, 118–132 (2017)
13. Meiling, J., Backlund, F., Johnsson, H.: Managing for continuous improvement in off-site construction. Evaluation of lean management principles. Eng. Constr. Archit. Manag. **19**(2), 141–158 (2012)
14. Bateman, N.: Sustainability: the elusive element of process improvement. Int. J. Oper. Prod. Manag. **25**(3–4), 261–276 (2005)
15. Patil, B.S., Ullagaddi, P.B., Jugati, D.G.: An investigation of factors impelling effective and continuous improvement of Indian construction industries quality management systems. In: IEEE-International Conference on Advances in Engineering, Science and Management (2012)
16. del Solar Serrano, P., del Rio merino, M., Saez, P.V.: Methodology for continuous improvement projects in housing constructions. Buildings **10**(199) (2020)
17. Lodgaard, E., Johannessen, L.P: Shop floor teams and motivating factors for continuous improvement. In: Wang, K., Wang Y., Strandhagen, J., Yu, T. (eds.) IWAMA 2018. Advanced Manufacturing and Automation, vol. 484, pp. 467-474. Springer, Singapore (2019). https://doi.org/10.1007/978-981-13-2375-1_59
18. Delgado-Hernandez, D.J., Aspinwall, E.M.: Improvements tools in the UK construction industry. Constr. Manag. Econ. **3**, 965–977 (2005)
19. Tillmann, P., Ballard, G., Tommelein, I.: A mentoring approach to implement lean construction. In: Proceeding IGLC-22 (2014)
20. Coughlan, P., Coghlan, D.: Action research for operation management. Int. J. Oper. Prod. Manag. **22**(2), 220–240 (2002)

Lean Translated from a Manufacturing Industry Context to Municipality Service Production: A Case Study

Eirin Lodgaard[1(✉)] and Maren Sogstad[2]

[1] SINTEF Manufacturing, 2830 Raufoss, Norway
eirin.lodgaard@sintef.no
[2] Norwegian University of Science and Technology, 2815 Gjøvik, Norway

Abstract. The expectations of Norwegian citizens with respect to the quality, availability, and effectiveness of public healthcare and welfare service provided by municipalities, are rising. Public resources are at the same time limited. This increased pressure is, however, an encouragement to optimize organizations and improve productivity and on delivering high-quality service. It is well known from the manufacturing industry that lean as an approach has achieved outstanding results regard to this. Thus, the aim of this study is to explore the translation process of new organizational idea of lean into the municipality, enabling to meet future increasing demand. The translation is carried out through a collaboration between a private actor and the municipality, where the private actor contributes with knowledge and skills about lean in a manufacturing industry context. An in-depth, exploratory single case study was carried out with the aim of understanding how the collaboration process takes place and the factors that drive the translation process. From how lean is rendered to fit the specific context to its meet in the recipient organization. The data material is made up of semi-structured interviews with key personnel from both actors and the analysis of written documentation from the collaboration process. The main drivers identified were being part of the same municipality, competent translation, internal translator to prepare the final translation, and solid anchoring and motivated leadership.

Keywords: Lean translation · Collaboration · Drivers · Municipality · Case study

1 Introduction

Citizens place increasing demands on governments to be responsive. Their expectations of the quality, availability and productivity of municipal services are also continually rising [1]. Resources are, however, limited. The motivation to close the gap between rising demands and constrained resources calls for improvement to manage forthcoming crisis in the municipality services sector.

Lean has been recently drawn from the manufacturing industry and adapted to sectors such as hospitals, healthcare and municipalities [2]. Lean is implemented in the municipality sector to create public value, increase quality and reduce waste in

© IFIP International Federation for Information Processing 2021
Published by Springer Nature Switzerland AG 2021
D. J. Powell et al. (Eds.): ELEC 2021, IFIP AICT 610, pp. 373–380, 2021.
https://doi.org/10.1007/978-3-030-92934-3_39

organizations. It is primarily applied through the use of implementation tools such as value stream mapping, team board and kaizen [2].

Lean interventions have been shown to have a high degree of variation, both in content, application, and outcome [3]. Successful implementation is not achieved through the blind adoption of the manufacturing version of lean, but through adapting the selected principles and tools to specific needs [4]. Thus, the aim of this study is to examines the interplay between a municipality and a private actor in translating the new organizational idea of lean into the municipality. Upon the collaboration and learning process between these actors a smooth translation process is of high value. Trying to avoid pitfalls regards a smooth translation process, it is a need to explore factors that determine what drives this. The fundamentals of our research question therefore relate to drivers as follows: *What are the drivers of the translation process when translation lean from an industrial context to municipality service production?* This will give us valuable knowledge about how to succeed with translation of lean from a manufacturing industry context into a municipality service context.

2 Lean Manufacturing

Lean's origin is an American interpretation of the Japanese approach to car manufacturing after the Second World War. Henry Ford's system of mass production had revolutionized the world of manufacturing. The Japanese engineer Eiji Toyoda however saw, when visiting the Ford factory in Detroit in 1950, the potential for improvements. This led to Eiji Toyoda and Taiichi Ohno developing the Toyota Production System, which is considered to be the most important innovation in manufacturing since Ford's mass production system [5]. Dr. James P Womac and colleagues defined, in a 1988 study from Massachusetts Institute of Technology, the Toyota organization as being "lean" and the phenomenon got its name.

Lean can be understood as being an organization trend, a management philosophy, a set of principles or a set of practices [6]. It is a systematic quality improvement approach that focuses on reducing non-value-adding activities in work processes. A lean organization uses less human capacity to conduct work, less material to produce a product, less time to develop it and less energy and space to produce it. Lean is simply the ability to do more with less [7].

3 Translation Theory

Transferring organizational ideas from one sector and organization to another has become widespread [8]. Translation is defined as being the process in which an organizational idea is transferred from one context and reinterpreted in a new context [9]. It is a relatively new perspective within organizational research [10]. The basis of this perspective is that ideas are constantly being processed and modified by the actors who receive them [10]. The carriers of the idea and the recipient organization collaborate to redefine or prioritize some practices and discard others. Røvik [10] claims that the translation theoretical perspective makes it possible to understand how organizational ideas are transferred between organizations and materialize and become a part of daily practice [10].

Translators are not passive recipients of organizational ideas. They are actively involved in the conversion process. Rørvik [10] distinguishes three degrees of conversion modes based on work in classical translation theory. The modes are the reproduction mode, the adding and subtracting mode and the radical mode. Degree of translation ranges from no conversion to significantly converted. This shows how important the translator's competence is, as the transformation of abstract presentations of an organizational idea into a local practice will always require interpretation. The translator's competence is therefore of great importance in achieving the desired organizational effectivity and effectiveness, which normally is the motivation for organizational change [11].

The de-contextualization of an organizational idea from one context and contextualization in another is not a straightforward process [9]. One challenge in the de-contextualization process is ensuring that the presentation contains all the information that is required to explain and understand how practices function in the source context. Translatability, the degree to which a practice can be translated to an abstract representation and the degree to which this can be achieved without excluding the elements that are required to define how it should function in the source context, is equally crucial [11]. The more complex the idea is, the more difficult it is to capture all of the factors that affect the way the practice works. Success depends on the translator having the freedom to add, subtract and convert the idea that is to be transferred. Some argue that translator should be as invisible as possible. Others argue that the translator should be as visible as possible. It is, however, almost impossible to reproduce accurately without the translator's own brand being overlaid onto the translated organization idea. Translation usually leads to new and unique versions [9, 12].

4 Research Design

Case research is a method that uses case studies as units of analysis. Case research is one of a number of ways of conducting social science and understanding complex social phenomena. It is used in many situations to contribute to our knowledge of groups, organizations and related phenomena within a real-life context [12].

A case study protocol was prepared to assure reliability. It included the research question, research methods, unit of analysis, procedure for data collection and data analysis guidelines. The unit of analysis was people from the municipality and the private actor that collaborated in the creation of an innovative solution for the municipality. We adopted a qualitative case-based enquiry because we wanted to understand how a non-public actor can become part of the municipal innovation system, what factors determine the successful transfer of an organizational idea.

The data collection process started with semi-structured interviews with key personnel involved in this collaboration process. Informants were recruited from both the municipality and the private actor. Informants from the municipality included the head of municipal administration, the head of lean implementation, three lean facilitators and three healthcare managers. The project manager was recruited as an informant from the private actor. The semi-structured interviews were conducted face to face to provide greater understanding and knowledge. The interviews were open-ended and conducted through informal conversation. The questions covered experience with the collaboration process, including the development of the innovation, the implementation phase and how to maintain the new way of working. All of the semi-structured interviews were tape recorded and transcribed.

The use of data triangulation was considered to be important in the deeper examination of data from the semi-structured interviews and in ensuring validation [13]. Data collection was therefore supplemented through informal conversations and through the analysis of written documents. The written documents include progress reports and working documents that describe the work process and results.

The analysis of data is probably the most challenging step in a research process. There are few standard recipes to guide this process [13]. The collected data was analyzed with the aim of exploring the drivers that facilitate the translation process. The data was coded and categorized into main and subcategories, to allow patterns and recurring themes to be found. The coded categories were then merged into common categories. The collected data was assembled into an array, the categories and the evidence from qualitative data from the in-depth study in columns. A within case analysis was then carried out in an attempt to identify patterns among the data in the array and identify preliminary conclusions [13].

5 The Case

The municipality in focus had, for some years, been following the implementation of lean as an approach to increase efficiency and quality. The approach had been applied in other municipalities in Norway and in other Nordic countries. The municipality's level of understanding of lean had increased through this. The municipality hosts one of Norway's largest industrial clusters, one third of turnover being derived from the development and production of automotive parts for a highly competitive global market. Structural changes in this cluster in the last decade have led to more connections and more frequent communication between the private actors in the cluster and the public sector. The municipality has, through this, gained an understanding of lean and how it can be used to achieve an organizational change and increase organizational effectivity and efficiency.

The private actor is a private research and development institution located in the case study municipality. The vision of this company is to create innovative, sustainable and effective solutions that provide competitive advantages to both manufacturing industry in Norway and to services provided by private and public actors.

This in-depth single case study played out in six pilot units. The private actor was mainly involved in the startup phase, in mediating lean as an organizational idea and customizing the idea locally. Pilot departments were selected by the municipality, lean being implemented in these departments over a time period of one year. The pilot implementation was to be used to make the final decision on whether lean would be used as a strategy for the entire organization. The municipality decided to start with their welfare departments where the potential impact was considered to be greatest and where the greatest organizational change and increases in organizational in efficiency and effectivity could be achieved.

The pilot phase mainly involved choosing principals and tools and testing those in six pilot departments. The phase was also used to build lean competence in the organization. The selection of lean principals and tools for tailoring lean was, as outlined by the private actor, influenced by experience from the local industrial park and other manufacturing companies in Norway and published research on the implementation of lean in Scandinavian municipalities and welfare and health care.

Each pilot department entering the implementation process was followed closely for a period of time. Representatives of each pilot department received basic training in lean before the implementation process began. This knowledge building was one of the private actor's main tasks. The manager of the pilot department selected the employees who would be involved in training courses held by the private actor. The aim of the courses was to create knowledge of lean and the chosen customized concept. They simultaneously held workshops and tried out the selected tools and practices in these. A further task was facilitating the pilot department in the pilot startup phase. The municipality administration also hired lean facilitators to take over the role of facilitator when pilot projects were up and running. Their role was to act as facilitators in the implementation of lean in new departments. They therefore participated in the pilot projects to gain knowledge of lean and to prepare for taking over as facilitators.

6 Findings and Discussion

A good starting point for the translation process was both actors realizing the need to translate the idea of lean from an industrial perspective into the municipal context [3, 10]. As one of the managers from the municipal management team stated *"Lean is a suitable approach and transferable into the public sector, but it has to be adjusted to the context, to the municipal service production. Production of municipal service at a nursing home is not the same as producing a component in the automotive industry".* Lean is therefore tailored to a local need but at the same time stays true to the origin of the concept [3, 8].

The impact of geographical, cognitive and social proximity between the public and the private actor has been shown by others to be an important influence for the start of collaboration [14]. Geographical proximity is where both actors are present in a municipality. The industry park that the private actor is part of is the largest workplace in the municipality, many of the actor's employees living in the municipality. The municipality's innovation was built on a long-term cooperation between the private actor and the municipality. This finding supports the study of Granovetter [15], who concludes that strong ties between actors are a powerful driver.

The private actor's high level of competence and experience in successful implementation of lean provided the motivation for the translation of the lean concept in municipal service production. The decision to start the pilot project by translating lean into the municipality context was made by the municipality before the active innovation process began. There had been informal meetings and discussions at management levels on the possibility of introducing lean into the municipality's practices. As stated by the head of the administration: *"The private actor suggested to use lean, we need time to mature this possibility, but after a while, we started to describe and elaborate it into a plan to check out the possibilities to implement lean as an organizational idea in our municipality."* The competence of the translators is a critical factor in the success of the transfer of an idea [11].

Both actors pointed out that humility and respect for each other's knowledge was an important driving factor. Inception was based on the municipality contributing their knowledge and experience within health and welfare service production and the private actor contributing, as consultants, their solid knowledge of lean and its application in industrial settings. Both actors realized that they were dependent on each other and on each other's knowledge and expertise for the successful translation of lean into the recipient organization. This is in line with the work performed by Rørvik [11], who stated that an important factor in the successful translation of ideas from one context to another is knowing the context from which the organizational idea is translated out of and the context into which it is to be translated into.

The private actor was an important player in the translation process in the trial pilot project period. The translator's main task is to make the new organization idea accessible and understandable. External translators can, however, impose their own influence on the process through their choice of translation. The private actor pointed out that there was a need to spend time together, particularly in the beginning, to gain knowledge of challenges and organizational opportunities in an unfamiliar sector. An important prerequisite is the acquisition of a thorough knowledge of the ideas and mechanisms of action of the organization which the idea is to be transferred from and that these conditions are replicated as closely as possible in the recipient organization [11]. The informants describe cooperation between actors as being good and built on trust, respect and sincere mutual interest. It can also be argued that showing humility for each other's knowledge combined with closed knitted relationships resulted in the development of the trust and confidence identified in previous research [16].

The public actor recognized the importance of being an active part of the translation of lean as opposed to just confining involvement to assigning this task to the private actor. The public actor highlighted the need for the lean model to be retranslated by the municipality. An external actor cannot gain sufficient knowledge of an organization as complex as a municipality to allow lean to be modified in the correct way. The public actor therefore took over as translator after the process of implementing in the pilot departments, adding some elements to the translation and rejecting others. A study by Andersen and Rørvik [3] concludes that a balance between tailoring lean to local needs and staying true to lean as an organizational idea for change is of importance.

A further observation was strong management commitment to the process and a well-anchored change process. This included the involvement of employees who adopt the innovation. The change process was strategically anchored at both the administrative and political top level in the municipality. The head of administration also had a strong positive reputation within the organization and played a central role both in and during the translation process. Participants interestingly expressed that it was an advantage that the head of administration was seen as being well-educated and driven by professional values and ambition. This study has shown it would have been difficult to achieve lasting innovation without this.

7 Concluding Remarks

The organizational idea studied in this work is the transfer of lean from the manufacturing sector and its reinterpretation in a municipality service context. The lean idea was, on its way through the municipality, translated into multiple iterations as identified in other studies [8]. Translation of lean included the incorporation of lean principles, the selection of tools and practices and the development of a customized implementation model that can ensure systematic implementation throughout the organization.

The drivers identified in this study relate to structural conditions and "being part of the same municipality", competence in "competent translation", internal translator to prepare the final translation, process factors in "solid anchoring" and "motivated leadership". These drivers may act as useful ideas for the practitioners.

The municipality's awareness of lean through proximity and strong ties with the industrial hub and the private actor seems to have enhanced its absorbing capacity, facilitating the innovation process. Together with strong management commitment from the receiving organization this was an important driver to achieve a lasting innovation. The crux of this study seems to pivot on the balance between knowledge of the new organizational idea and knowledge of the recipient organization. This study indicates that knowledge of the organizational idea is mandatory and is essential if the process is to stay true to lean and that core elements of the organizational idea are not washed out when tailoring new contexts. Knowledge of how a municipality is organized and works in practice is also essential if lean is to be successfully adapted to local context and needs. A mutual humility among the partners and for each other's differences in knowledge furthermore enabled a good collaboration process. This is reported by Dittmer et al. [17] as being an important driver. There are grounds for believing that organizations that need to collaborate with an external actor to obtain knowledge of new organizational ideas, will need to develop the final version due to the external translator not having the knowledge held by the recipient organization. This indicates that optimal success requires the internal translator to prepare the final translation, as external actors may miss recipient context essentials and consequently not fully adapt tailored lean-to local needs.

Acknowledgment. The authors would like to thank to all participants who took part of this study and the research project SFI Manufacturing, latter funded by the Research Council of Norway.

References

1. Sørensen, E., Torfing, J.: Enhancing collaborative innovation in the public sector. Adm. Soc. **43**(8), 842–868 (2011)
2. Arlbjørn, J.S., Freytag, P.V., De Haas, H.: Service supply chain management. A survey of lean apllication in the municipal sector. Int. J. Phys. Distrib. Logist. Manag. **41**(3), 277–295 (2011)
3. Andersen, H., Røvik, K.A.: Lost in translation: a sase-study of the travel of lean thinking in a hospital. BMC Health Serv. Res. **15** (2015)

4. Radnor, Z., Walley, P.: Learning to walk before we try to run: adapting lean for the public sector. Public Money Manag. **28**(1), 13–20 (2008)
5. Womack, J.P., Jones, D.T., Roos, D.: The Machine That Change the World. R. Associates, New York (1990)
6. Ingvaldsen, J., Ringen, G., Rolfsen, M.: Lean på global vandring. Fagbokforlaget, Bergen (2014)
7. Womack, J.P., Jones, D.T.: Lean Thinking Banish Waste and Create Wealth in Your Corporation. Free Press, New York (2003)
8. Morris, T., Lancaster, Z.: Translating management ideas. Organ. Stud. **27**(2), 207–233 (2005)
9. Czarniawska, B., Sevon, G.: Translation Organizational Change. De Gruyter, Berlin (1996)
10. Røvik, K.A.: Trender og translasjoner. Ideer som former det 21. århundrets organisasjon. Universitetsforlaget, Oslo (2007)
11. Røvik, K.A.: Knowledge transfer as translation: review and elements of an instrumental theory. Int. J. Manag. Rev. **18**, 290–310 (2016)
12. Madsen, D.Ø., Risvik, S., Stenheim, T.: The diffusion of lean in the Norwegian municipality sector: an exploratory survey. Cogent Bus. Manag. **4**, 1–25 (2017)
13. Yin, R.K.: Case Study Research. Design and Methods. Fourth edn. Sage Publications, Beverly Hills (2009)
14. Boschma, R.A.: Proximity and innovation: a critical assessment. Reg. Stud. **39**(1), 61–74 (2005)
15. Granovetter, M.: The Strenght of Strong Ties, in Networks in the Knowledge Economy. Oxford University Press, Oxford (2003)
16. Curtis, T., Herbst, J., Gumkovska, M.: The social economy of trust: social entrepreneurship experience in Poland. Soc. Enterp. J. **6**(3), 194–209 (2010)
17. Dittmer, P., Christiansen, C., Kierkegaard, G.F.: Public private partnership for innovation in Denmark. Eur. Public Priv. Partnersh. Law Rev. **4**(4), 240–242 (2008)

Challenges in Prototyping a Problem-Solving Practice

Christina Villefrance Møller$^{(\boxtimes)}$ (iD)

Technical University of Denmark, 2750 Ballerup, Denmark
chrvim@dtu.dk

Abstract. The purpose of this paper is to describe challenges in collaborative research in practice. The research object is a development process where a medium-sized and large manufacturer applied prototypes of a problem-solving practice on practical problems in product realization. These applications provided the manufacturers with insights into practical problems in product realization and their problem-solving capabilities. In addition, the application of the prototypes generated data for a case study about challenges in collaborative research. The findings reveal three challenges in collaborative research. The first challenges relates to scoping problems and action. The second challenge relates to practitioner's immediate outcome from applying the problem-solving practice. The third challenge addresses practitioner's commitment for change. Practitioners gained insight by applying prototypes; however, implications for practitioners emerge when mutual stakeholders have ambiguous or conflicting objectives for participating in research. Implications for research include tension between researchers and practitioners in the organization. Therefore, researchers, practitioners and consultants can benefit from considering these challenges when engaging in collaborative research.

Keywords: Problem-solving · Collaborative research · Product realization

1 Introduction

Research continuously explores the possibilities for collaboration, interaction, and co-creation of knowledge with practitioners [1, 2]. Researchers [2, 3] are occupied by research quality, epistemological discussions, and whether a trade-off exists between contributions to practice and research. The challenge for researchers is delivering knowledge with practical relevance, ensuring the scientific quality of research [3], linking theory, practice, and collaboration, and capturing differences while sustaining collaboration and managing quality [4].

Research provide guidance for researchers and practitioners collaborating in research within Operations Management. Engaging in collaborative research require that practitioners develop confidence in the language and process. For researchers collaborative research require skills such as listening to, motivating and convincing people, involving various areas of competence, designing and conducting change management. The research process engage researchers and practitioners in social interaction and collective inquiry into the research topic [4]. Even though collaborative

© IFIP International Federation for Information Processing 2021
Published by Springer Nature Switzerland AG 2021
D. J. Powell et al. (Eds.): ELEC 2021, IFIP AICT 610, pp. 381–390, 2021.
https://doi.org/10.1007/978-3-030-92934-3_40

research is common within Operations Management and research provide guidance, empirical case studies of the challenges in conducting collaborative research are rare.

This paper aims to describe the challenges in collaborative research in practice and contribute to the collaborative research stream within operations management. The research object is a development process where a researcher collaborate with practitioners in applying prototypes of a problem-solving practice for product realization. Practitioners in two manufacturing enterprises applied prototypes of a problem-solving practice twice before applying a final version. The mutual objective was to gain insight into practical problems in product realization while developing a problem-solving practice. Practitioners and the researcher gained insight at an early stage of the development process, which allowed them to refine the problem-solving practice. In addition, early insight gained by practitioners targeted their need for short-term gains, whereas the researcher improved applicability of the problem-solving practice.

The remainder of the paper presents the literature on collaborative research gains and prototyping a problem-solving practice. Subsequent sections describes the applied action research methodology. The empirical context is a development process of a problem-solving practice that practitioners can apply to improve product realization. The findings indicate three challenges: scoping problems and action, learning from outcome, and committing to change. Finally, the paper discuss authority relations and learning in collaborative research, and concludes by proposing future research directions.

2   Literature

2.1   Collaborative Research Gains

Collaborative research serves a dual purpose for practice and theory. For practitioners, the purpose is to generate actionable knowledge to pursuit practical solutions to issues of pressing concern to people for worthwhile human purposes [3, 5]. For research, the purpose is to generate a theoretical understanding that pertain to the academic field of management and organizational studies [4].

Collaborative research is a true partnership between researchers and members of a living system [3] grounded in a participatory worldview and growing out of a concern for the flourishing of individuals and their communities [5]. Collaboration between research and practice can be mutually beneficial as researchers gain access to real-life data and practitioners gain access to applicable knowledge [2, 4, 6, 7]. In addition, researchers and practitioners benefit from helping each other while practitioners help researchers understand practical problems and researchers help practitioners explicate learnings about the effects of their actions [4].

Practitioners claim that collaborative research creates an arena for reflection and conceptualization comprising the application of newly created knowledge to achieve competitiveness [2–4]. Researchers gain value by creating new knowledge and sharing findings with fellow researchers (e.g., journal papers [4]).

The process of collaborative research is a methodological approach characterized as a participatory and democratic process that reconnects action and reflection, and theory and practice [2, 5]. Further, collaborative research is an emergent and systematic inquiry process that helps design and implement appropriate management tools, and procedures in the field based on defined transformation projects [3]. As such, the practice system is a site for learning for both practitioners and researchers who needs each other to generate shared understanding, conceptualizations and interpretations [2, 4]. However, the time frame of creating value within research is a longitudinal and resource-demanding process where practitioners expect short-term gains [2, 3].

2.2 Prototyping a Problem-Solving Practice

Literature within operations management promotes conceptualizations of product realizations such as lean product and process development [8], and agile stage gate [9]. However, Benner and Tushman [10] suggested that researchers take a more problem-focused approach to developing theories on innovation and organizations. Implementing such universal programs can potentially deliver unexpected outcomes or even harm organizations [10]. Therefore, the problem-solving practice for product realization is developed through a prototyping process in collaboration with practitioners.

Developers create prototypes to gain customer or internal stakeholder feedback at an early stage of development [11]. Conventional product development applies prototyping as a practice to initiate a dialog with manufacturing concerning manufacturability and marketing about customer needs [12]. Designers in product development create prototypes of product concepts. Engineers create prototypes of the production designs, and software developers create prototypes of software for beta testing.

Collaborative prototyping acts as a boundary object across functional, hierarchical, and organizational boundaries in new product development, thus improving the overall prototyping process [13]. In addition, collaborative prototyping provides an approach to problem-solving that continuously iterates the prototype, improving functionality and usability through design changes [13]. As such, prototyping is a learning process for both users and designers [11].

This case study consider a problem-solving practice in product realization to be the product and two manufacturers to be the users applying the prototypes of problem-solving practice. Similar to prototypes of products, the objective was applying prototypes of a problem-solving practice to generate insight into improvements of product realization for practitioners and provide feedback about the problem-solving practice to the designer (the researcher in this case).

3 Research Method

A pragmatic position is taken and a practice–based perspective is applied on the research topic [14]. This paper study the challenges in collaborative research and apply action research in two manufacturing enterprises. Applying action research enabled a mutual learning process and collaborative partnership between practitioners and the researcher. Developing a problem-solving practice is a complex problem that requires a reflective and questioning process [3]. Such a process requires actions like "finding a problem; finding a group; identify their questions, reflections, and insights; how to build their (and the researcher's) commitments; and helping them while being open to their help" [4, p. 1682]. Breaking down the research process into smaller steps enables mutual reflection and builds creative confidence, benefiting practitioners and the researcher [4].

Following the recommendations from collaborative research [2–4, 6, 15], this case study takes small steps in a reflective and collaborative research process aimed at determining and designing solutions for real-life organizational problems. A problem-solving practice for product realization was designed through four action research cycles. Each cycle comprise four steps: developing a construct, planning action, taking action, and evaluating action. The first of the four action research cycles aimed to scope the activities with management representatives in two manufacturing enterprises to develop a mutual understanding of the relevant issues to address using the problem-solving practice. In the second and third action research cycle, appointed practitioners applied prototypes of the problem-solving practice to the issues within product realization. After completing each research cycle, management representatives decided application focus for the next cycle. The researcher then revised the problem-solving practice and planned the next step with the practitioners. Finally, in the fourth action research cycle, the practitioners applied a revised problem-solving practice on similar issues within product realization.

The research process implied that the researcher managed clarifying the problems in product realization, negotiating acceptance for applying prototypes of the problem-solving practice, and managing stakeholders in the research program. To paying attention to the first person voice [15, 16], the researcher kept a personal journal, observation notes and recorded research activities with participants. In addition, resúmés and transcribed interviews was shared with participants. Reflection-on-action by participants' self-evaluation and sharing insights and reflections with management representatives provided a second person voice to the collaborative research [15, 16]. Activities in the four action research cycles generated data multiple types of data; transcribed interviews, observations at workplaces and meetings, résumés from applying the problem-solving practice, chronicle workshops, and field data such as notes and pictures.

4   Empirical Findings

4.1   Applying Prototypes of a Problem-Solving Practice

Two manufacturers participated in the development process. Manufacturer A was a large global manufacturer delivering make-to-stock products with medium product variation and complexity in high quantities. It was essential for Manufacturer A to be flexible enough to adapt new technologies and simultaneously improve quality and lead-time to introduce new products and processes. These issues required collaboration across development and manufacturing functions. Manufacturer B was a medium sized manufacturer delivering engineered-to-order projects with high product variation and complexity in limited quantities. Manufacturer B changed strategic focus from single stand-alone projects to a small, customized series of projects, where engineering increasingly reused designs from previous projects. In addition, Manufacturer B found it necessary not to tie designers into a bureaucracy that could hamper their creativity in design and capability to meet customer needs. These two manufacturers are typical cases illustrating the challenges in product realization.

Each action research cycle was planned in collaboration with the two manufacturers relating the activities to specific situations with expressed purpose. Table 1 and Table 2 presents the activities, situations, purposes, outcomes and learnings for the two manufacturers. These activities occurred from February 2015 to June 2017 (further details are available [17]).

4.2   Outcome from Applying Prototypes of a Problem-Solving Practice

From the practitioner's perspective, the immediate outcome of the collaborative development process was applying the problem-solving practice onto practical problems in product realization. The findings indicate that management in the two manufacturers gained insight from scoping and evaluating feedback. Participants taking part of the problem-solving practice also gained insight into creating and implementing new organizational practices addressing the practical problems in product realization. As such, the manufacturers experienced practical and short-term outcomes by taking part in the collaborative research.

Similarly, the researcher conducting this collaborative research gained insight into the challenges in product realization and the applicability of the problem-solving practice through the development process. Learning from applying the prototypes and evaluating the application through refinements led to a final version of the problem-solving practice. The researcher gained early insight into the context and problem-solving practices, and collected data for further research. In this way, the collaboration also met the long-term objective for research.

4.3   Challenges in Applying Prototypes of a Problem-Solving Practice

Three challenges emerged from analyzing data from the development process: scoping problems and action, learning from outcome, and committing to change. Despite the similarities in challenges, differences also exist between the two manufacturers, and between the manufacturers and the researcher.

The challenge of scoping problems and activities arise when practitioners and the researcher collaborate on framing the purpose of solving problems, negotiate access and organizational involvement, and evaluate the outcome. Changing contact persons twice at Manufacturer A caused a discontinuity in the research process. Aligning expectations under such circumstances was a challenge at Manufacturer A, where the responsibility for accepting access was less clear and mutual stakeholders had different expectations. For the researcher adapting wording to internal terminology eased the negotiating process. Consequently, negotiations with Manufacturer A postponed actions applying prototypes of the problem-solving practice until late in the research project. In contrast, Manufacturer B initiated actions to apply a prototype of the problem-solving practice instantly.

Table 1. Activities, situation, purpose, outcome, and learning from four action research cycles in Manufacturer A.

Action research cycle	Situation	Purpose	Outcome	Learning
1. Scoping	A1 Production system	To identify challenges in product realization	SIPOC for the production system	Identified challenges in developing a future production system
2. First prototype application	A2 Challenge	To test the equipment capacity and train designated employees	Insight into situations that reduce the equipment capacity	Identified problems that reduced equipment capacity, were not used in further problem-solving activities
	A2 Breakdown	To get the production line up and running after a breakdown	Insight into and shared understanding of the problem causing stops on the production line	Identified problems and their causes not used in problem-solving activities to prevent similar situations
3. Second prototype application	A3 Analysis function	To improve the exchange of failing prints and fixtures for testers	Insight into the problem on the board and structures for integrating coordination between two functions into work practices	Participants focused on making a task list and completing tasks more than understanding what caused the problems
4. Testing the practice	A4 Analysis function	To reduce lead-time and failures from repair (FPY) To ensure new types of failures are detected and analyzed	Technicians gained insight into each other's work practices as a prerequisite for mutual adjustments and coding knowledge into procedures. New layouts supported integrating new work practices into the process	Instant collective action in changing the layout supported collective thinking and experimentation with prototypes of suggested solutions
	A5 Composite	To reduce lead-time by reducing the number of problem-solving loops or shortening the loops. To ensure that all relevant stakeholder knowledge is considered in the process	Shared understanding of the insufficiencies in the development process and suggestion to develop and integrate the functional-activity list as a new work practice.	Lack of facts and limited experimentation restrained the challenges of cognitive maps, and few suggestions for solutions were developed

Table 2. Activities, situation, purpose, outcome, and learning from four action research cycles in Manufacturer B.

Action research cycle	Situation	Purpose	Outcome	Learning
1. Scoping	B1 engineered-to-order project	To identify challenges in an engineered-to-order project	Management team chose to focus on knowledge sharing across customized projects	Identified challenges related to resource planning, knowledge sharing and diverse business units
2. First prototype application	B2 Knowledge sharing I	To improve knowledge sharing between projects in a project group	A board visualizing knowledge in the project was integrated into work practice	Time constraints made designers cut corners and left limited time for improving their work practices
3. Second prototype application	B3 Knowledge sharing II	To prevent deviations from recurring in successive projects	Insight into and shared understanding of the problem	The problem highlighted issues of ambiguous priorities that management representatives needed to be involved in
4. Testing the practice	B4 Knowledge sharing III	To prevent deviations from recurring in successive projects	Insights into each others work practices and solutions into the existing systems and procedures	Frequent fact-checking challenged participants' cognitive maps and thus provided new insight Experimentation with prototypes revealed how suggestions could be integrated into existing systems and procedures

Differences between the two manufacturers also became apparent as Manufacturer A had extensive experience collaborating with researchers whereas this experience at Manufacturer B was limited. These differences influenced the building of trust, and contributed to ambiguous and conflicting objectives. Especially ambiguous and conflicting objectives challenged scoping problems and actions for both practitioners and the researcher.

The challenge of learning from outcome addresses primarily local learning for practitioners participating in problem-solving activities. Practitioners emphasized improved collaboration and mutual understanding as important outcomes along with solving the problems. Despite that solved problems in product realization provided short-term outcomes and learning, Manufacturer B continuously asked, "What do we get out of this". At Manufacturer B, disagreements occurred regarding the counter-measures selected to solve the problem. These disagreements had roots in the management team and, as such, were out of scope for the research process. Management at Manufacturer A lacked attention and commitment to the problem-solving activities and outcome, which resulted in limited learning for the rest of the organization. Furthermore, ownership of the problem-solving practice remained for the researcher to share externally. Learnings based on analyzing the development process was shared with the two manufacturers but gained limited attention.

The challenge of committing to change arise in activities where practitioners and the researcher evaluated application of the problem-solving practice. The problem-solving practices include confronting assumptions and visualizing insight, which required active participation as a critical resource. Despite a scoping process with management, resources for such activities were scarce. However, participants were committed to solving problems relevant to their daily work. The problems that were mitigated and further improvements had to be followed up on by team management. As the researcher withdraws, there is a risk that solving the specific problem and the problem-solving practice will be discontinued. In this case, the contact persons at the manufacturers were not given a role. Therefore, commitment for change relied on the practitioners individually and not on the organization.

5   Discussion

Researchers have proposed various approaches to mitigating challenges in collaborative research [2, 4]. However, the efforts to meet these proposals do not necessarily lead to success for all stakeholders taking part in collaborative research. When scoping problems and action, practitioners and researchers aimed at developing mutual understanding of the research objective and negotiated practical application of the problem-solving practice. Collaborative management research is based on true partnership among the individuals "encompassing the dynamics and equality of integrated collaboration, emergent and systematic inquiry" [4, p. 1682]. However, establishing a true partnership based on equality requires attention to the authority relations between managers on different levels, such as manufacturers of different sizes and stakeholders within a research program. Authority relations and other intra-organizational activities, such as organizational changes that are out of the scope for the research project also influence the required development of mutual trust. Practitioners do have ambiguous or conflicting objectives for participating in research.

Practitioners and researchers focus on the task to learn from outcome, and commit to change. Ellström [2] suggest that a research model address a three-fold task aimed at practical concerns, creating scientifically acceptable knowledge and enhancing the competencies of the involved parties. The first two tasks were accomplished through the action research cycles. As for the learning task, researcher, especially Ph.D. students, accept the precondition of entering the learning process. Especially young researchers need skills to help practitioners design and co-create solutions to complex problems within operations management, and to improve the accessibility of research for practitioners by clarifying implications [4]. The researcher conducting this case study was a former consultant experienced in process consultation, scoping, negotiating and facilitating problem-solving processes but was inexperienced in enhancing practitioners' competencies in participating in collaborative research. Consequently, attention draws to the role of the supervisor.

Practitioners gained insight by applying prototypes of a problem-solving practice; however, implications emerge for practitioners when mutual stakeholders have ambiguous or conflicting objectives for participating in research. Implications for research draw attention to the tension between researchers and practitioners in the organization. Therefore, researchers, practitioners and consultants can benefit from considering these challenges when engaging in collaborative research.

6   Conclusion and Future Research Directions

This case study describes three challenges in collaborative research applying prototypes of a problem-solving practice. The first challenge relates to scoping problems and action collaboratively between the researcher and management representatives. The second challenge addresses practitioner's immediate learning from outcome of the problem-solving practice. Finally, the third challenge addresses practitioner's commitment to change. The findings exemplify challenges in applying prototypes of a problem-solving practice better than providing an exhaustive list of challenges in collaborative research. As such, collaborative research for this paper provided value for both practice and research. Practitioners gained insights from applying the prototypes and solving real problems in product realization. For the researcher, the application of prototypes generated data for research about the challenges in collaborative research.

Topics for further research could include authority relations between stakeholders such as a contact person as an insider, a researcher, and management representatives in collaborative research. In what way does the interplay of organizational systems with ambiguous or conflicting objectives influence the management of collaborative research? Furthermore, findings draw attention to describing the supervisors' role in a research program with more than one research project, researchers from more than one university. Collaborative research is based on the assumption that students have strong advocacy of the supervisors. However, what does that imply?

References

1. Svensson, L., Nielsen, K.A.: Action and Interactive Research : Beyond Practice and Theory. Shaker Pub (2006)
2. Ellström, P.: Knowledge Creation Through Interactive Research: A Learning Perspective1 (2015)
3. Adler, N., Shani, A.B.R., Styhre, A.: Collaborative Research in Organizations: Foundations for Learning, Change, and Theoretical Development. Sage Publications Inc, Thousand Oaks (2004)
4. Coughlan, P., Draaijer, D., Godsell, J., Boer, H.: Operations and supply chain management. Int. J. Oper. Prod. Manag. **36**(12), 1673–1695 (2016)

5. Bradbury, H., Reason, P.: Action research: an opportunity for revitalizing research purpose and practices. Qual. Soc. Work **2**(2), 155–175 (2003)
6. Coghlan, D., Brannick, T.: Doing Action Research in your Own Organization, 4th edn. Sage Publications, London (2014)
7. Karlsson, C.: Research Methods for Operations Management (2016)
8. Morgan, J.M., Liker, J.K.: The Toyota Product Development System : Integrating People, Process, and Technology. Productivity Press (2006)
9. Cooper, R.G.: Agile–stage-gate hybrids. Res. Manage. **59**(1), 21–29 (2016)
10. Benner, M.J., Tushman, M.L.: Reflections on the 2013 decade award–exploitation, exploration, and process management: the productivity dilemma revisited ten years later. Acad. Manage. Rev. **40**(4), 497–514 (2015)
11. Ulrich, K.T., Eppinger, S.D.: Product Design and Development. McGraw-Hill/Irwin (2012)
12. Cole, R.: From continuous improvement to continuous innovation. Total Qual. Manage. **13**(8), 1051–1056 (2002)
13. Bogers, M., Horst, W.: Collaborative prototyping: cross-fertilization of knowledge in prototype-driven problem solving. J. Prod. Innov. Manage. **31**(4), 744–764 (2014)
14. Saunders, M., Lewis, P., Thornhill, A.: Research Methods for Business Students (2012)
15. Coughlan, P., Coghlan, D.: Action research for operations management. Int. J. Oper. Prod. Manage. **22**(2), 220–240 (2002)
16. Argyris, C., Schon, D.: Organizational Learning II. Method and Practice. Addison-Wesley Publishing Company, Boston, Theory (1996)
17. Møller, C.V.: Organizational learning perspective on continuous improvement and innovation in product realization - DTU Findit. Echnical University of Denmark (DTU) (2018)

Leonardo da Vinci: Lean Educator or Lean Sensei?

Gianpaolo Perlongo[✉] and Monica Rossi[✉]

Department of Industrial Engineering and Management, Politecnico di Milano,
20133 Milano, MI, Italy
{gianpaolo.perlongo,monica.rossi}@polimi.it

Abstract. The inspiration behind the research conducted in this conference paper roots back to the 6th European Lean Educator Conference (ELEC 2019) held in Milan. Named "The Lean Educator and Practitioner Mashup", the conference was meant to explore the latest academics and industrial contribution to lean education, embracing the figure of Leonardo da Vinci as a conceptual leitmotif.

Both formal and informal events were paramount in the inspirational process leading to this research, highlighting some relations between lean thinking principles and the character of Leonardo da Vinci.

In this regard, analogies have been analysed to study if and how Leonardo's tools, theories and techniques could describe him as one of the contemporary figures of the Lean Educator or Lean Sensei.

The focus of the research was on Leonardo's tendency to outline the method and process through which his genius developed and manifested, to point out similarities and differences with the contemporary lean thinking and practice and to compare his figure with the one of a Lean Educator first and of the Lean Sensei afterwards.

Keywords: Leonardo da Vinci · Lean thinking · Lean Educator · Lean Sensei

1 Introduction on Lean Thinking and Leonardo da Vinci – LEANardo

The inspiration behind the research led in this work roots back to November 2019, during the 6th European Lean Educator Conference (ELEC 2019) held in Milan. Named "The Lean Educator and Practitioner Mashup", the conference was meant to explore the latest academics and industrial contribution to lean education, embracing the figure of Leonardo da Vinci as a conceptual leitmotif, in celebration of the 500th anniversary of his death.

Several influent academics and professionals in lean management took the stage to share latest studies and experiences led in this context. Speech after speech, it was built an enriching and stimulating environment, enhanced by more convivial meetings around the city in order to deepen interesting aspects about the genius of Leonardo da Vinci.

Both formal and informal events were key in the inspirational process leading to this research, highlighting some relations between the lean thinking principles and the character of Leonardo da Vinci.

© IFIP International Federation for Information Processing 2021
Published by Springer Nature Switzerland AG 2021
D. J. Powell et al. (Eds.): ELEC 2021, IFIP AICT 610, pp. 391–399, 2021.
https://doi.org/10.1007/978-3-030-92934-3_41

In this regard, analogies have been analysed in order to study if and how Leonardo's tools, theories and techniques could describe him as one of the contemporary figures of the Lean Educator or Lean Sensei.

The research was first general and concerned insights about principles and practices related to lean thinking and their relevance with Leonardo da Vinci. Just in a second moment, the focus was shifted towards Leonardo's tendency to outline the process through which his genius developed and manifested, to point out similarities and differences with the contemporary lean thinking and practice and to compare his figure with the one of a Lean Educator first and of the Lean Sensei afterwards.

To answer these questions, it was first conducted a literature review regarding lean thinking and practice to highlight the relevance of the five lean principles and the most common traits and tools to lean thinking. Followingly, another literature review was carried out, to understand the possible contributions that Leonardo's method and attitude towards life could have had, unconsciously or not, on lean thinking and practice, bringing to life the concept of LEANardo, followingly defined more in detail.

The first work of literature on lean thinking and practice produced several insights, which will be presented in a brief overview followingly.

As a matter of fact, Toyota represents in the common imaginary the best application of lean thinking, the place where everything started. For this reason, its production system has been first studied and then replicated all over the world thanks to the formalization by Womack and Jones of The Five Lean Principles, representing important pillars to be accurately followed in the implementation of Lean practices [1, 2].

Basically, Womack and Jones tried to convey that it is implicit in the interaction of all these principles that the never-ending improvement process it is aimed at with lean thinking can be summarized in a continuous pursuit for perfection, which is represented by a continuous boost for value, exposing hidden Muda in the value stream [3].

To do it, some tools and techniques can be used in a lean thinking perspective, both related to the TPS (along with TPD, TQM and Hoshin Kanri [4]) and more oriented to Product Development as *Set-based Concurrent Engineering* [3–5]. Among others:

– The *5S*, used to eliminate *Muda* and improve discipline and standards [1, 2],
– JIT (*Just In Time*), about producing just what needed at a specific time [1, 2, 6],
– *Kanban*, a tool through which the JIT is made possible [1, 2],
– *Kaizen*, related to the concept of continuous improvement [1, 2],
– *Zero Defects*, dedicated to the elimination of waste and defective parts [1, 2, 6],
– *Andon*, a control device reporting the status of a machine, line or process [7].

2   Context and Background

This paragraph will be mainly a literature research based on Leonardo da Vinci's method and tools, to highlight his "lean attitude" ante litteram through a comparison between The Five Lean Principles and *The Seven Da Vincian principles* [8].

This translates, in a second moment, into the research of the traits of a Lean Educator and a Lean Sensei in Leonardo's way of approaching life.

2.1 The Seven Da Vincian Principles

The Seven Da Vincian Principles, formalized by Gelb in the homonymous book, are considered the pillars of Leonardo's methodology and attitude towards life [8]. The objective of this analysis is actually to give an idea of his major and most common traits, the ones who made him the recognized genius that is known today, in order to make afterwards some reflections about his affinities with lean thinking and practice.

Given that, it is now time to introduce the seven principles:

1. Curiosità (Curiosity), curious approach towards life and research for continuous learning,
2. Dimostrazione (Demonstration), propension to test knowledge through experience and commitment, together with the willingness to learn from mistakes,
3. Sensazione (Sensation), experience the world through the five senses,
4. Sfumato, willingness to embrace ambiguity and paradox,
5. Arte e Scienza (Art and Science), continuous balance between art and science, logic and imagination,
6. Corporalità (Corporality), propension towards grace and ambidexterity,
7. Connessione (Connection), awareness of the importance of interconnections among all things and phenomena [8].

Curiosità can be defined as Leonardo's desire to know more about the world surrounding him, its dynamics and processes. He was extremely fascinated by nature and passionate in his research for truth and beauty. This attitude towards life "fueled the wellspring of his genius throughout his adult life, as he was able to transmute his passion into inquisitiveness" [8]. This continuous desire to learn is the cause of the depth of his studies and of the range of the topics considered.

The second principle is related to the concept of Dimostrazione, i.e., the human propension to absorb the most out of an experience. As a matter of fact, "Leonardo's practical orientation, penetrating intelligence, curiosity and independent spirit led him to question much of the accepted theory and dogma of his time" [8]. This is one of the reasons why he was so ahead of time, he was driven by experience, which was the first tool he used to question the status quo and common knowledge. The learning process was based on experience and new knowledge was created a mistake upon another.

The third principle, Sensazione, is strictly related to the second one, as the best way to make experience is through the five senses, especially sight. "*Saper vedere* (knowing how to see) was one of Leonardo's mottoes and the cornerstone of his artistic and scientific work" [8]. By improving his senses, he meant to improve his mind and the experiences accordingly.

Sfumato, literally "going up in smoke", is related to the concept of paradox and unknown. "Keeping your mind open in the face of uncertainty is the single most powerful secret of unleashing your creative potential. And the principle of sfumato is the key to that openness" [8, 9]. Accordingly, paradox was a common trait in his research for the truth, as learning more and more about everything, he was dragged deeper into ambiguity and towards the unknown.

Arte e Scienza is the principle expressing his inborn dichotomy. Art and Science were two traits influencing each other, but that had to be continuously balanced. For this reason, he is labelled by Gelb "the supreme whole-brain thinker", able to see the world in all its facets and to find connections among them to understand its intrinsic dynamics [8].

The principle of Corporalità is related to the Latin saying "Mens sana in corpore sano". Giorgio Vasari, artist and art historian of the same period of Leonardo, can be helpful in this regard reporting about "his great physical beauty and more than infinite grace in every action. His great personal strength was joined to dexterity" [8, 10]. This is reflected in his obsession for the human body and interest in anatomy.

Finally, the seventh and last principle, Connessione, is probably the one that better gives a measure of Leonardo's genius. He recognized an intrinsic relation among all things and phenomena of the world surrounding him and found in these connections a way to create new knowledge out of their interactions. Gelb says that "one secret of Leonardo's unparalleled creativity is his lifelong practice of combining and connecting disparate elements to form new patterns" [8].

2.2   Lean Sensei vs Lean Educator

Before discussing the concept of LEANardo more in detail, analyzing the common traits of lean thinking and Leonardo's way, it is necessary to better define what a lean educator and a lean sensei is. As a matter of fact, this research aims at taking unusual and unique insights on Leonardo da Vinci in relation to his lean attitude towards his research and creation processes.

The Cambridge Academic Content Dictionary defines an educator as "a person whose work is teaching others, or one who is an authority on methods or theories of teaching" [11]. Moreover, according to Doscatsch, a general educator is "one who teaches or educates others with no indication of mentoring or working with preservice or fellow teachers" [12]. Therefore, a lean educator can be defined as an authority on methods or theories of teaching, who educates their students with no indication of mentoring, applying the lean methods depicted through the five lean principles. This figure, then, is defined by their pronounced attitudes towards dedicating into the shaping of others' education, through a direct approach.

On the other hand, a lean sensei "is not a title you can take for yourself. It is given to you as a sign of respect from the people who want to learn from you because they recognize your mastery, such as it is. This mastery requires a blend of theoretical knowledge, practical skills, and good judgement from experience, as well as a genuine intention to help – even when the lessons can be on the brutal side" [13].

So, the lean sensei is not internally moved by the same motivations of a lean educator, he is basically claimed to convey his mastery and knowledge by people recognizing his authority and who are eager to learn [14].

3   Discussion – Is Leonardo a Lean Educator or a Lean Sensei?

3.1   Analogies: The Concept of LEANardo

In the previous paragraph, it was shown what can be concisely called "Leonardo's way", considered as an attempt to universally describe his method and attitude towards life. While reading it, it's impossible not to catch some analogies with the Five Lean Principles. Such analogies will be now better shown in this paragraph; so, some considerations will be drawn to give a measure of how cutting-edge and innovative he was in this topic, and to give an explanation to the concept of LEANardo.

To accurately apply Lean Thinking and Practice, attention must be paid to the specification of value, that has to be defined in the customer's perspective, so a product (or good or service) is valuable when it is capable to satisfy customers' needs and these are willing to pay for it. In the same extent, Leonardo for most of his life worked for commissioners, usually important patrons of the major cities, and had to satisfy their desires. His customers had expectations that had to be met in order to keep benefiting of their protection and sponsorship, even if he didn't believe in what he was asked to do, or his genius would have suggested him to dedicate to other projects [15]. Basically, everything revolved around the customers and their needs, however he was not willing to renounce to projects he was very passionate about and sometimes persuaded his customers, satisfying needs they didn't actually have.

Another analogy can be found in Leonardo's tendency to make several sketches of concepts he wanted to develop. For instance, Fig. 1 shows his attention in the development of models for chains, his concept is enriched and improved at every attempt in order to draw on paper exactly what he had in mind, seeking perfection in the execution [16].

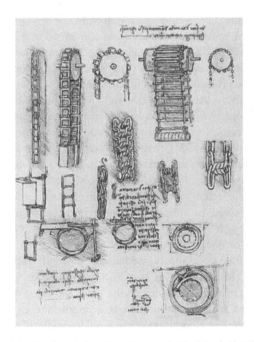

Fig. 1. Models for chains from Leonardo da Vinci (details) [16]

This vision is strictly connected to the lean concept of *Kaizen*, which is related to a research of perfection through continuous improvements. Moreover, sketches in Fig. 1 don't have to be intended just as ways to visualize and improve a concept, but also as a way to test it both as a whole and at components level, investigating different alternatives in terms of mechanics, connections or materials. This could be interpreted as an evidence for another analogy: his propension for Rapid Prototyping, typical of Lean Product Development applications. This concept has already been relatively addressed by Tarelko in his work "Leonardo da Vinci: precursor of engineering design" and by Sampaio in "Historical evolution of technical drawings in engineering", where it is discussed Leonardo's tendency to study all components in detail, evaluating different alternatives to understand, through models on paper, which was the best suitable for his purposes [17, 18]. In this way, the testing phase for his concepts was diluted throughout the all design, the two phases were conducted in parallel and their results influenced each other, in an ancestral and simplified version of *Set-based Concurrent Engineering*.

The last thing to be said regarding his sketches, is that he used to write notes on the side. Notes were about everything that could come to his mind while thinking of his concept, so some lines are reminders regarding how to develop the physical prototype once the concept was defined, others were about details that shouldn't have been forgotten, or about insights for other possible works [17]. This was basically what in Lean Thinking and Practice is now called *Job instruction and Standardization*, it is to say defining strict guidelines to be followed by whomever is in charge of doing the job, once he is carefully trained for it. In the same extent, Leonardo described the procedure step by step, in order to be clear and make it simple.

In addition to this, Leonardo's projects were conceived in a lean "pull" logic. As a matter of fact, in Lean Thinking and Practice, activities must be organized in a "pull" logic, so you realize an activity just when the downstream process requires it. In the same way, Leonardo's projects derived from a demand to satisfy, both internal (for instance his desire to fly in the sky) or external (works of military engineering).

In any case, his projects were driven by experience. As a matter of fact, as he was an illegitimate son, he couldn't be instructed as every child in the upper middle-class of the time, he never studied Latin or Greek, so he had no chance to read all the classical and technical literature of previous times. For this reason, he focused on experience, without relying much on theory, he developed his own necessary encyclopedia through what he was able to experience in his life. He didn't learn by spending hours studying literature, he rather learned by doing, a mistake upon another, understanding what hadn't worked and why [N8]. The analogy with the lean concepts of *Learn by doing* and *Make mistake faster* goes without saying.

Moreover, as just stated previously, Leonardo constantly felt the urgency to know more about the world, in a process of continuous learning driven by experience and commitment. These concepts are basically the pillars of lean thinking and practice, which finds its strength in the commitment that people are expected to put in their work, which is conceived as a continuous circle of learning to create new knowledge.

However, it is in this context that one of the analogies driving this work of thesis arises. As a matter of fact, Leonardo's learning process was totalizing and embracing different fields, so what made him so ahead of time was his strong multidisciplinarity, it is to say his interest in the most disparate fields, reflecting in his achievements in quite every domain of human knowledge.

In the same extent, *Set-based Concurrent Engineering* promotes the development of multidisciplinarity in project teams, in a powerful attempt of denial for *silos-thinking*, in order to create synergistic knowledge [3].

Finally, to wrap up, LEANardo is basically a concise way to express everything that has been figured out during the development of this literature research, it is the result of an accurate analysis of literature that highlighted the relevance of the indirect influence of Leonardo da Vinci on lean thinking, graphically summarized again in Table 1.

Table 1. Analogies between *Leonardo's way* and lean thinking and practice in keywords

Analogies in Keywords	
Leonardo's Way	**Lean Thinking and Practice**
His projects had to meet commissioners' requirements in order to keep their sponsorship.	Value defined in the customer's perspective. A product (or good or service) is valuable if it satisfies customers' needs.
Customer-centric approach	Customer is the starting point of the process
Several sketches to continuously improve his concepts	*Kaizen*: seeking perfection through continuous improvement
Several sketches to perform testing of components during the design	*Rapid Prototyping* in *Set-based Concurrent Engineering*: Test - Design - Build
Process described step by step through notes	*Job Instruction and Standardization*
Projects arise from a demand to satisfy	Lean "pull" logic
Learn through direct experience and a mistake upon another	*Learn by doing* *Make mistake faster*
Multidisciplinarity Interdisciplinarity	Denial for *silos-thinking* in *Set-based concurrent Engineering*

3.2   Lean Educator or Lean Sensei?

Given all the points previously discussed, Leonardo da Vinci as a lean educator could be considered as a forced analogy. As a matter of fact, he was a lean educator just potentially, he seldomly expressed and transmitted his potential directly to his contemporaries. Future generations throughout the centuries learned from Leonardo's attitude and method, but always indirectly and according to their interests.

For this reason, more than an educator, he could be considered as a lean sensei. As a matter of fact, this figure is more appropriate to describe Leonardo's figure and the authority that it constituted all over the centuries. Indeed, as previously mentioned, this title "is given to you as a sign of respect from the people who want to learn from you because they recognize your mastery, such as it is. This mastery requires a blend of theoretical knowledge, practical skills, and good judgement from experience, as well as a genuine intention to help – even when the lessons can be on the brutal side" [13].

4   Conclusions

The main purpose of this work was to analyse the method and attitude towards life of Leonardo da Vinci, in order to depict the analogies detected through the literature with lean thinking and lean education.

In this regard, all the analogies were schematized and studied to prove if Leonardo da Vinci can be defined a lean educator first and a lean sensei afterwards.

The results show that considering Leonardo da Vinci as a lean educator could be not totally pertinent, as he rarely got in contact with students or directed transmitted his knowledge to his contemporaries. He has been regarded as a genius since ever, as an unreachable figure you can only learn from. His authority is undeniable, for this reason he can be regarded more as a lean sensei towards the centuries.

References

1. Womack, J.P., Jones, D.T., Roos, D.: The Machine that Changed the World. Rawson Associates, New York (1990)
2. Womack, J.P., Jones, D.T.: Lean Thinking: Banish Waste and Create Wealth in your Corporation, 1st edn. Free Press, New York (1996)
3. Rossi, M., Morgan, J., Shook, J.: Lean Product and Process Development. The Routledge Companion to Lean Management, Chapter 6 (2017)
4. Leuschel, S.R.: Sensei Secrets: Mentoring at Toyota Georgetown. Align Kaizen (2020
5. Kerga, E., Akaberi, A., Tasich, M., Rossi, M., Terzi, S.: Lean product development: serious game and evaluation of the learning outcomes. In: Emmanouilidis, C., Taisch, M., Kiritsis, D. (eds.) APMS 2012. IAICT, vol. 397, pp. 590–597. Springer, Heidelberg (2013). https://doi.org/10.1007/978-3-642-40352-1_74
6. Ríos, J., Bernard, A., Bouras, A., Foufou, S. (eds.): PLM 2017. IAICT, vol. 517. Springer, Cham (2017). https://doi.org/10.1007/978-3-319-72905-3
7. Iuga, M.V., Kifor, C.V.: Lean manufacturing and its transfer to Non-Japanese organizations. Calitatea: Access Success, 15(139) (2014)
8. Gelb, M.J.: How to think like Leonardo da Vinci: Seven Steps to Genius every Day. New York, Delta Trade Paperback, reissue edition, June 2004
9. Davidson, S.J.: Complex responsive processes: a new lens for leadership in twenty-first-century health care. Nurs. Forum 45(2), 108–117 (2010)

10. Bondanella, J.C., Bondanella, P.: Giorgio Vasari – The Lives of the Artists. Oxford University Press, New York (1998)
11. Cambridge Academic Content Dictionary. Cambridge University Press. https://dictionary.cambridge.org/it/dizionario/inglese/educator
12. Doskatsch, I.: Perceptions and perplexities of the faculty-librarian partnership: an Australian perspective. Ref. Serv. Rev. **31**(2), 111–121 (2003)
13. Ballé, M., Chartier, N., Coignet, P., Olivencia, S., Powell, D., Reke, E.: The Lean Sensei. Go See Challenge. Lean Enterprise Institute, Boston (2019)
14. Reke, E., Powell, D., Olivencia, S., Coignet, P., Chartier, N., Ballé, M.: Recapturing the spirit of lean: the role of the sensei in developing lean leaders. In: Rossi, M., Rossini, M., Terzi, S. (eds.) ELEC 2019. LNNS, vol. 122, pp. 117–125. Springer, Cham (2020). https://doi.org/10.1007/978-3-030-41429-0_12
15. Koestler-Grack, R.A.: Leonardo da Vinci: Artist. Philadelphia, Chelsea House Publisher, Inventor and Renaissance Man (2006)
16. Innocenzi, P.: The Innovators Behind Leonardo: The True Story of the Scientific and Technological Renaissance. Springer International Publishing AG (2019)
17. Tarelko, W.: Leonardo da Vinci – Precursor of Engineering Design. International Design Conference – Design 2006, Dubrovnik (2006)
18. Sampaio, A.Z.: Historical Evolution of Technical Drawing in Engineering. IST University of Lisbon, Portugal, Lisbon (2018)

Lean Contribution to the Companies' Sustainability

Pedro Teixeira[1], José Carlos Sá[1,2(✉)], Francisco José Silva[1,2],
Gilberto Santos[3], Pedro Fontoura[4], and Arnaldo Coelho[4]

[1] School of Engineering, Polytechnic of Porto, Porto, Portugal
{1141259, cvs, fgs}@isep.ipp.pt
[2] INEGI, Porto, Portugal
[3] Design School, Polytechnic Institute Cavado Ave, Barcelos, Portugal
gsantos@ipca.pt
[4] Faculty of Economics, University of Coimbra, Coimbra, Portugal
uc2011122684@student.uc.pt, acoelho@fe.uc.pt

Abstract. In recent years, there has been an increase in the adoption of lean manufacturing principles, as well as tools, in multiple industrial sectors. In fact, this management philosophy, which is already well-established, has been applied successfully in other various contexts than just production. However, companies are under pressure to manage their activities, considering their effects in social and environmental terms. In this sense, this study aims to characterize the acceptance of lean by Portuguese companies, as well as the results obtained both in terms of Environmental, Social, and Governance (ESG), as well as in terms of organizational competitiveness (COM) by companies with lean implemented. To this end, a questionnaire was distributed to 3957 companies operating in Portugal, having obtained a total of 373 responses, of which 201 were validated, i.e., a rate of 53.89% valid responses. The results of this study show that companies with lean adopted, also adhere to green management practices (GMP), managing to improve their performance relative to each one of the dimensions of the triple bottom line (TBL), i.e., their ESG outcomes, as well as their COM.

Keywords: Lean · Sustainability · Competitiveness · Triple bottom line · Survey

1 Introduction

Lean, also known as lean manufacturing, or lean thinking, seeks to continuously improve production processes, through the removal of what does not bring value for the product [1] and/or service.

In fact, for lean philosophy, avoiding the occurrence of waste is a key point [2]. In this sense, given the success in this chapter, its tools and principles can currently be noticed in various sectors of the industry, showing interesting outcomes [1], as well as in contexts beyond the industrial one [3]. Nevertheless, there is an increased pressure for companies to handle their activities in a responsibly way with regard to their effects on the environment and on society [4]. In addition, there is still little empirical studies

D. J. Powell et al. (Eds.): ELEC 2021, IFIP AICT 610, pp. 400–408, 2021.
https://doi.org/10.1007/978-3-030-92934-3_42

about the impacts of lean on the environmental pillar [5, 6], as well as on the social one [7], so there is a lack of clarification of these aspects [5]. Thus, the present study aims to assess the degree of adherence to lean by the companies that operate in Portugal and what results have been obtained, by the ones who adopted lean, in relation to its ESG, as well as in terms of its organizational COM. In this sense, a questionnaire was distributed to 3957 companies operating in Portugal to assess if they have lean adopted and, if so, what results they have obtained in terms of the ESG components and its organizational COM.

2   Literature Review

2.1   Lean Connection with Triple Bottom Line

Lean and its methodologies are recognized on a global scale for the gains they bring to companies [8], since it emerged as an alternative for the management and organization of manufacturing companies [9], above all. According to the literature, which addressed the adoption of this type of practices, most of them report gains related to cost reduction and increases in the quality of products and/or services, as well as gains in productivity [10, 11]. As such, considering the eight types of waste that lean seeks to eliminate, i.e., defects, inadequate processing, overproduction, transportation, unnecessary inventory and movement [1], besides the waste of workers skills [2], this indicates that it has also potential for help companies in achieving a production system more environmentally sustainable and socially responsible. In other words, better outcomes in terms of ESG [6]. However, as mentioned before, not many studies have listed lean contributions to the environmental pillar [12], and, mainly, to the social one [9].

In fact, given the pressure from companies' stakeholders, it is necessary today to have a positive performance in all components of the TBL [13]. This led to the introduction of green management term, which translates into a set of environmental strategies and methods with the aim of making companies more ecologically efficient [6, 14]. In this sense, there is some literature that points out that lean enables green concept to be adopted [15]. As such, with the joint adopting both lean and GMP, companies can potentially enhance their cost and environmental risk reduction, enabling them to obtain a better social image among their stakeholders [16]. This indicates that both concepts (lean & green) are concerned with the efficient use of resources, i.e., removing waste [17, 18] and by that way, sharing a common point with sustainability concept [19].

2.2   Lean & Green Techniques and Sustainable Outcomes

There is a significant amount of literature that points out lean positive effects on economic/operational dimension, namely through its practices and tools. According to Rodrigues et al. [20] study in a metalworking organization, through the implementation of ten lean tools (e.g. 5S and daily kaizen), it was possible to obtain "quick-wins", i.e., fast positive gains, namely in terms of operational efficiency and increased workers motivation. However, even though lean has already proven to help obtaining positive results, mainly in operational terms, there is a lack of holistic studies about its impacts on the TBL in an integral way [21]. In this sense, there are a few studies that have reported positive benefits to the other components of sustainability, including through its combination with green concept. For instance, in the case study of Cheung et al. [22], at a plastic injection molding company, through the application of a model which integrated lean tools (e.g. Kanban and 5s) and green tools (life cycle assessment), it was possible to reduce the organization ecological impact by 40%. This allowed also to decrease its pollutant emissions (40%) and its electricity costs (41%).

In terms of social impact, Cordeiro et al. [9] found, through a case study in a hospital maintenance services organization, that lean tools (e.g. 5S and One Point Lesson), induces better working conditions (e.g. 80% decrease in the time spent in routes for picking materials and tools). As for Gonçalves et al. [8] study, these researchers created the Safety Stream Mapping (SSM) tool, in order to assess the safety risk level of a textile organization. This came from the combination of two tools (Value Stream Mapping and Waste Identification Diagram), which allowed to pinpoint this company most critical areas and by that way, to contribute to its social pillar.

3   Method

This investigation is based on an online questionnaire that was addressed to companies from multiple sectors of activity in Portugal, namely 3957, gathered from Portuguese Institute of Accreditation (IPAC) annual list. Its distribution, done through Google Forms followed by a short message, took place during the four final months of the second semester of 2020 until the middle of the first semester of 2021. The contacted firms were asked to forward the questionnaire, preferably, to people who held leadership positions and who had knowledge about lean and/or continuous improvement.

The criterion used for validating the responses was that the contacted companies had adopted lean and, if they did not have it, the questionnaire was then automatically closed. The questionnaire was divided into 10 sections. The first one with a short presentation of the investigation aim. The second one, related to demographic questions about the organization and the third one about the respondent profile. The fourth one, determined if the respondent could progress to the following sections. This advance depended on whether the organization had (or not) lean adopted, and for this the respondent was asked about which lean management practices (LMP) and tools were in place and how long ago. From the fifth to the tenth section, the respondent was asked to score, according to a Likert scale from 1 ("strongly disagree") to 7 ("strongly agree") a set of different items. Fifth section aimed to assess the degree of implementation of LMP, through a set of 15 items that sought to capture the different types of waste that lean aims to eliminate. Sixth section aimed at capturing whether the companies also applied GMP, i.e., they also incorporate the green concept, through 13 items. Sections seven to nine, sought to evaluate the results in terms of companies' TBL, i.e., operational performance (OPE), environmental performance (EPE) and health and safety performance (HSP), considering their last year, but without the effect caused by the covid-19 pandemic, through a set of 7, 14 and 10 items, respectively. Finally, tenth section sought to understand how respondents scored their companies in relation to their organizational COM, through a set of 9 items. Moreover, it is worth mentioning that respondents were instructed to respond to sections seven to ten, based on their responses in the previous two sections. A total of 373 responses were obtained, but only 201 were validated, i.e., 53.89% valid responses, since only these were from firms with lean adopted, these being this study sample.

3.1   Sample and Data Collection

In this section, this study sample is described. The responses obtained demonstrate that most of the companies have their headquarters in Portugal (98%) with only 2% being based abroad, namely in Spain, France, Brazil and the United States of America (Table 1). In terms of its regions' distribution throughout Portugal (Table 1), considering the Nomenclature of Territorial Units for Statistical Purposes (NUTS II), it is verified that most companies' headquarters are based in the North and Center regions (79.5%).

Table 1. Distribution of companies' headquarters by country and region (based on NUTS II)

Country	Total	%	Region of Portugal	Total	%
Portugal	197	98%	North	89	45%
Brazil	1	0.5%	Center	68	34.5%
France	1	0.5%	Lisbon Metropolitan Area	37	19%
Spain	1	0.5%	Alentejo	2	1
United States of America	1	0.5%	Algarve	1	0.5
Total	201	100%	Total	197	100%

Regarding the sector of activity (Fig. 1), most of the companies operate in the industrial sector (62%). Within this sector, the metallurgical one is the most representative (30%), followed by the textiles and clothing sector alongside with the automotive one, both representing 12% of all the companies that compose the industrial sector percentage.

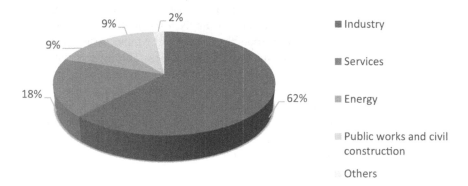

Fig. 1. Distribution of the companies for sectors of activity

As for the companies' size (Table 2), there is a higher representation of small and medium-sized companies (SMEs) (74%), i.e., companies with less than 250 employees. In this sense, these data show that, in this sample, SMEs are predominant in relation to large companies (26%).

Table 2. Distribution of companies by size

Companies' dimension	Total	%
SMEs	149	74%
Large companies	52	26%
Total	201	100%

Table 3 shows the position held by each respondent in their company, which shows there is a greater predominance of department managers (47.8%).

Table 3. Respondent job title

Title	Total	%
Department manager	96	47.8%
CEO	33	16.4%
Quality manager	27	13.4%
Health and safety manager	5	2.5%
Sustainability manager	5	2.5%
Other	35	17.4%
Total	201	100%

4   Results

In this section, the results obtained through the questionnaire are reported. According to the LMP construct results (Table 4), which had a global score of 80.4%, it appears that indeed most of the companies that declared to apply lean, have in place practices to reduce improper processing, overproduction, transport, unnecessary stocks and movement, as well as the waste of the workers capacities. The aspects with highest score were the promotion of 5S (83.2%) and error proofing systems (i.e., poka-yoke) (82.8%). Similarly, the same happens for the GMP construct (Table 4), which had a global score of 77.0%, showing that companies that have lean adopted are also committed to the application of GMP, mostly in the design of environmentally sustainable products/services, which had a score of 80.9%, as well in promoting a policy of reusing equipment's and materials, which had a score of 80.6%.

Table 4. Questionnaire results for LMP and GMP

Measure	Construct	
	LMP	GMP
Score (%)	80.4	77.0
Average	5.6	5.4
Standard deviation	1.3	1.5

As for the respondents' assessment of their companies TBL performance (Table 5), it appears that, in each of its dimensions (financial, represented by OPE, environmental, represented by EPE and social, represented by HSP), there has been an improvement in the past year, which means that TBL was enhanced as well. In terms of the OPE construct, which had a global score of 78.0%, the aspects with the highest score were compliance with established commitments (80.8%), such as delivery deadlines, and the increase in the quality of products/services (82.4%), considering the perspective of its stakeholders. Regarding EPE construct, which had a global score of 76.0%, the aspects with the highest score were the reduction in the number of accidents and/or near-environmental accidents (78.9%), besides the compliance with the applicable environmental legislation (85.64%). In the HSP construct, which had a global score of 82.5%, the aspects with the highest score were compliance with the applicable safety regulations (87.0%), employees worked hours being lower than the maximum value stipulated by law (83.6%) and the reduction in the rate of occurrence of high-severity accidents (84.7%). Regarding the COM construct (Table 5), this was the one that got the highest global score (84.0%). In this sense, the aspects that contributed the most to this result were the score attributed to best social image, among their stakeholders (87.8%), and the highest degree of customer loyalty (87.1%), both in relation to their competition.

Table 5. Questionnaire results for sustainability performance and competitiveness

Measure	Construct			
	OPE	EPE	HSP	COM
Score (%)	78.0	76.0	82.5	84.0
Average	5.5	5.3	5.8	5.9
Standard deviation	1.2	1.5	1.3	1.2

5 Discussion

The results of this study show that the lean philosophy is increasingly adopted in companies from different sectors of activity, that operate in Portugal. Moreover, it shows that firms where LMP are implemented, GMP are also being considered, given the relatively high overall score that the GMP construct obtained (Table 4), which may be due to the lack of studies that report (positive) effects of LMP on environmental [12] and, mainly, on social terms [9]. Regarding the results related to the TBL performance of companies (Table 5), these suggest that both LMP and also GMP can promote its improvement in each of its three dimensions. In this sense, stands out the HSP (82.5%), namely not only through the aforementioned compliance with safety rules and working hours limits plus the decrease in the rate of high-severity accidents, but also in improving working conditions and the capacity for detecting safety problems, which had, respectively a score of 82.9% and 79.9%. This way, these results are aligned with Cordeiro et al. [10] study, who found that lean tools can lead to better working conditions. In addition, regarding the EPE (76.0%), it was found that the aspect linked to the reduction of polluting emissions was also scored high (78.2%). This corroborates the findings of Cheung et al. [22] study. In terms of the results related to the OPE (78.0%), as expected, they were also positive, emphasizing, besides the previously mentioned aspects, the aspect related to the improvement in the management of available resources with a score of 78.1%. In this sense, these results corroborate the study of Rodrigues et al. [20]. Finally, it is interesting to note that the COM construct is where companies' respondents self-assessed themselves with a higher global score (84.0%), standing out, besides the aspects already mentioned, the aspect related to the greater reliability of their products and/or services, relatively to its competition, with a score of 85.9%.

6   Conclusion

The analysis carried out in this study, using data collected through a questionnaire, addressed to companies operating in Portugal, indicates that LMP, as well as GMP, can help to respond to stakeholders requests regarding the TBL performance by contributing (positively) to its improvement as a whole [13], i.e., ESG outcomes, in addition to the organizational COM. In addition, this study results also suggest that LMP and GMP may be compatible, allowing to achieve positive performance, not only in operational terms, but also in environmental and health and safety terms, conjunctly. As for the fact that the organizations surveyed have also assumed that they have GMP in place, this suggests that these organizations, where lean is adopted, seek to boost their TBL performance and, in turn, their organizational COM, by combining LMP with GMP. Future studies can focus on comparing the TBL performance, as well as organizational COM, between lean versus non-lean organizations, in order to verify if there are significant differences, since this study focused only on lean organizations. Another interesting possibility for future researches is to seek to analyze the effects of lean on sustainability and COM, considering the size of companies', namely the SMEs. Finally, future studies can consider the number of years that companies have adopted lean, in order to try validate the results obtained and to identify if there are significant differences.

References

1. Lopes, R., Teixeira, L., Ferreira, C.: Lean thinking across the company: successful cases in the manufacturing industry. In: Silva, F., Ferreira, L. (eds.) Lean Manufacturing: Implementation, Opportunities and Challenges, Nova Science Publishers Inc., pp. 1–31 (2019)

2. Brito, M., Ramos, A.L., Carneiro, P., Gonçalves, M.: The eight waste: non-utilized talent. In: Lean Manufacturing: Implementation, Opportunities and Challenges, Nova Science Publishers Inc., pp. 151–164 (2019)
3. Lopes, R., Freitas, F., Sousa, I.: Application of lean manufacturing tools in the food and beverage industries. J. Technol. Manag. Innov. **10**, 120–130 (2015). https://doi.org/10.4067/S0718-27242015000300013
4. Leon, H.C.M., Calvo-Amodio, J.: Towards lean for sustainability: understanding the interrelationships between lean and sustainability from a systems thinking perspective. J. Clean. Prod. **142**, 4384–4402 (2017). https://doi.org/10.1016/j.jclepro.2016.11.132
5. Garza-Reyes, J.A., Kumar, V., Chaikittisilp, S., Tan, K.H.: The effect of lean methods and tools on the environmental performance of manufacturing organisations. Int. J. Prod. Econ. **200**, 170–180 (2018). https://doi.org/10.1016/j.ijpe.2018.03.030
6. Silva, F.J.G., Gouveia, R.M.: Cleaner Production – Toward a Better Future. Springer Nature Publishing, Cham, Switzerland (2020). ISBN: 978-3-030-23164-4
7. Dey, P.K., Malesios, C., De, D., Chowdhury, S., Ben Abdelaziz, F.: The impact of lean management practices and sustainably-oriented innovation on sustainability performance of small and medium-sized enterprises: empirical evidence from the UK. Br. J. Manag. **31**(1), 141–161 (2020). https://doi.org/10.1111/1467-8551.12388
8. Gonçalves, I., Sá, J.C., Santos, G., Gonçalves, M.: Safety stream mapping–a new tool applied to the textile company as a case study. In: Arezes, P.M., et al. (eds.) Occupational and Environmental Safety and Health. SSDC, vol. 202, pp. 71–79. Springer, Cham (2019). https://doi.org/10.1007/978-3-030-14730-3_8
9. Powell, D., Coughlan, P.: Corporate lean programs: practical insights and implications for learning and continuous improvement. Procedia CIRP **93**, 820–825 (2020). https://doi.org/10.1016/j.procir.2020.03.072
10. Cordeiro, P., Sá, J.C., Pata, A., Gonçalves, M., Santos, G., Silva, F.J.G.: The impact of lean tools on safety–case study. In: Arezes, P.M., et al. (eds.) Occupational and Environmental Safety and Health II. SSDC, vol. 277, pp. 151–159. Springer, Cham (2020). https://doi.org/10.1007/978-3-030-41486-3_17
11. Silva, F.J.G., Ferreira, L.P.: Lean Manufacturing–Implementation, Opportunities and Challenges. Nova Science Publishers, NY, U.S.A. (2019). ISBN: 978-1-53615-725-3
12. Dieste, M., Panizzolo, R., Garza-Reyes, J.A.: Evaluating the impact of lean practices on environmental performance: evidences from five manufacturing companies. Prod. Plan. Control (2019). https://doi.org/10.1080/09537287.2019.1681535
13. Das, K.: Integrating lean systems in the design of a sustainable supply chain model. Int. J. Prod. Econ. **198**, 177–190 (2018). https://doi.org/10.1016/j.ijpe.2018.01.003
14. Toke, L.K., Kalpande, S.D.: Critical success factors of green manufacturing for achieving sustainability in Indian context. Int. J. Sustain. Eng. **12**(6), 415–422 (2019). https://doi.org/10.1080/19397038.2019.1660731
15. Dües, C.M., Tan, K.H., Lim, M.: Green as the new lean: how to use Lean practices as a catalyst to greening your supply chain. J. Clean. Prod. **40**, 93–100 (2013). https://doi.org/10.1016/j.jclepro.2011.12.023
16. Silva, S., Sá, J.C., Silva, F.J.G., Ferreira, L.P., Santos, G.: Lean green–the importance of integrating environment into lean philosophy–a case study. In: Rossi, M., Rossini, M., Terzi, S. (eds.) Proceedings of the 6th European Lean Educator Conference. ELEC 2019. Lecture Notes in Networks and Systems, vol. 122, pp. 211–219. Springer, Cham (2020). https://doi.org/10.1007/978-3-030-41429-0_21
17. Verrier, B., Rose, B., Caillaud, E., Remita, H.: Combining organizational performance with sustainable development issues: the lean and green project benchmarking repository. J. Clean. Prod. **85**, 83–93 (2014). https://doi.org/10.1016/j.jclepro.2013.12.023

18. Correia, D., Silva, F.J.G., Gouveia, R.M., Pereira, T., Ferreira, L.P.: Improving manual assembly lines devoted to complex electronic devices by applying lean tools. Procedia Manuf. **17**, 663–671 (2018). https://doi.org/10.1016/j.promfg.2018.10.115
19. de Carvalho, A., Granja, A., da Silva, V.: A systematic literature review on integrative lean and sustainability synergies over a building's lifecycle. Sustainability **9**(7), 1156 (2017). https://doi.org/10.3390/su9071156
20. Rodrigues, J., Sá, J.C., Silva, F.J.G., Ferreira, L.P., Jimenez, G., Santos, G.: A rapid improvement process through 'quick-win' lean tools: a case study. Systems **8**(4), 55 (2020). https://doi.org/10.3390/systems8040055
21. Henao, R., Sarache, W., Gómez, I.: Lean manufacturing and sustainable performance: trends and future challenges. J. Clean. Prod. **208**, 99–116 (2019). https://doi.org/10.1016/j.jclepro.2018.10.116
22. Cheung, W.M., Leong, J.T., Vichare, P.: Incorporating lean thinking and life cycle assessment to reduce environmental impacts of plastic injection moulded products. J. Clean. Prod. **167**, 759–775 (2017). https://doi.org/10.1016/j.jclepro.2017.08.208

Author Index

Printed in the United States
by Baker & Taylor Publisher Services